Power Systems

For further volumes:
http://www.springer.com/series/4622

Ulf Häger · Christian Rehtanz
Nikolai Voropai
Editors

Monitoring, Control and Protection of Interconnected Power Systems

Springer

Editors
Ulf Häger
Christian Rehtanz
Institute of Energy Systems, Energy Efficiency and Energy Economics
TU Dortmund University
Dortmund
Germany

Nikolai Voropai
Energy Systems Institute
Irkutsk
Russia

ISSN 1612-1287 ISSN 1860-4676 (electronic)
ISBN 978-3-642-53847-6 ISBN 978-3-642-53848-3 (eBook)
DOI 10.1007/978-3-642-53848-3
Springer Heidelberg New York Dordrecht London

Library of Congress Control Number: 2014931999

Printed on acid-free paper

Springer is part of Springer Science+Business Media (www.springer.com)

Foreword

Electric energy systems have undergone two major paradigm changes during the last 20 years:

- Liberalization of the electricity market by unbundling generation and energy market on one hand and transmission/distribution systems on the other hand.
- Transition from the existing fossil fuel and nuclear generation to an increasing share with respect to renewable energy sources.

To effectively support the aims of these changes, the transmission systems have to be regulated because they are a natural monopoly. Planning and operation of modern transmission systems become more and more complex for the following reasons:

- The location of generation sites depends on the availability of renewable energies and not on the location of the major customers.
- The nature of renewable energy sources creates a substantial volatility in the transmission network.
- Dispatchable power sources are replaced by stochastic generation patterns which need complex and sophisticated prediction tools.
- The acceptance for new transmission lines by the public is a major barrier for the adequate expansion of the transmission system.
- The key for the successful transformation from fossil fuels/nuclear generation to renewable energies is a powerful transmission system tailored to its needs.

The continuous growth of interconnected transmission systems is an immense challenge because it leads to the most complex technical system ever built by engineers. Although the frontiers for the size of future interconnected systems are continuously expanding there is a conjecture concerning the optimal size of a transmission system. This conjecture cannot be confirmed without looking into the details of monitoring, control and protection of interconnected power systems keeping the above mentioned paradigm changes in mind.

The expansion of the existing interconnected electric power transmission systems offers significant advantages with respect to operational security, integration of renewable energy, as well as energy trading. On the other hand, the complexity of operational problems significantly increases and hence large R&D

efforts are urgently required in order to make full use of recent technological innovations with respect to new power system components like wide area monitoring, control, and protection equipment, as well as advanced network controllers such as flexible AC transmission systems (FACTS) and HVDC systems. Furthermore, power system disturbances may result in major blackouts if monitoring, operation, and control of interconnected power systems is not based on efficient and innovative information technologies. The objectives of the FP7 project Intelligent Coordination of Operation and Emergency Control of EU and Russian Power Grids (ICOEUR) sponsored by the EU commission are directly linked to the aspects of a secure and economic operation of large interconnected power transmission systems, the integration of renewable energy power generation, and the efficient system handling under emergency conditions.

To achieve these objectives, a close and trustful cooperation between many experts with a wide range of expertise is an important prerequisite for creating a large impact on the future development of interconnected power systems. Leading experts in all the relevant fields successfully completed the ICOEUR project. It is most welcome that the important results and insights obtained in this successful project are documented in this book. It supports the dissemination efforts of the ICOEUR research consortium in order to adopt the accomplished results concerning innovative monitoring, simulation and control concepts, experience with tools and equipment, and the implementation of the results.

This book offers a systematic approach in looking for the optimal size of a large interconnected system from a technical point of view. A suitable basis for tackling the related problems is a well-defined basis consisting of system models and systematic description of relevant dynamic phenomena. This leads to a holistic simulation approach indispensable for the thorough understanding of the future energy system. Monitoring aspects based on state estimation and wide area monitoring deal with the reliable and complete assessment of the current operational state of the power system. Without detailed knowledge of the actual system state, a secure and economic operation is unthinkable. Since the first efforts in the 1960s of the last century much technological process significantly contributes to new effective solutions for the secure assessment of the current system state.

In view of the aforementioned increasing volatility due to renewable generation, the dynamic control of interconnected power systems is of increasing importance. These control aspects are based on new operational equipment such as FACTS and HVDC as well as modern information technologies such as multiagent control systems. Dynamic control intrinsically is related to system stability and the associated protection techniques. The reliable assessment of a stable operating point with respect to voltage and frequency is a great challenge for modern information systems. Interesting contributions have been achieved based on neural network approaches and artificial intelligence techniques. Although being powerful, all these methods have limits and suitable methods are required for stabilizing the power system under emergency conditions. Under-frequency load shedding is a suitable method to guarantee a stable system state even under extreme conditions. All these methodological approaches lead to an important answer to the aforementioned conjecture related to possible limits of future interconnected transmission systems. The close interaction between energy and information technologies is under all circumstances an important prerequisite for successfully tackling the future challenges.

This book is a comprehensive description of all aspects related to modern power transmission systems. As it is the result of a fruitful cooperation under the FP7 program of the European Commission in collaboration with the Russian Federal Agency for Science and Innovation, it is an impressive contribution across national boundaries based on a successful cooperation between scientists and engineers. May this book be an important reference for all those responsible for the future electric energy transmission systems.

Germany 2013 Edmund Handschin

Acknowledgments

This book has arisen out of the results of the joint research project Intelligent Coordination of Operation and Emergency Control of EU and Russian Power Grids (ICOEUR) with partners from Europe and Russia. We express our deepest thank to the European Commission (under the 7th Framework Program) and the Russian Federal Agency of Science and Innovation for supporting this work.

The editors would like to thank all authors from the ICOEUR project for their very valuable contributions to this book. Furthermore, a special thanks is given to Sven Christian Müller, Hanno Georg and Prof. Christian Wietfeld from TU Dortmund University for their contribution to Chap. 19 supported by the German Research Foundation DFG as part of research unit FOR1511 "Protection and Control Systems for Reliable and Secure Operation of Electrical Transmission Systems".

We express a special gratitude to Prof. Edmund Handschin for providing his very valuable support to the ICOEUR project. Also we would like to thank all members of the ICOEUR stakeholder committee for discussions and practical advices on project results.

The challenging task of writing and editing this book was made possible by the excellent cooperation of the team of authors together with a number of colleagues and friends. Our sincere thanks to all contributors, proofreaders, the publisher and our families for making this book project happen.

Contents

Part VI Integrated System

Part I
Introduction

Chapter 1
Requirements for Monitoring, Control and Operation

Christian Rehtanz, Nikolai Voropai and Ulf Häger

1.1 Introduction

The interstate integration of power grids provides multiple advantages concerning operational security as well as energy trading. Due to these facts the Central European Power System (ENTSO-E-CE, formerly UCTE) expands continually since its establishment and the ties to other interconnected systems like NORDEL grow stronger and further new ties with neighboring countries are either being constructed or planned. The recent interconnection of Turkey underlines this trend.

Similar developments are valid in the IPS/UPS system of Russia and its neighboring countries. Particular consideration has to be given to different scenarios of joint operation of the ENTSO-E- and IPS/UPS-systems.

With large scale deployment of renewable generation throughout Europe, in particular large scale wind farms and future solar power plants, interstate interconnections are of growing importance to secure energy supply. They optimize the utilization of energy sources within larger areas, promote electricity trading between different regions, and meet the requirements of economic development. IPS/UPS can achieve similar benefits (so called system effects), by initiating joint operation with ENTSO-E using interstate interconnections.

The major benefits that motivate TSOs to build up interconnections to neighboring transmission systems are:

- Optimization of the use of installed capacities
- Reliability improvements reducing the economic cost of power outages

C. Rehtanz (✉) · U. Häger
TU Dortmund University, Dortmund, Germany
e-mail: Christian.Rehtanz@tu-dortmund.de

U. Häger
e-mail: ulf.haeger@tu-dortmund.de

N. Voropai
Energy Systems Institute, Irkutsk, Russia
e-mail: voropai@isem.sei.irk.ru

U. Häger et al. (eds.), *Monitoring, Control and Protection of Interconnected Power Systems*, Power Systems, DOI: 10.1007/978-3-642-53848-3_1,

- Improved control of system frequency to minimize major disturbances
- Sharing reserve capacities and reducing the level of reserves required
- Providing mutual support for the interconnected systems in case of emergency
- Improved energy market conditions in better integrated large scale systems
- Facilitating large scale integration of renewable energies due to higher flexibility of the interstate network operations.

Due to the fact that both systems of Europe and Russia alone and especially with interconnections are second to none in the world in terms of the scale and distance of the interconnection and number of countries involved, strong R&D and innovations are urgently required along with the recent development of technologies. Presently, there are numerous enlargement projects of ENTSO-E and IPS/UPS under consideration and investigation:

- interconnection of Turkey was recently established,
- interconnection to northern Africa (Tunisia, Libya, Morocco, etc.),
- interconnection to P.R. China,
- and, most significantly, the interconnection of the two largest systems
- ENTSO-E and IPS/UPS.

It has to be mentioned that there are actually several system bottlenecks identified within the networks of EU and Russia. These congestions have to be considered as well and need to be relieved with the right technologies strengthening the interconnected power systems.

The realization of an interconnection of bulk power systems, which differ in their technical characteristics, is not trivial and its technical and economical efficiency depends on the chosen technology as well as its impact on system operational security. Currently there are multiple transmission technologies with miscellaneous technical properties available: i.e. cost efficient and well proven HVAC technologies, with the disadvantage of direct disturbance extension between interconnected systems or more sophisticated HVDC transmission systems with better controllability but high investments. In order to improve system stability, to control load flow, to facilitate electricity trading and to optimize the utilization of energy resources in interconnected power systems Flexible AC transmission Systems (FACTS) and HVDC as well as other innovative compensation or control devices can be used. Due to that complexity as a first step the technically and economically optimal realization of future large scale interconnected power systems have to be investigated regarding interconnection technologies. The beneficial integration of appropriately selected technologies is a precondition for the future development of large scale interconnected power systems.

However, bulk power grids may encounter major blackouts, often with catastrophic consequences for system and consumers. Some of such severe blackouts occurred for instance in Europe and Russia in 2003, 2005 and 2006, respectively. Among the main factors leading to occurrence and development of such emergencies, researchers call complication in operating conditions of the power grids and their control in a market environment as well as insufficient coordination of control at an interstate level. The latter particularly manifested itself during the 2006 European blackout. Therefore the possible future extension of power system interconnections requires elaborating methods for monitoring, control and operation of large scale systems and especially for the support of their interconnections. Besides, the possible future interconnection between the Pan-European and Russian electricity transmission systems would be greatly simplified if common/compatible software tools, hard ware equipment and operational procedures are adopted by all TSOs involved. The joint development of these tools and equipment will promote their adoption. The presentation of recent developments and their prototypically demonstration is the major goal of this book. The operability of the results is demonstrated based on extensive network simulations using realistic network data.

1.2 Large-Scale Interconnected Power Systems

All investigations in this book are related to real power system requirements. As examples the interconnected power systems of Europe (ENTSO-E) and Russia (IPS/UPS) are considered.

1.2.1 General Characteristics of IPS/UPS

The Interconnected Power Systems/Unified Power Systems (IPS/UPS) is a power union presently comprising synchronously operated power systems of 14 countries: Azerbaijan, Belarus, Kazakhstan, Kyrgyzstan, Moldova, Mongolia, Russia, Tajikistan, Ukraine, and Uzbekistan belonging to the Commonwealth of Independent States (CIS) and Estonia, Georgia, Latvia, Lithuania as unbelonging th the CIS. The system is actually based on the former USSR Unified Power Systems originated in the mid 1950s of the last century and being continuously developed over the last 50 years.

Synchronous operation of the power systems of these countries is coordinated by the Electric Power Council of the CIS (EPC CIS). Within the framework of the EPC CIS the Commission on Operational-Technological Coordination of parallel operation of the power systems of the CIS and Baltic countries (COTC) establishes recommendatory principles of technical interaction and develops corresponding documents.

The cooperation of the Baltic power systems with the power systems of the CIS countries is performed within the framework of the BRELL-Committee established on the base of multilateral international agreement between TSO's of Belarus, Russia, Estonia, Latvia and Lithuania signed in 2002.

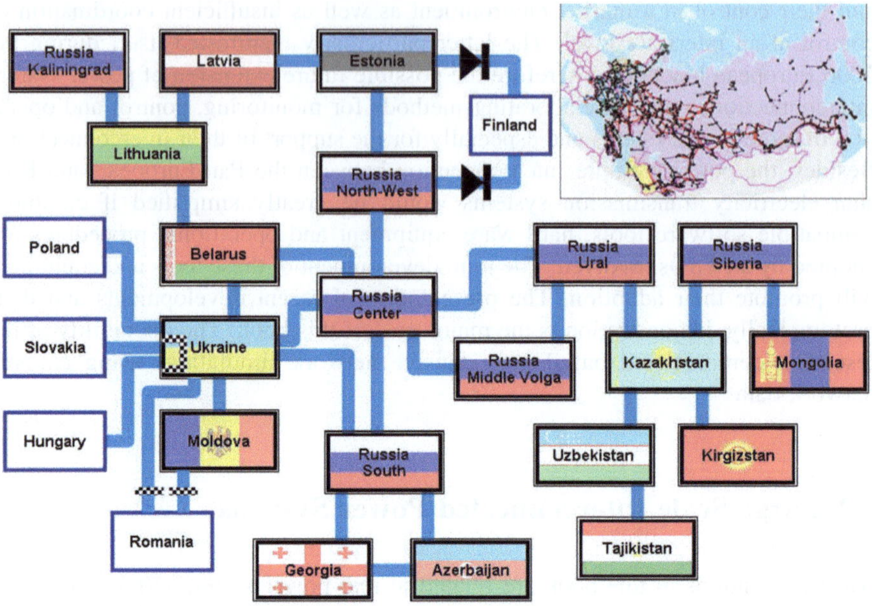

Fig. 1.1 Structure of interconnected power system (IPS/UPS) of Russia and its neighboring countries

At the moment, with 335 GW of installed capacity IPS/UPS annually supplies about 1,200 TWh to more than 280 million consumers. This is the world's most geographically extended power system, spanning over 8 time zones. Such vast territory impedes certain specific features of the power system:

- Comprises of internally almost balanced regional power systems interconnected in most of the cases by congested links;
- Extensive use of long-distance extra high voltage transmission lines (up to 1,150 kV);
- Use of automatic emergency control systems (in certain cases the N-1 criterion is only satisfied with the automatic emergency control system);
- All power systems composing IPS/UPS are structurally allocated to 14 power regions (see Fig. 1.1):
  - 6 IPS in Russia (North-West, Centre, Middle Volga, South, Ural, and Siberia),
  - Baltic States (Estonia, Latvia, and Lithuania);
  - Ukraine and Moldova;
  - Central Asia (Kirghizstan, Tajikistan, and Uzbekistan);
  - 5 individual powers systems of other countries (Azerbaijan, Belarus, Georgia, Kazakhstan, Mongolia).

In addition to Fig. 1.1, IPS/UPS has weak cross-border AC interconnections with Norway, Turkey, Iran, Afghanistan and China.

Each power system regulates the active power balance with or without frequency deviation correction, with or without automatic systems. The UPS of Russia controls frequency in the whole synchronous zone.

North-Western IPS of Russia has a DC-link with Finland. Another DC-link between Estonia and Finland (ESTLINK) was commissioned in December 2006. AC radial operation of near generation is operated with Finland, too.

Two scales of nominal voltages are used in the IPS/UPS: 750-330-220-110 kV and 1150-500-220-110 kV (now 1,150 kV equipment operates at 500 kV). The backbone network of 220–1150 kV performs power transmission; while the lower voltage lines form distribution grids.

In fact many electric ties in IPS turn out to be underloaded for a long time and their transfer capabilities are even below the limits determined by the standard margins.

The European part of Russia's UPS including Ural, has a rather developed closed structure of the main network. It encompasses relatively weak extended transmission lines between and within IPSs, which cause problems of irregular power fluctuations and angle stability (small signal and transient). The Asian Part of the UPS of Russia is characterized by lengthier transmission lines that are mostly extended in latitudinal direction. In West-Siberia they are mostly extended structure of the main network and have a chain-like structure in East Siberia and Trans-Baikalia. The problems of irregular power fluctuations in transmission lines and small signal and transient stability with respect to angle are also pressing here. Until recently voltage stability problems arose mainly at local nodes that contained large amount of asynchronous loads and appeared in the centers of oscillations during emergencies.

In the last decades development of large cities and megapolises has changed the main network structure. This resulted in formation of rather large highly meshed zones with relatively short transmission lines between substations. For such zones the problem of system voltage stability is getting urgent. This problem was shown by the blackout in Moscow and adjacent area in May 2005.

The following technical regulations are now in force in the synchronous area of IPS/UPS but, however, not fully confirmed by other countries than Russia:

- IPS/UPS Intergovernmental standard 1516.3-96 "Electrical equipment for a.c. voltages from 1 to 750 kV. Requirements for dielectric strength of insulation" [1];
- Methodical Guidelines for Power System Stability of RF Ministry of Energy [2];
- Guidelines of technical maintenance of Power Plants and Grids of Russian Federation [3];
- IPS/UPS Intergovernmental standard 14209-97 "Loading guide for oil-immersed power transformers" [4].

Fig. 1.2 Geographical representation of synchronous areas in Europe (ENTSO-E and IPS/UPS) (Map: © Lutum+Tappert)

1.2.2 ENTSO-E CE

The "European Network of Transmission System Operators for Electricity" (ENTSO-E) is the association of transmission system operators in Europe. The sub-region for Continental Europe the former UCTE is named ENTSO-E-CE in the following. The organization aims to provide a reliable market base by coordinating the operation of power system interconnections over the entire European mainland.

The transmission networks of the ENTSO-E-CE members supply electricity to about 450 million people with an annual consumption of approximately 2,500 TWh. The ENTSO-E-CE system covers 23 European countries with some 220,000 km of 400-kV- and 220-kV-lines, thus being by far the largest interconnected system in Europe. The annual peak load in 2006 was about 390 GW. Figure 1.2 gives a geographical overview about the synchronous areas of ENTSO-E-CE, IPS/UPS and other synchronous areas in Europe [5].

Over the 2nd half of the 20th century the ENTSO-E-CE interconnected system was designed in order to implement principles of solidarity and economy. The ENTSO-E-CE system developed progressively into the highly meshed network that provides routes for electricity from the generation in-feed to the consumption and allows getting missing power from a neighboring control area through the available reserves of partners. Building on the essential principle of solidarity, the reliability, adequacy and quality of supply were continuously improved.

Today, TSOs are in charge of managing the security of the operation of their own networks in a subsidiary way based on the ENTSO-E Operation Handbook. Individual TSOs are responsible for procedures of reliable operation in their control area from the planning period as in view of the real-time conditions, with contingency and emergency conditions. The coordination between TSOs contributes to enhancing the shared solidarity to cope with operational risks inherent to interconnected systems, to prevent disturbances, to provide assistance in the event of failures with a view to reducing their impact and to provide re-setting strategies and coordinated actions after a collapse.

sparely HVDC transmission system. The used transmission lines have short distances and the resulting grids are highly meshed.

In contrast to this, an interconnection of EU and Russian networks is an interconnection of power systems with significant differences. The IPS/UPS network includes high rates of HVAC links, due to the fact that large distances have to be covered between generation and load centers. Therefore the interconnection design for such power systems has to consider the specific features of both networks.

Another motivation for new innovations is the fact that the AC technology has already reached its limits. Due to this, previously minor issues for the European interconnected grids are now raised in their importance, i.e. voltage stability problems that previously presumed to be a problem of weak, sparse meshed or large distance grids are considered as raising problem in the ENTSO-E CE network. In particular, the creation of very large synchronously interconnected electrical systems is a potential source of stability problems, e.g. with regard to inter-area oscillations.

Such phenomena, if not adequately controlled and damped, can cause unexpected and cascading tripping of critical cross-sections of the system, thereafter determining unmeshed system operation conditions and outages, also in absence of significant system disturbances. This can be solved only by enhancement of existing power networks and making them more effective with regard to interconnected operation, by means of proper controls.

Further challenges to the interconnected power grids result from new large changing power flow scenarios due to liberalized electricity markets and a growing share of renewable generation.

Due to this previously adequate designed networks and interconnections are actually operated towards their limits. Therefore measures for wide area grid and interconnection enhancement are required.

1.3.2 State Estimation of Large-Scale Interconnected Systems

State estimation is normally applied for internal TSO control areas and considers only steady state network behavior. New sources of generation, like large-scale wind power penetration, and smart transmission devices, like HVDC and FACTS, are either not considered at all or not modeled adequately. The current generation of state estimators assumes steady state behavior within their calculation intervals of some tens of seconds. During alert or emergency situations, when the system state changes fast, the accuracy of state estimators deteriorates drastically. In addition, inaccurate information about the state of the neighboring systems may create a false sense of security and hence affect the effectiveness of security controls taken in case of large disturbances.

Monitoring the state of a transmission system is already a very complex task, and it is becoming more and more complex because of the increasing degree of interaction between the various TSO control areas. Therefore an insight into the state of the interconnected system is essential to maintain secure operation of the interconnected as well as of individual control areas. The increasing number of blackouts in recent years caused amongst others through insufficient knowledge of the interconnected system state highlights the necessity of a new generation of state estimation tools, which will help to solve this problem. The vision of growing interconnected systems suggest that for the first step also the fundamental step for monitoring and control of large scale interconnected systems, new state estimation tools with consideration of the development in innovative power system measurement and control technologies will give much more appropriate and accurate system state information of its own system and the interconnected system, hence provide better basis for taking appropriate control actions against large system faults or disturbances.

The need of innovations in the field of state estimation strengthens with the increasing implementation of FACTS within grid enhancement measures. Appropriate models of these smart FACTS devices and their integration in the new generation of state estimation tools are needed to provide intelligent monitoring tools with satisfactory functionality for system operation and security.

Improved network security can be achieved through increased collaboration and exchange of information between transmission system operators.

The new generation of state estimators has to provide TSOs with an accurate snapshot and robust indicators of the system state even during emergency situations. The key technologies to achieve this improved estimator performance are Wide Area Monitoring System (WAMS) technology for time-synchronized measurements combined with a central data concentrator and distributed state estimators.

Progress is envisioned in three specific directions as follows:

- The methodology for distributed state estimation to be developed provides each TSO with an accurate snapshot of the state of its part of the transmission system in relation to the overall state of the interconnected transmission system. This methodology will be based on the mutual exchange of system information and on-line measurement data between interconnected networks. The exchange of time-synchronized phasor measurements between TSO control areas with mutual access by the interconnected parties will be incorporated. The information exchange will be used for the modeling of neighboring network areas in such a way that the amount of data to be exchanged between individual state estimators is minimized.
- In comparison to the current state estimation processing principle based on steady-state models, the approach here is a fully dynamic state estimator. Dynamic phenomena like oscillations affecting the entire system will be identified online. With the new system state information obtained from the dynamic state estimator, the operators can initiate security mitigation actions. The

dynamic state estimator is able to follow the system changes during an alert or emergency state with much higher accuracy than that using existing technology.

- The method for online parameter identification to be developed provides system operators of system simulation tools with far more accurate dynamic models and parameters for the transmission systems and their components than that currently available. Accurate dynamic models and model parameters are needed for operational security assessment but also during the system operational planning stage.

1.3.3 Protection Functions Securing Stable Operation of Interconnected Networks

The security standards applied by European TSOs are based on the conventional (n−1)-criterion. In security analysis this implies that the security and stability of the transmission system is evaluated for a predefined list of events (mainly single contingencies). The use of this well defined deterministic approach to security assessment has in the past resulted in a very high reliability of supply for both national and transnational transmission systems.

The European interconnected power system implements the cooperation of TSOs for maintaining frequency stability. Within this cooperation all TSOs mutually provide primary reserve, which is activated automatically in case of power imbalances. Additionally each TSO activates secondary reserve to displace primary reserve and to control the frequency in its own control area. Such cooperation exists actually only for maintaining frequency stability, which is considered today as a solved problem. Similar TSO cooperation in the field of angle or voltage stability or concerning instability due to inter area oscillations does not exist, what leads to wide propagation of such disturbances affecting neighboring interconnected grids.

Some limited relief is provided by locally activated protection systems, which have very short reaction times and response to local measured parameters. These allow protecting important devices or limited network areas to be affected by the disturbance. Long history of blackouts shows that although protecting their local objects protection mechanisms often aggravate small disturbance and contribute to its large propagation.

History shows very clearly that most large scale blackouts occurred though the (n−1)-criterion was already used in the period of scheduling/operation planning or during a series of several disturbances. This highlights the importance of complementary methods and mechanisms for the maintenance of system stability. These might include innovative cooperation rules regarding angle and voltage stability and inter-area oscillations. This possible TSO cooperation requires actually not existing methods for identification of eminent stability loss. In case of interconnection of large scale power systems as that of Pan-European and Russian networks, priority will be given to total system stability as well as to the avoidance of large scale blackouts affecting several power systems with interconnections between them and to the maintenance of each individual network reliability. This requires new protection mechanisms which are activated systematically and protect the power systems against improper states. Such protection systems must consider smart devices as FACTS integrated in the protected system.

Significant progress is to be made in fast stability assessment methods dealing with the challenge of identification of instable modes of large scale power systems due to improper voltage or angle deviations as well as inter-area oscillations. Realization of such a system is possible under combined utilization of strategic placed PMUs, centralized data collection and efficient stability assessment algorithms.

Protection systems securing individual systems must be activated selectively, systematically and coordinated by a central entity based on not only local but also global information. They have to regard partial as well as total system stability issues trying to maintain stability of the interconnected wide area system as wide as possible.

1.4 Structure of the Book

The book presents results targeting on the required and envisioned innovations specified above.

For investigating large scale interconnected systems, models of the respective networks are required. Such models for static and transient investigations are presented in Part 2 of this book. These models are based purely on public available sources and can be used as reference cases for research purposes. It has to be stated clearly that the focus is both on the internal needs of the network development as well as their extensions and interconnections with surrounding systems.

A basic requirement for future large scale power systems is a most modern monitoring technology beyond the nowadays state of the art. Nowadays, regional control centers get steady state estimation information of their respective system only. With increasing interconnections and size of the system, system wide information of the entire interconnected system as well as dynamic information is needed to prevent large scale disturbances.

Therefore the development of methods and tools for monitoring of large scale power systems is a key requirement and objective. The communication and data exchange between control centers as well as system state estimation based on wide area system monitoring are innovative approaches. Use will be made of the latest developments in synchronized wide area measurements, in information and communication technology, and in system identification. New developments need to privilege the interconnection concepts and technical solutions that offer flexibility and minimize the impact on the power system operational organization to permit a progressive and modular extension of the electrical system to be interconnected with the pan-European system. Proposals for innovative solutions in this area are presented in Part 3 of this book. Wide area monitoring systems build the basis for further developments. Distributed state estimation and dynamic state estimation are proposed promising to increase the quality of system information significantly.

Beyond the better monitoring, new control approaches are required. The basic technologies are wide area control mechanisms which help for coordinated and automated control schemes. New control technologies and control schemes are presented in Part 4. The focus is on coordinated power flow control, control of interconnections and optimization of interconnections.

Another required innovation for interconnected power systems are new protection mechanisms. Efficient protection functions for secure operation of large scale systems of EU and Russia in both isolated operation mode and interconnected mode are needed. In this context methods for quantifying the operational risk and for assessing the stability of a large transmission system have to be developed. Strategies for keeping system stability as well as for certain disconnection of parts of the systems in case of imminent stability loss of internal and external interconnections have to be defined. Part 5 of this book focuses on stability and protection techniques. Protection technologies are analyzed. Dynamic security assessment and risk estimation is investigated and the containment of disturbances is discussed.

References

1. IPS/UPS Intergovernmental standard 1516.3-96, *Electrical Equipment for A.C. Voltages from 1 to 750 kV. Requirements for Dielectric Strength of Insulation*, http://www.eprussia.ru/lib/base/gost15163962003.rar (in Russian)
2. RF Ministry of Energy, *Methodical Guidelines for Power System Stability*, http://energorisc.ru/content/ukaz-30062003.pdf (in Russian)
3. RF Ministry of Energy, *Guidelines of technical maintenance of Power Plants and Grids of Russian Federation*, http://www.tehdoc.ru/repository/industrialsafety/1988443631.zip (in Russian)
4. IPS/UPS Intergovernmental standard 14209-97, *Loading Guide for Oil-immersed Power Transformers*, http://www.ruscable.ru/doc/docgost/files/21819903408.doc (in Russian)
5. UCTE IPS/UPS study, *Synchronous Interconnection of the IPS/UPS with the UCTE: Summary of Investigations and Conclusions*, Dec 2008. Available at: http://www.ucteipsups.org/
6. WG SYSTINT (with contributions from WG SYSTMED experts), *European, CIS and Mediterranean Interconnection: State of play 2004. 2nd SYSTINT Report* (EURELECTRIC, Brussels, Feb 2005). Ref: 2004-030-0105

Part II
System Model and Dynamic Phenomena

Chapter 2
Load Flow and Dynamic Model

André Seack, Johannes Schwippe, Ulf Häger and Daniil Panasetsky

2.1 Introduction

Based on the general requirements presented in Chap. 1, particular consideration has to be given to different scenarios of joint operation of the ENTSO-E- and IPS/UPS-systems. To provide a basis for the following investigations in this book, this chapter focuses on the development of an aggregated network model of these power systems which will be used as general test case in the following chapters.

Detailed information of the network topology and the used equipment are confidential data of TSOs which is difficult to access. Furthermore, if confidential data is used, publication of results is difficult. Therefore, a reduced network model was created by using only public available data. A core of this power system model consisting of the CWE region and a few more surrounding countries was already available at the Institute of Energy Systems, Energy Economics and Energy Efficiency (ie^3) at TU Dortmund University [1].

2.2 Aggregated Network Model

In the following sub-sections the general modeling method of the grid, the unit commitment and the enhancement for dynamic simulations are described.

A. Seack (✉) · J. Schwippe · U. Häger
TU Dortmund University, Dortmund, Germany
e-mail: andre.seack@tu-dortmund.de

J. Schwippe
e-mail: johannes.schwippe@tu-dortmund.de

U. Häger
e-mail: ulf.haeger@tu-dortmund.de

D. Panasetsky
Energy Systems Institute, Irkutsk, Russia
e-mail: panasetsky@gmail.com

U. Häger et al. (eds.), *Monitoring, Control and Protection of Interconnected Power Systems*, Power Systems, DOI: 10.1007/978-3-642-53848-3_2,

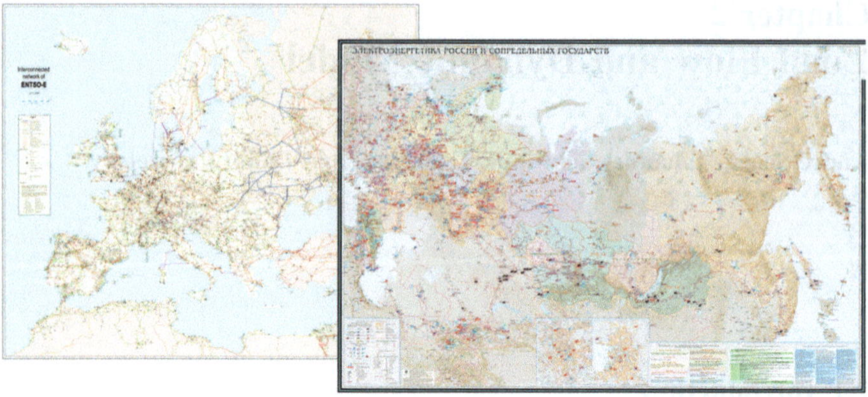

Fig. 2.1 Interconnected network of ENTSO-E and IPS/UPS [2, 3]

The aggregation method was for the first time introduced in [1], when the ENTSO-E CWE part of the model was developed. The methodology of the aggregation process is divided into four steps:

- Identification of important network areas to reduce the number of network nodes.
- Aggregation of transmission lines in the considered network area to the reduced number of network nodes.
- Allocation of loads and generation to the aggregated nodes and the unit commitment by a Merit-order model.
- System enhancement with dynamic data of the power stations to allow detailed analysis of the power system response to disturbances and innovative network assets.

The geographical maps of the European and Russian transmission network with the interconnections are available from public sources [2, 3]. These maps are outlined in Fig. 2.1.

The geographic location and voltage levels of nodes in the aggregated grid model are determined according to these grid maps. Additionally, the length and the number of transmission lines between these nodes are estimated according to these sources. Not public documented ties between these areas are neglected.

The specific electrical parameters and the nominal current correspond to typical transmission lines according to [4, 5]. Regional differences of these parameters are considered whereby the voltage level of the transmission lines is mainly 380 kV in the European part. Due to the longitudinal structure, the Russian network part is dominated by transmission lines with the voltage level of 330, 500 and 750 kV.

2.2.1 Aggregation of Detailed Network Nodes

In order to build a representative network an aggregation technique which maintains the electrical characteristic of the real transmission network was used. Based

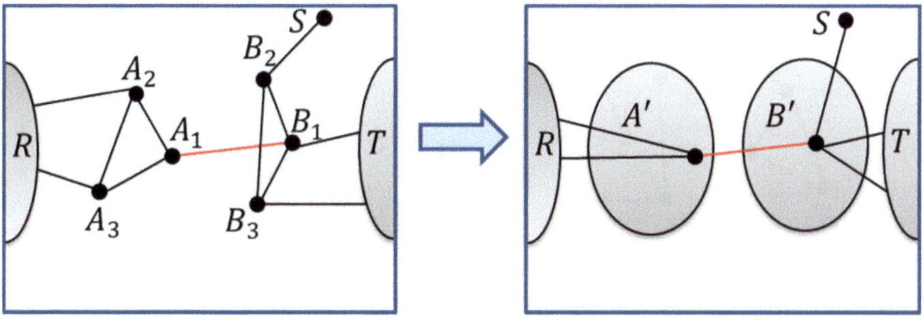

Fig. 2.2 Congestion limited node aggregation (*left* detailed network, *right* aggregated network)

on public available grid maps [2, 3] important network areas are identified, which typically represent load or feed-in areas.

Real nodes, which build a highly meshed grid in contrast to the surrounding network, represent one node in the aggregated network model (Fig. 2.2). The internal connection capacity $TC^{XY}_{d,\mathrm{int,m}}$ of this area has to be higher than the transfer capacity between two potentially aggregated nodes $TC^{XY}_{d,\mathrm{ext,n}}$ of the resulting areas for every connection.

$$TC^{XY}_{d,\mathrm{ext,n}} \leq TC^{XY}_{d,\mathrm{int,m}} \tag{2.1}$$

If the restriction is not met, congestion may occur and additional nodes are required to achieve a similar behavior as the original network.

Figure 2.2 shows a typical situation where an area cannot be aggregated because of insufficient internal transfer capacity between A_1, A_2, A_3 and B_1, B_2, B_3. The internal connection capacity in the detailed network TC^{AB}_{d} does not exceed the connection capacity of the nodes to the surrounding network R, S and T.

$$\begin{aligned} TC^{AB}_{d} &\leq TC^{AR}_{d} \\ TC^{AB}_{d} &= TC^{AS}_{d} \\ TC^{AB}_{d} &= TC^{AS}_{d} \end{aligned} \tag{2.2}$$

Even though an aggregation of the network areas (A' and B') would be suitable in terms of feed-in and load concentration, the network restricts the aggregation process to the substituted nodes A' and B'.

The opposite situation is shown in Fig. 2.3. The internal connection capacity exceeds or is equal to the external connection capacity for every connection. This enables the aggregation of the individual nodes A_1, A_2, A_3 and B_1, B_2, B_3 to the aggregated network node A'. For this particular example six nodes of the detailed network can be simplified to one node while maintaining the system transmission characteristic under these conditions:

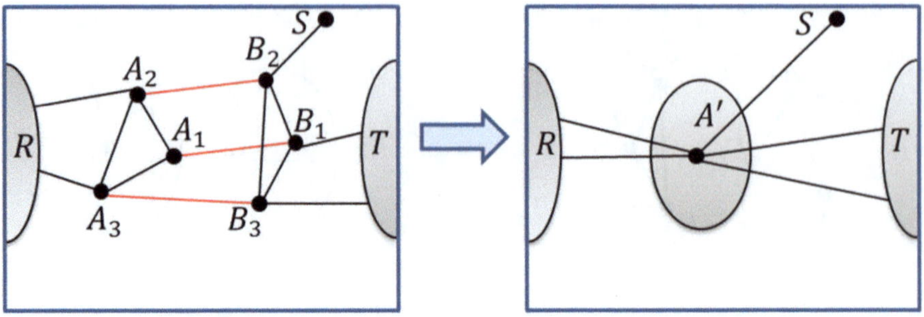

Fig. 2.3 Node aggregation process (*left* detailed network, *right* aggregated network)

$$TC_d^{AB} = TC_d^{AR}$$
$$TC_d^{AB} > TC_d^{BS} \quad (2.3)$$
$$TC_d^{AB} > TC_d^{BT}$$

With the described methodology the original network data of the ENTSO-E and IPS/UPS network with an overall complexity of more than 8,000 nodes is reduced to 545 aggregated nodes, while maintaining the transmission characteristic of the interconnected network.

2.2.2 Aggregation of Transmission Lines

To meet the reduced number of nodes in the aggregated network model the connecting transmission lines have to be matched to the aggregated network configuration. The voltage level, length and number of transmission lines between the aggregated nodes is estimated according to publicly available data. Transmission capacity TC_a^{KI} between node K and I of the aggregated network is calculated for n paths in the detailed network with all sections of a path as Ψ^{KI} according to:

$$TC_a^{KI} = \min\{TC_{d,n} | TC_{d,n} \subset \Psi^{KI}\} \quad (2.4)$$

It states, that the limiting factor between two aggregated nodes is the connection with the lowest transfer capacity.

Figure 2.4 shows a simple example where node A and node D are identified as aggregated nodes. Node B and C are removed in the aggregation process due to insignificance. With $TC_d^{AB} = 2$, $TC_d^{BC} = 1$ and $TC_d^{CD} = 1$ the resulting transmission capacity for the aggregated network interconnection is limited by either TC_d^{BC}

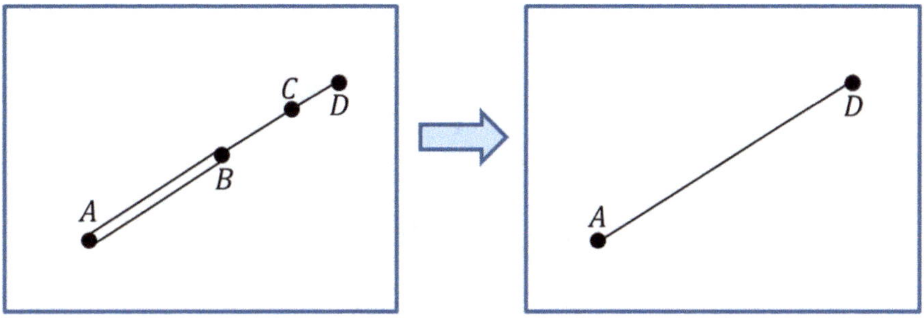

Fig. 2.4 Aggregation of transmission line path sections

or TC_d^{CD}. In this example the resulting transfer capacity for the aggregated connection is $TC_a^{AD} = 1$ limited by $TC_d^{BC} = 1$ and $TC_d^{CD} = 1$.

$$\begin{aligned} TC_d^{AB} &= 2 \\ TC_d^{BC} &= 1 \\ TC_d^{CD} &= 1 \end{aligned} \tag{2.5}$$

The aggregated path length between aggregated nodes K and I is the sum of all path sections m in the detailed network.

$$l_a^{\prime KI} = \sum\nolimits_{n=1}^{m} l_{d,n}^{KI} \tag{2.6}$$

In the particular example, shown in Fig. 2.4 the resulting path length is:

$$l_a^{\prime AD} = l_d^{AB} + l_d^{BC} + l_d^{CD} \tag{2.7}$$

Parallel transmission lines on the same path between aggregated nodes are not simplified and reduced to an alternate impedance. As illustrated in Fig. 2.5 every system is modeled as a single line in the aggregated network. A transmission line can be disabled independently without disabling the whole transmission path between aggregated nodes. This is essential for considering n − 1 situations.

The transmission lines in the aggregated network are modeled as typical overhead lines. In the Western European part of the network model the conductor type 253/40 Al/St is used as 2-wire bundle for the 220 kV-level and as 4-wire bundle for 380 kV-level. The corresponding specific parameters of these transmission lines are shown in Table 2.1.

In the Western European part the voltage level of 380 kV is dominant whereby a detailed model of the 220 kV-level is neglected. The electrical characteristics of the 220 kV-level are transferred to the voltage level of 380 kV. The remaining voltage layers in IPS/UPS network are modeled with similar line parameters.

The aggregated network models level of detail for the Central Europe region is summarized in Table 2.2. Interconnection lines between neighboring countries are

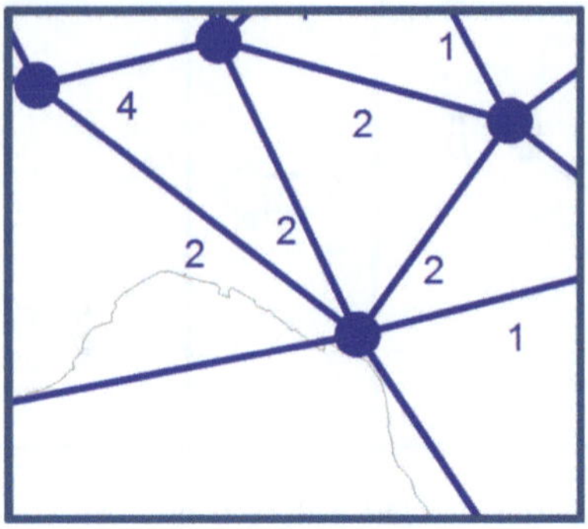

Fig. 2.5 Parallel lines on aggregated transmission system paths

Table 2.1 Exemplary aggregated line parameters for 220 and 380 kV systems [4]

Conductor configuration	TC [kA]	$R'_L\left[\frac{\Omega}{\text{km}}\right]$	$X'_L\left[\frac{\Omega}{\text{km}}\right]$	$C'_L\left[\frac{\mu S}{\text{km}}\right]$
220 kV overhead line (2 × 240/40 Al/St)	1.28	0.08	0.32	3.8
380 kV overhead line (4 × 240/40 Al/St)	2.56	0.03	0.26	4.3

Table 2.2 Level of detail of the Central Europe Region

Country	Number of aggregated nodes	Number of aggregated lines
Austria	8	44
Belgium	4	15
Czech Republic	7	22
Denmark	5	17
France	13	79
Germany	31	159
Poland	13	42
Netherlands	10	28
Switzerland	7	53

not assigned to one country. The level of detail of every country is based on the system complexity of the original network. For example, important transfer corridors as well as load and feed-in centers in the Belgium system are easier to be identified as in the German system. This results in a significant difference in terms of node and line count in the aggregated network model.

The IPS/UPS network model includes two parts: the transmission grid of Russia, Belorussia and Baltic countries and the transmission grid of the Ukraine and Moldavia.

Table 2.3 Existing interface lines

ENTSO-E RG CE		IPS/UPS		Voltage (kV)	Length (km)	Capacity (MVA)
Substation	Country	Substation	Country			
Rzeszow	Poland	Khmelnitska NPP	Ukraine	750	395	2,600
Vel'ke Kapusany	Slovakia	Mukachevo	Ukraine	400	51	900
Sajoszeged	Hungary	Mukachevo	Ukraine	400	142	692
Albertirsa	Hungary	Zakhidnoukrainska	Ukraine	750	479	2,600
Kisvarda	Hungary	Mukachevo	Ukraine	220	96	312
Tiszalok	Hungary	Mukachevo	Ukraine	220	54	310
Rosiori	Romania	Mukachevo	Ukraine	400	115	831
Isaccea	Romania	Pivdennoukrainska	Ukraine	750	409	2,600
Isaccea	Romania	Vulkaneshty	Moldova	400	59.7	955

For the transmission grid of Russia, Belorussia and Baltic countries the node count is not directly associated with the real network complexity. A diversity of voltage layers was taken into account and modeled as separate network nodes for every transmission system station.

The transmission grid of Ukraine and Moldavia is modeled in detail, as these countries include the interconnection between the European grid and IPS/UPS. These interconnection lines have been modeled based on the data of the existing but not operational lines between the systems. A list of these lines is presented in Table 2.3. These transmission lines were operated as an integrated part of the IPS/UPS and power system "Mir" until 1995 when Poland, Hungary, Slovakia and Czech Republic were synchronously interconnected to UCTE.

2.2.3 Allocation of Load and Generation

Every node of the aggregated network model is equipped with a load and a generation cluster representing the total load and generation of the aggregated area.

It is assumed, that the load is proportional to the population. To determine the load at the aggregated network nodes, statistical data on the population of districts from the European statistic database Eurostat [6] is analyzed and the population is added to the nearest node of the aggregated network model. Finally all data of one country is normalized to the total population of the country. The result is a nodal load distribution factor for every country, which can be scaled by the country specific residual load for every time frame determined by public available load statistics [7].

Figure 2.6 shows the exemplary allocation of seven communities to three nodes of the aggregated network model. The geographic distance between area centers and the aggregated nodes is determined for the allocation to the nearest node and

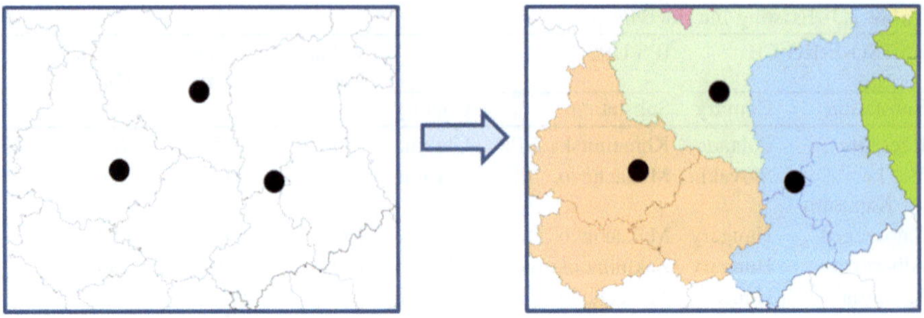

Fig. 2.6 Allocation of population per district

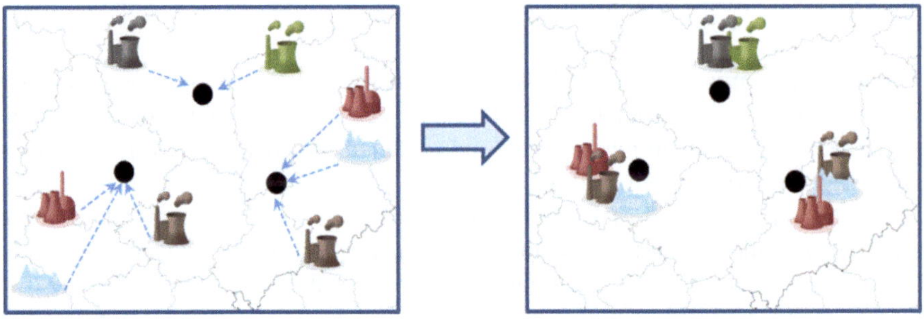

Fig. 2.7 Geographical allocation of power stations

the population is summed up. If an area contains several network nodes the load is distributed pro rata.

Analog to the load allocation all power generating units, taken from the open database [8] were assigned to the geographically nearest node in the corresponding country. Figure 2.7 illustrates the allocation process for three exemplary nodes of the network. In contrast to load allocation the individuality of every power station is maintained to enable detailed dynamic simulations in the network model. All unit specific information, fuel type, year of constriction and block size are kept for the unit commitment in the market model.

2.2.4 Unit Commitment

Input data for each market scenario is the load, non-controllable generation of each country, exported or imported power and the detailed power station data (location, fuel type, year of construction) [8]. To reduce the computation burden only power stations with an installed capacity of more than 50 MW are taken into consideration. The result of the market simulation is the unit commitment of the power

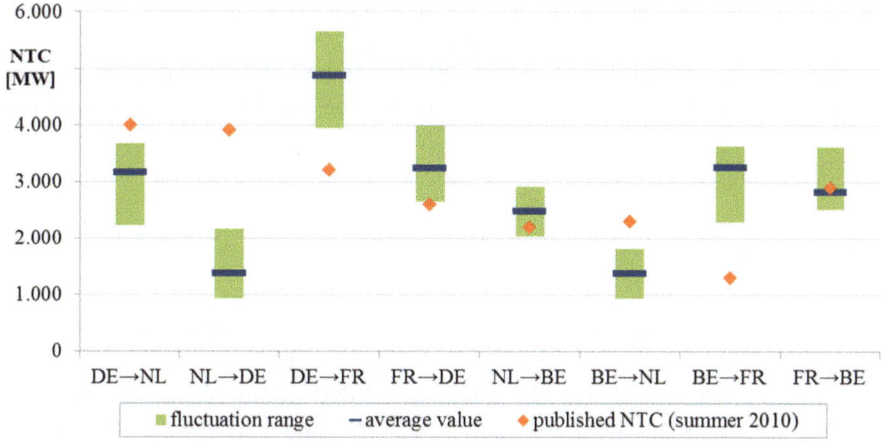

Fig. 2.8 Validation of NTC capacities of the aggregated network model

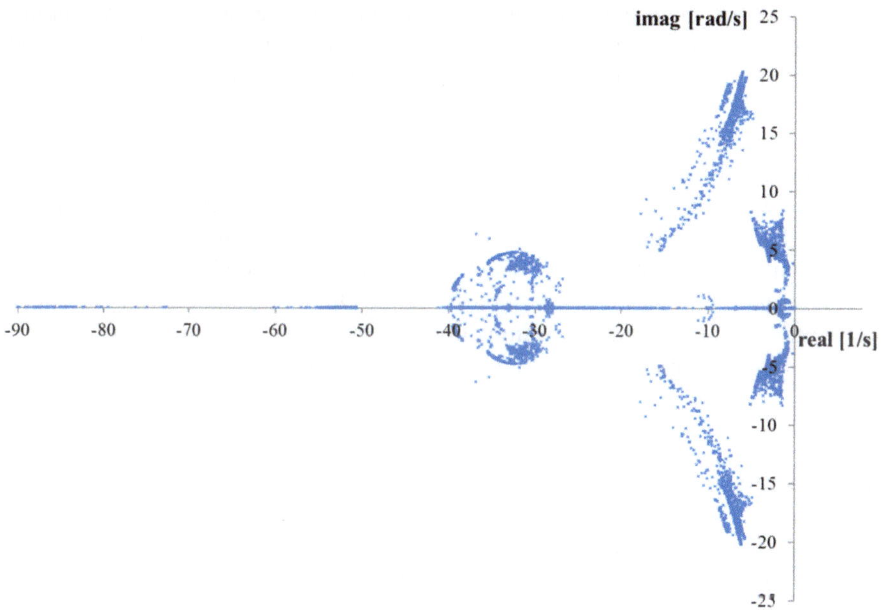

Fig. 2.9 Modal analysis results for one dynamic model

each power station and no typically parameter sets are available. As approximation to the real system calibration, the instable nodes are identified and the parameters of the involved power stations are modified to reduce the instability. With the new parameter set a new modal analysis is carried out to verify the modification. The result of this iterative process is a large scaled dynamic and stable network model. The modes of the final modal analysis for one dynamic scenario are shown in

Fig. 2.9. It can be concluded, that the dynamic model is stable but tends to oscillate more than the real transmission system.

Additionally, typical disturbances in the interconnected network are simulated and the system shows reasonable reaction on line trippings, shortcuts and unit outages.

2.4 Conclusions

The presented network model provides a first realisation of an interconnected European and Russian static and dynamic network model with a market based unit commitment. The model has 168 nodes and 1,860 power plants in continental Europe, 368 nodes and 499 power plants in the Russian part and is based only on public available data (The aggregated network model of the CWE-region can be download from http://www.icoeur.eu). The motivation for developing the reduced model is that detailed data on the network topology and the used equipment are confidential. Furthermore, the aggregation approach reduces the computation burden of the unit commitment and dynamic simulations in such large scale transmission systems and it allows for publishing results, including the network data.

Validation of the network model has shown, that the static and dynamic characteristics of the aggregated network model are comparable to the real network. Network extensions and system instabilities can be related to geographic regions. To transfer the simulation results to the real transmission network, more detailed input data is necessary, which is not publicly available.

Additionally, the aggregation process reduces the computational burden of the unit commitment and dynamic simulations dramatically that causes of instability or necessary network extensions are more likely to be identified.

References

1. D. Waniek, *Lastflussbasierte Bewertung von Engpässen im elektrischen Energieübertragungsnetz*, Ph.D. dissertation, TU Dortmund University, 2010
2. ENTSO-E, *Interconnected network of ENTSO-E*, July 2011 [Online] available: http://www.entsoe.eu
3. IPS/UPS, *Elektroenergetika Rossii I Sopredel'nyh Gosudarstv (2009)*, Associated Cartographic Center [Online] available: http://www.carta.ru
4. ABB, *ABB Switchgear Manual*, 11th revised edn. (Cornelsen Verlag Scriptor GmbH & Co. KG, Berlin, 2006)
5. P. Kundur, *Power System Stability and Control* (McGraw-Hill Inc., U.S.A, 1994)
6. EUROSTAT, *Bevölkerung nach NUTS-2-Regionen*, July 2010, [Online] available: http://epp.eurostat.ec.europa.eu
7. ENTSO-E, *Statistical Yearbook*, July 2011, [Online] available: http://www.entsoe.eu

8. VGE, *Jahrbuch der Europäischen Energie- und Rohstoffwirtschaft 2010*, (VGE-Verlag, Essen, 2010)
9. ENTSO-E, *Annual Report*, July 2011 [Online] available: http://www.entsoe.eu
10. IEEE, *Recommended Practice for Excitation System Models for Power System Stability Studies*, IEEE Standard 421.5-1992
11. ENTSO-E, *Indicative values for Net Transfer Capacities (NTC) in Europe*, July 2010 [Online] available: http://www.entsoe.eu
12. SO-UPS, *The report on the functioning of UPS of Russia in 2010*, Dec 2010 [Online] available: http://www.so-ups.ru
13. BMdJ, *Energieleitungsausbaugesetz—EnLAG*, Berlin, August 2009
14. DENA, *Netzstudie 1 -Energiewirtschaftliche Planung für die Netzintegration von Windenergie in Deutschland an Land und Offshore bis zum Jahr 2030*, (2005)

Chapter 3
Power System Dynamic Phenomena

Enrico Gaglioti and Adriano Iaria

3.1 Introduction

To prepare the reader for the following parts of the book on monitoring, control and protection of interconnected power systems, this chapter provides a general overview of the power system dynamic phenomena: the purpose is to provide general concepts of the power system stability focusing on interconnection between bulk power systems and in particular for large interconnected systems such as the perspective one consequent to the possible interconnection between the Continental Europe "CE" synchronous system, including Turkey, and the IPS/UPS system of the Russian Confederation.

After a classification of various types of phenomena mainly based on their typical time ranges, the main concerns are highlighted. With the aim of optimally focus the model, an analysis of the relevant parameters is presented. Finally various assessment methods, depending on the phenomena type, and the corresponding countermeasures are described.

The huge dimension of the investigated power system stresses the phenomenon of interarea power oscillations, characterized by extremely low frequency and involving new problems: therefore the overview is mainly focused on this type of phenomena.

The following subparagraphs present a general description of the power system dynamic phenomena including classification, main concerns for large interconnected systems, relevant parameters, methods for the phenomena assessment and the corresponding countermeasures.

E. Gaglioti (✉) · A. Iaria
RSE S.p.A, Milano, Italy
e-mail: Enrico.Gaglioti@rse-web.it

A. Iaria
e-mail: Adriano.Iaria@rse-web.it

U. Häger et al. (eds.), *Monitoring, Control and Protection of Interconnected Power Systems*, Power Systems, DOI: 10.1007/978-3-642-53848-3_3,

3.2 Classification

Power systems are very complex and are characterized by dynamic phenomena spanning over a large time range, starting from milliseconds [electromagnetic transients mainly driven by resistive/inductive/capacitive (RLC) characteristics of lines and transformers and by subtransient inductance of synchronous machines], over seconds (electromechanical transients mainly driven by inertia characteristics of synchronous machines, automatic voltage regulators, governors and primary frequency regulation), until tenths or hundreds of seconds [long term dynamics driven by Automatic Generation Control (AGC), boiler response, secondary voltage control, on load-tap changers, over and under excitation limits of generators].

Electromagnetic transients, ranging from tenths of microseconds to tenths of milliseconds, impact mainly on the sizing and the design of protective electrical devices like circuit breakers, surge arresters and so on: in fact the main considered phenomena are the lightning overvoltages, switching overvoltages and transient recovery voltages. As regards the power system as a whole complex system to be monitored and controlled, the main concerns derive by the medium-term dynamics (electromechanical transients) and by the long-term dynamics: this is why in the following the focus will be on medium and long term dynamics.

The power system stability is in general the ability of the power system to maintain an equilibrium state overcoming possible transient disturbances that can be large or small. Stability may concern the rotor angle of the synchronous machines (angle stability) or the network voltages (voltage stability). The first ones are driven by the generators while the second ones are driven by the loads.

Large disturbances such as short-circuits, loss of transmission lines or auto-transformers, loss of generation or load, may jeopardize the system operation, leading to loss of synchronism between generator (angle instability), power unbalance with under or over frequency transients (with consequent automatic or manual load shedding or generation tripping) as well as overloading of lines or transformers with possible tripping and voltage instability. The transient stability is in particular the ability of a power system to maintain synchronism between its areas when subjected to a large disturbance of particular severity. The voltage stability is the ability of a power system to maintain acceptable voltages at all buses in the system under normal operating conditions and overcoming possible disturbances, load variations and changes in the system condition.

Small disturbances occur frequently in the system because of unbalances in load/generation and do not generally cause problems unless power system is particularly weak (structurally or for contingency) or stressed by the power flows. The small-signal stability is the ability of a power system to face small disturbances without the loss of synchronism among the machines and without persistency of electromechanical oscillations. Also the voltage may be unstable for small disturbances: this may happen in a weakened system owing to contingency and/or to increased flows in the system.

3.3 Main Concerns for Large Interconnected Power Systems

Large alternating-current AC systems, such as the possible interconnection between the Continental Europe "CE" synchronous system, including Turkey, and the IPS/UPS system of the Russian Confederation, could provide multiple advantages in terms of operational security. Reasons are diversification type and location of energy sources, mutual assistance between subsystems and integration of renewable energy as well as energy trading. On the other hand increasing dimensions may cause technical problems such as transient and small-signal stability, voltage stability, especially in case of long distance transmission and synchronous interconnections.

Inter area power oscillations, related to small-signal stability, are one of the main undesired phenomena: they may be critical (poorly damped or undamped) on weak/long distance systems and/or highly exploited systems. Oscillations could be triggered also by small disturbances like normal variations of load and/or generation and mainly consist of fluctuation of power flow on critical corridors with possible serious consequences like system separation in islanded areas, load shedding, up to the possibility of a black-out. These phenomena are emphasized in large interconnected power systems, in particular in case of a high utilization of critical corridors also due to market liberalization.

Conventional thermal units may improve damping, while hydro ones worsen it, because of penstock water dynamics. For hydro units, the penstock water time constant, proportional to generated power, and the transient frequency droop (normally greater than permanent one in order to mitigate the worsening due to hydro response) are the sensitive parameters; dead-bands in terms of frequency deviations or, in extreme situations, blocking off the governor may provide some relief.

In the past, when the dimensions of synchronously interconnected power systems were smaller, power oscillations were characterized by frequencies generally higher than 0.6–0.7 Hz; governors, whose dynamic is slower, did not act on the phenomenon. When dealing with modern interconnected system of increasing size, governor dynamics interact with power oscillations: conventional thermal units may improve damping, while hydro ones worsen it, because of penstock water dynamics, especially if equipped with not properly tuned regulators. Dead-bands in terms of frequency deviations or, in extreme situations, blocking off the governor may provide some relief.

Another concern for very large systems arises from the large number of synchronous machines participating to the interarea oscillations, especially for "global oscillation modes" involving large peripheral areas. Conventional countermeasure consisting in Power System stabilizers (PSS) applied to generator voltage regulators become less effective since the total effectiveness is "shared" by a large number of units: in order to get a significant performance improvement too many generators would have to be equipped with PSS. In this case an additional and more effective improvement may be obtained by equipping some transmission devices like High Voltage Direct Current links (HVDC) and Flexible Alternating Current Transmission System devices (FACTS) with properly designed PSS, especially if located along critical highly exploited transmission corridors.

Transient instability leads to loss of synchronism of synchronous machines after a large disturbance, normally at first swing; it generally involves only one machine, which goes out of step respect to the remainder of the system, but it may also interest several generators, leading to loss of synchronism between areas.

Disturbances causing transient instability can have different degrees of severity; usually the system is designed to face a defined set of contingencies such as short circuits. In this case the fault has to be cleared by a well-timed opening of circuit breakers. In fact transient stability studies are aimed to assess the critical clearing time; when too small critical clearing times are needed, the main concern is to assess the maximum transferable active power along a critical corridor preserving stability with reasonable clearing times, leading to restriction of active power generation dispatching. The utilization of (very fast) special protection schemes (SPS), like the partial rejection of generation following the disturbance, may help limiting the above explained dispatching constraints.

Load voltage collapse consists of a low-voltage profile in a load widespread area supplied by a large system; it's the final consequence of a sequence of local voltage instabilities. Many factors contribute to the process of system collapse because of voltage instability: weakness of the transmission system, high power-transfer, low power factor load characteristics, generator reactive power capability limits and low availability of reactive power compensating devices. Prevention of voltage collapse can in general be achieved by means of reactive power compensators or by a secondary voltage regulation strategy, which monitors and controls the voltage of some "significant" busses of the system acting on generators reactive power production. Another option is blocking the transformer tap-changers close to a low voltage area.

3.4 Relevant Parameters for Transient and Small Signal Stability

The reliability of power system studies results will depend on the accuracy of the system model and data. In case of very large power systems the amount of data is huge and therefore a priority list is advisable. Depending on the power system analysis type, a certain model is required and, inside this model, a subset of parameters may be particularly relevant for the accuracy of the results. Dealing with different types of power system analysis it is therefore a good practice to focus the attention on certain parameter values. In order to state the main important parameters it is expedient to highlight the main physical aspects of the phenomena.

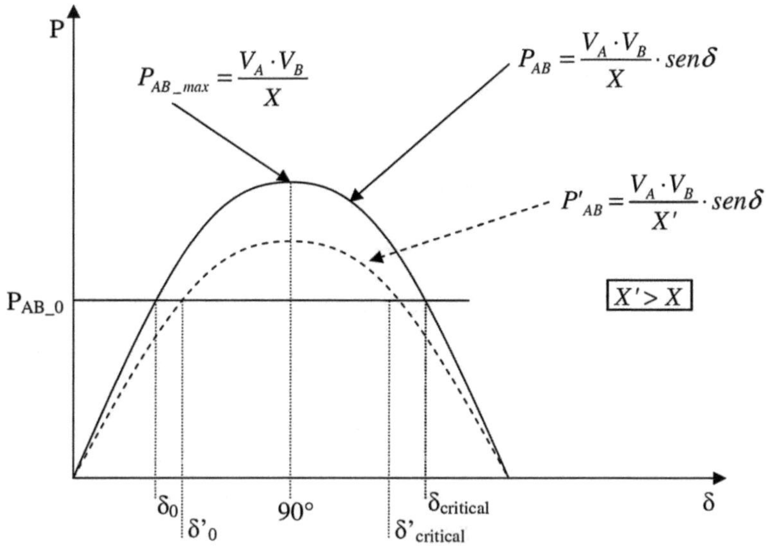

Fig. 3.1 Power transfer characteristics with two different link reactances

As well known, the active power transferred from a bus A to a bus B is:

$$P_{AB} = \frac{V_A \cdot V_B}{X} \cdot sen\,\delta \tag{3.1}$$

with

X	Reactance of the electrical corridor between the buses A and B
V_A, V_B	Modules of the voltage phasors in the two buses
δ	Angle between the two voltage phasors.

The higher the reactance of the considered link, the lower is the maximum transmittable power. The following Fig. 3.1 depicts the comparison between the two transfer power characteristics; the stability margin is defined by the area between the transmittable power and the level of actually transmitted power P_{AB_0} at the equilibrium starting point "0". The lower the reactance, the lower is the angle (power flow) and the higher the voltages, the higher is the stability margin.

For example, as regards the power flow, a higher value of transmitted power P'_{AB_0} at the starting point, involving a higher angle, reduces the stability margin as shown in Fig. 3.2 where it's possible to note the reduced stability area (dark grey corresponding to P'_{AB}).

Large dimensions of interconnected power systems, nowadays growing at high rate, means high link reactances X between the peripheral system areas. Due also to the even more stressed utilization of critical corridors, the stability margin is reduced concerning the transmittable power. These two aspects play a key role for the occurrence of inter-area oscillations. A particular attention must be paid to the

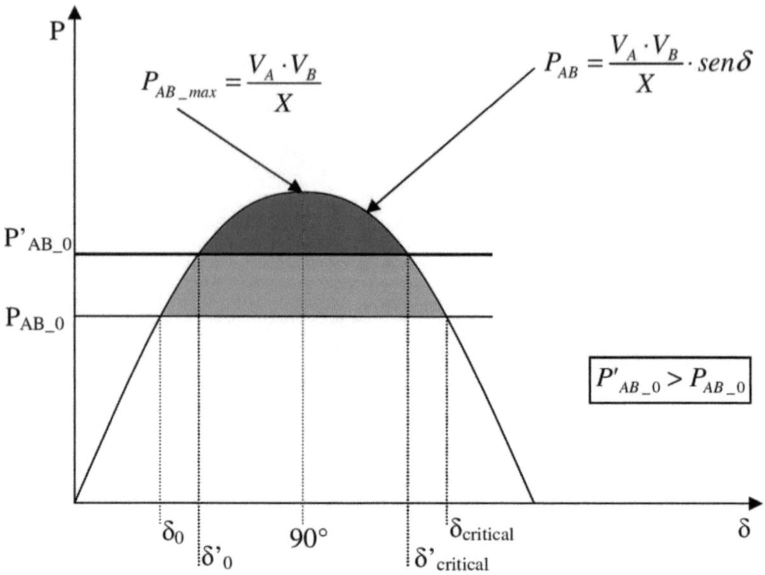

Fig. 3.2 Power transfer characteristics with two different level of power transfer

system configuration (transmission links out of service) and to *active power dispatching* since instability in the system may occur depending on the above mentioned factors.

Since the generators terminal voltages depend on internal excitation voltages which in turn depend on the *generators* reactances, a particular attention must be paid on the above described parameters and the excitation system realized by the *Automatic Voltage Regulators* (AVRs). Among the synchronous machine reactances, the most relevant ones impacting the frequency and the damping of electromechanical oscillations, are the direct axis transient reactance (X'_d) and the subtransient reactances (X''_d and X''_q, normally very similar).

With particular reference to small signal stability *governors* must be modeled: in order to simulate the effects of governors and prime movers on electromechanical oscillations simple linearized governor models are sufficient; one or two state variables allow a good linearized model. For hydro units, the penstock water time constant, which is proportional to the water flow rate and then to the generated power, and the transient frequency droop (normally larger than the permanent one in order to mitigate the worsening due to hydro response) with its time constant are the sensitive parameters. For conventional thermal units, the percentage of high-pressure turbine stage power with the steam-chest time constant, and the transient frequency droop (normally equal to the permanent one) have to be well stated. For simple/combined cycle turbogas, the physical process dynamics is very fast, so only frequency droop is required; the normally positive impact may be reduced by a gas valve opening rate limiter, depending on the oscillation amplitude.

3.5 Assessment Methods

Depending on the kind of problem to be analysed, different assessment methods have been developed in the past. Power system dynamic behaviour is inherently non-linear, so requiring a non-linear model, but in case of small disturbances analysis the linearized behaviour around the initial steady-state operating point can be considered.

A non-linear approach has to be carried out for transient and large-disturbance voltage stability analyses. Many types of approaches have been utilized to study the transient stability of a power system, but the most important and common approach is the time-domain simulation. Other types of approaches could be used (Lyapunov's or direct method based, for example, on the energy function), that don't need an extensive time domain simulation, but normally require a drastic approximation and reduction in the studied power system, so that the relevant evaluation of the stability might be not accurate enough. On the other side, time domain simulations based on a deterministic approach (a restricted number of reasonable starting points, fixed perturbations) has the following disadvantages:

- it gives an evaluation of power system stability under specific conditions and disturbances, but does not give any information about the stability degree, i.e. stability margin;
- its effectiveness is very dependent on experience of people who decide the system configuration and the disturbances.

Dynamic time domain simulations are needed lasting over a period of few seconds (for transient stability) to some minutes for voltage stability simulating dynamics of on-load transformer tap-changers, under and over excitation controls of generators and loads. For slow dynamics capturing voltage-stability problems, simplified dynamic models are advisable, allowing greater integration time steps. Time domain simulation can be performed by explicit integration methods (separated state variable integration and algebraic system solution) and implicit ones (unique and contemporary solution of a differential algebraic system including state variables); the explicit approach requires a prefixed time step while the implicit one may automatically recognize the passing from fast dynamics to slow ones, adapting the time step.

Any method different from the time domain simulation is too much approximated; moreover it must be noted that, owing to several switching actions during a large perturbation, the studied system changes many times in its configuration making the evaluation of system stability more difficult.

A linear approach is used for small-signal stability analysis allowing to asses inter-area oscillations by means of the system dynamic model linearization around an operating point and the successive eigenvalues calculation. This methodology yields complete information about the oscillation modes in terms of frequency, damping, observability, controllability and consequent control design. Time domain simulations are then recommended to validate the accuracy of the modal analysis results.

A modal analysis tool, suitable for the power system small signal analysis, realizes a linearization of the algebraic-differential equations describing the power system oscillation, a set-up of the corresponding reduced state matrix and the calculation of eigenvalues (frequency and damping) and eigenvectors (mode's observability, controllability, participation factors and sensitivities). If the tool has to be applied to very large interconnected systems, it has to include, besides the classical components like synchronous machines and passive network components (lines/transformers), the models of modern devices for the interconnection, like FACTS and HVDC. Since these last ones may contribute to improve the small signal performance, also their PSS control has to be considered.

A Thyristor Controlled Series Capacitor (TCSC) model must simulate a regulated series reactance for keeping constant the active power flow. A Static Var Compensator (SVC) model must simulate a modulated shunt susceptance in order to keep the bus voltage quasi constant. The thyristor firing angle control for the modulation of the TCSC reactance and of the SVC susceptance is considered to be instantaneous since it is very fast compared to power oscillations. Normally a constant gain is enough to simulate this control.

Regarding HVDC links, dynamic transients are generated by the variation of firing and extinction angles following changes of power order and/or AC voltages. The control of the firing and extinction angles are considered instantaneous since they are very fast, like the dynamic response of resistive/inductive/capacitive DC lines, therefore they can be considered as steady state. On the other hand the converter transformer ratios, which are changed in order to maintain the angles at their nominal values, are considered constant since they are very slow. As a consequence the classical thyristor based HVDC model is substantially represented by algebraic relations expressing the active and reactive powers at the converter stations as a function of the converters voltage at the AC side [1]. The power order is generally considered constant (subject to the operator): it may be modulated by the frequency control and by the PSS, if any. Frequency control is suitable only for weak AC systems asynchronously connected via HVDC to a strong one, like the connection between Sardinia and the Italian Continent. Modeling of HVDC based on Voltage Source Converter VSC differs from the classical thyristor based HVDC for the dependence on the AC voltage of the reactive power: classical HVDC implies a very low dependence, while VSC HVDC, being able to control AC voltage by means of reactive power Q at the converters, implies a Q strong dependence on the AC voltage if the control strategy implement this opportunity. It must however be noted that the dependence on the AC voltage of Q has not much effect on power oscillations, being the active power the main factor.

3.6 Improvement Methods

Remedial actions to be taken in order to face the undesired dynamic phenomena can be classified in *preventive* and *corrective/curative*. Preventive actions are taken in advance whereas corrective actions are implemented in real-time

Table 3.1 Preventive and corrective actions in case of small signal instability

Dynamic phenomenon	Corrective actions	Preventive actions
Small signal instability (inter-area electromechanical oscillations)	When electromechanical oscillations appear: PSS intervention (neglecting dead bands); governors blocking, especially of not properly tuned controllers of hydro unit generators	Implementation of additional PSS on generators (traditional solution)
	Only in case of large excursion of generators rotor angles between incoherent areas of oscillation: Areas separation by line/s tripping, eventually followed by power balancing (generation or load shedding)	Adaptation to damping preservation criteria of hydro power plant speed governors having detrimental impact (setting of a transient frequency droop higher then the permanent one, frequency dead-band)
		Use of suitable FACTS devices (e.g. SVC, TCSC) and HVDC links with additional stabilizing PSS controls designed with the help of specialised tools for critical oscillations assessment
		Decreasing of power flow on critical corridors (uneconomic because of market constraints)
		Strengthening of the extra high voltage EHV system by new links, reducing link reactances (expensive, environmental impact)

when a certain undesired event or dynamic phenomenon occurs. Preventive and corrective actions vary according to the nature of the occurring dynamic phenomenon; in the following (Tables 3.1, 3.2, 3.3), the actions are summarized on the basis of the instability type that has to be faced.

Analyzing more in detail **interarea oscillations**, they could be hampered by decreasing of the power flow on critical corridors and strengthening of transmission system by new links. However, the first solution is uneconomic and against the market constraints while the second one is very challenging and expensive. The best solution, in terms of improvement of the oscillation behaviour with low related costs, is the implementation of suitable damping controls PSS on FACTS devices and HVDC links, in addition to conventional PSS applied on automatic voltage regulators of generators. HVDC links have been introduced in the second half of the 20th century and became a reliable technology for long distance transmission. FACTS devices have been introduced to improve the system's performance; in particular, several operating experiences are available world-wide concerning two nowadays reliable FACTS devices: TCSC devoted to active power control and SVC dedicated to the voltage control.

Table 3.2 Preventive and corrective actions in case of transient instability

Dynamic phenomenon	Corrective actions	Preventive actions
Transient instability	Adoption of circuit breakers clearing times less than critical values	Use of Automatic Voltage Regulator (AVR) with high speed first response and very high excitation voltage ceilings in order to limit terminal voltage drop during first swing (decisive for stability maintenance)
	Intervention of special protection schemes (SPS), if present (i.e. partial rejection of generation following the disturbance in order to preserving the remaining generation)	When corrective actions are not sufficient (needed clearing times smaller than technological lower limits), preventive restriction of active power generation dispatching by assessment of the maximum transferable active power along a critical corridor preserving stability with feasible clearing times

Table 3.3 Preventive and corrective actions in case of voltage instability

Dynamic phenomenon	Corrective actions	Preventive actions
Voltage instability (collapse)	Blocking off the transformers tap-changers close to the low voltage area when the voltage sags	Implementation of reactive power compensators with adequate control range
	Load shedding (as last option)	Implementation of secondary voltage regulation strategy on generators, monitoring and controlling the voltage of some pilot nodes of the system

Controllability of inter-area modes is not straightforward; it depends on the PSS location, types of generator exciters, as well as location and characteristic of the loads. Especially the damping device location plays a key role on the ability of this device to stabilize a swing mode. While a generator, equipped by a PSS, can have a local damping effect (many generators with PSS are needed to damp widespread interarea modes), HVDC and FACTS devices, if sited in the middle of transmission system, may have a high observability of the oscillations and therefore, if equipped by properly designed PSS, can contribute to stabilise inter-area modes.

In large interconnected systems, the most effective stabilization of poorly damped inter-area modes by HVDC and FACTS requires a proper assessment of their location and their control design. As concerns the location, it has to be underlined that, while TCSC and SVC devices may be installed (and designed) specifically to improve the power system stability, HVDC transmission links are sited where new or additional power transfer capability is needed (in accordance with positive cost-benefit assessment) and not for network stabilization.

The analysis of low-frequency oscillations and the development of proper countermeasures by PSS design on HVDC and FACTS can be done by means of small-signal stability analysis (modal analysis for the eigenvalues calculation). Modal analysis has to be carried out on a detailed dynamic grid model in order to accurately represent the power system dynamic behavior.

Choosing local input signals for PSSs avoids the use of remote information with consequent additional costs and reliability issues. Being FACTS devices typically located far from the generating synchronous machines and close to the middle of the oscillations, the active power flow is preferred as input to PSSs controls since it has a higher modal observability than the frequency. For HVDC links between synchronously interconnected areas, the power flow, kept constant by DC control, loses the observability; the remaining available local signal (frequency) as PSS input, is poorly effective. Only the use of remote information, like frequency close to the generators or power flows on parallel corridors, may improve HVDC damping performance, with consequent additional costs and reliability issues. An accurate and reliable Wide Area Measurement Systems (WAMS) should allow for obtaining the maximum enhancement of the oscillatory system behaviour by HVDC's PSSs. An example of the improvement that can be reached by a HVDC link with PSS fed by reliable remote data from another critical corridor is shown in [2].

In case of DC interconnection between asynchronous areas, the power oscillations are substantially confined inside each subsystem and the situation is equivalent to the separated systems. It must be noted that this type of solution can supply a full control on power exchange between the different subsystems and does not require keeping a transmission capacity margin to face transients following a generation tripping. The HVDC PSSs based on appropriate remote signals, like a properly chosen power flow, can significantly improve the damping of oscillations inside the subsystems [2].

Dealing with very large interconnected systems, turbines and speed governors dynamics interact with power oscillations. Conventional thermal units may improve damping because of their relatively fast response of mechanical power produced in the high-pressure turbine stage, while hydro ones worsen it, because of penstock water dynamics, as explained in the following.

The detrimental effect of the hydro power plants is mainly caused by the negative first response in terms of mechanical power to a valve opening: the prevalent instantaneous effect is an energy reduction at the penstock gate, because of the penstock water dynamics, while the water flow increases with a certain delay related to the water starting time constant T_W of the hydraulic prime mover. This effect is more evident for Francis turbines because of higher water flow rates and lower hydraulic heads with respect to Pelton turbines.

The starting time T_W of the hydraulic prime mover is [3]:

$$T_W = \frac{Q \times L}{g \times H \times S} \tag{3.2}$$

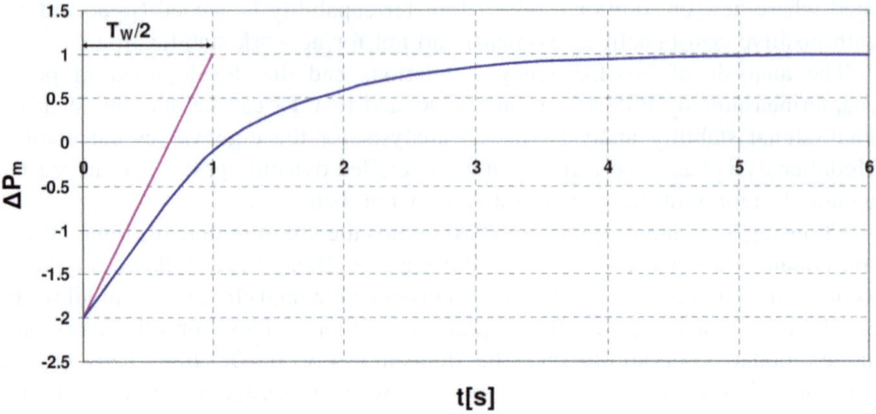

Fig. 3.3 Time response with negative zero due to water dynamics (typical case with Francis turbine and $T_W = 2$ s)

where:

Q (m^3/s)	Is the water flow rate
L (m)	Length of penstock
G (m/s^2)	Gravity acceleration
H (m)	Hydraulic head at the penstock gate
S (m^2)	Penstock cross section

The Fig. 3.3 shows the negative response, in terms of mechanical power P_m, of a hydraulic prime mover system with a Francis turbine and a starting time of $T_W = 2$ s, in case of an unitary step in the valve opening. The transfer function is:

$$\frac{\Delta P_m}{\Delta Y} = \frac{1 - 2s}{1 + s} \tag{3.3}$$

For example a recent analysis [4], performed on the Continental European interconnected system, emphasized the need for the adaptation of hydro power plant controllers in Turkey (most of them concerning Francis turbines). This countermeasure is mainly indicated to face the possible occurrence of a slow inter-area mode involving Turkey-Southern Balkans-Romania against the Iberian Peninsula. This East–West mode was also highlighted in the small signal stability analysis carried out in the recent ICOEUR project [5]; the calculated modal frequency is 0.17 Hz and the geographical shape is depicted in the following Fig. 3.4.

In order to improve oscillations damping, one extreme solution can be the governor blocking off in case of very slow oscillations (frequency less than 0.3 Hz). The main drawback of blocking off is the loss of requirements for frequency primary control; a less drastic solution mitigating the worsening due to hydro response may be the adoption of a transient frequency droop higher than the permanent one (in other words a transient gain lower than the permanent one).

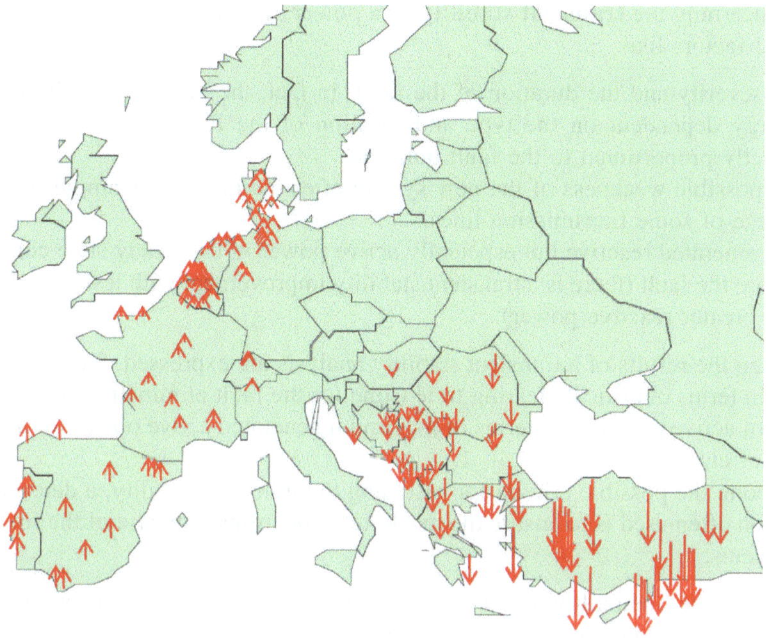

Fig. 3.4 Geographical shape of the East–West mode in CE-Turkey interconnected system (*source* [5])

Also dead-bands in terms of frequency deviations may provide some relief. For hydro units, the penstock water time constant, proportional to generated power, and the transient frequency droop are the sensitive parameters.

With the aim to get an idea about the strong influence on damping of the PSS and of the turbines and the speed governors dynamics, some extreme modeling hypothesis were investigated within the ICOEUR project. The modal analysis carried out on the merged systems CE-IPS/UPS, in case of synchronous interconnection, highlighted the following results concerning some modes spanning across the interconnection and having frequencies in the range 0.1–0.6 Hz.

1. All units modeled as conventional thermal with 5 % permanent droop involve a damping increase by 20–30 % with respect to constant mechanical power (governor blocking off).
2. All units modeled as hydro with 25 % transient droop and 2 s water time constant involve a damping decrease by 5–15 % with respect to constant mechanical power.
3. All units equipped by PSS with standard parameters historically utilized by the Italian utility ENEL (control gains of rotor speed $K_{\Omega} = 5$ and of active power $K_P = 0.3$) involve a damping increase by 10–25 % with respect to no PSS.

Concerning the **transient stability** of a power system, it may be influenced by several factors like:

- the severity and the duration of the fault; in fact, the generators gain a kinetic energy dependent on the type and location of the fault (severity) and about directly proportional to the fault duration.
- the possible weakness of the new system after the fault for example in case of outage of some transmission lines.
- the generated reactive but especially active power in the steady state conditions before the fault (there is a transient stability improvement with less active power and greater reactive power).

Often the results of a transient stability analysis are expressed, for each type of fault, in terms of critical clearing time (time for the fault elimination) for a certain value of active power or in terms of maximum generated active power for a certain value of clearing time.

Among the possible criteria for assessing the transient stability, a deterministic criterion often used is to ensure the machines synchronism in case of the following two events:

- three-phase line fault with unsuccessful automatic reclosing with sound system.
- single-phase line fault with successful automatic reclosing with the system already in first contingency.

Significant improvements of power systems transient stability performances can be achieved through the following means:

- High-speed fault-clearing by means of properly tuned line protections and highly performing circuit breakers.
- Use of Automatic Voltage Regulator (AVR) with high initial-response exciters and very high excitation voltage ceilings in order to limit terminal voltage drop during first swing (decisive for stability maintenance). Use of power system stabilizers on AVR controls is needed in order to deal the negative damping introduced by higher-response exciters involving local plant modes of oscillation.
- Series capacitors, especially for prevailing radial systems, in order to reduce electrical distance between the generators, so improving the synchronizing torque among the machines, even if they could arise some subsynchronous oscillations problems. Fast increase of capacitance during first swing, obtained by a thyristor control, may further improve stability.
- Special protection schemes (i.e. immediate partial rejection of generation after a disturbance close to a power plant in order to preserve the remaining generation and avoiding the loss of synchronism of the whole power plant).
- To utilize speed governors characterized by a fast response; sometimes this can lead to an out-of-step at the first swing and a resynchronizing at second or

successive swings (on this type of behavior there is not an unanimous judgment but however the synchronism is not lost).

- Furthermore a 'fast valving' device can improve first swing transient stability but can also arise some overpressure problems in reheater and moreover is harmful for not very serious perturbations. In order to avoid the overpressure a by-pass system can be used to discharge the steam before passing through the turbine, but in this case it needs some minutes to come back to the initial steam flow in the turbine.
- By providing an adequate voltage support at intermediate buses of important connection systems (for example by means of shunt reactors and capacitors, synchronous condenser, static var compensator).
- A higher generated reactive power in the pre-fault steady state condition; in fact the stability margin increases with a smaller angle between the machines that can be obtained, for a given active power, increasing the injected reactive power that implies a higher voltage level.
- If above measures are not enough, resulting a critical clearing times smaller than technological lower limits, an extreme solution can be the preventive restriction of active power generation dispatching by assessment of the maximum transferable active power along a critical corridor preserving stability with feasible clearing times.

As regards the **voltage stability**, voltage collapse may be avoided by means of the following countermeasures ranging from system protection design to real-time operation:

- properly regulated reactive power compensation having a sufficient control capability; in case of fast availability of reactive power, like fast switching capacitors and Static Var Compensators, also the short-term voltage instability related to the risk of induction motors stalling may be faced;
- a rescheduling of the reactive power injected by generators in order to maximize each contribution to the voltage support: the reactive rescheduling can be actuated by the operator or may be implemented by a secondary outer loop control of generators excitation (slowest than the primary regulation and with response time not less then 10 s in order to avoid an adverse interaction between the two loops);
- the adoption of a Under Load Tap Changing strategy based on knowledge of the distribution network characteristics (industrial or domestic prevalent load, i.e. how much the load is sensitive to the voltage);
- an adequate reduction of active and reactive load power consumption, reducing the voltage on the distribution-side by means of load tap-changers (LTCs): this reduction can be obtained by reducing LTCs set point, by bringing back LTCs to a predetermined position or by blocking them;
- as an extreme solution, the under-voltage load shedding analogous to the under-frequency ones.

References

1. R. Marconato, *Electric Power Systems*, vol. 1. Background and Basic Components (CEI edition, 2002)
2. E. Gaglioti, A. Iaria, G. Giannuzzi, A. Gubernali, D. Panasetsky, N. Voropai, Inter-area oscillations in the CE/Turkey and IPS/UPS power systems, in *CIGRÈ Symposium*, Bologna (2011)
3. Working Group on Prime Mover and Energy Supply Models for System Dynamic Performance Studies, Hydraulic turbine and turbine control models for system dynamic studies. IEEE Trans. Power Syst. **7**(1) (1992)
4. E. Grebe, J. Kabouris, S. López Barba, W. Sattinger, W. Winter, Low frequency oscillations in the interconnected system of continental Europe. Power and energy society general meeting, IEEE (2010)
5. E. Gaglioti, A. Iaria, A. Gatti, Tool for oscillatory analysis of large-scale power systems. FP7 ICOEUR research project (2011)

Part III
Monitoring of Interconnected Power Systems

Chapter 4
Monitoring Technologies

Kay Görner, Ettore Bompard, Tao Huang and Michael Kleemann

4.1 Introduction

This chapter provides an introduction to the third part of this book on monitoring of interconnected power systems, by introducing monitoring technologies. It serves as a basis for several modern monitoring applications presented in Chaps. 5, 6, 7 and 8.

Proven technology such as conventional measurement devices of current, voltage and power provide snapshots of the power system state. Some measurements are synchronized with different accuracy of timestamp. Quite often data is not complete, consistent or lost at all. The state estimation algorithm enables detection of wrong data and provides consistent system state. State Estimation results are the input for applications such as load flow calculation. System Integrity Protection Schemes (SIPS) receive measurement data from selected nodes and react according to the implemented algorithm in order to maintain secure power system state [1].

Measurement equipment and data evaluation, its algorithms together with communication networks are generally known as '***Monitoring Technology***'.

Current conventional Monitoring Technology exchanges static snapshots of measurement and status messages with a time-cycle of several seconds. However analysis of blackouts and disturbances in the past revealed the need for new

K. Görner (✉) · M. Kleemann
TU Dortmund University, Dortmund, Germany
e-mail: kay.goerner@tu-dortmund.de

M. Kleemann
e-mail: Michael.Kleemann@tu-dortmund.de

E. Bompard · T. Huang
Politecnico di Torino, Torino, Italy
e-mail: ettore.bompard@polito.it

T. Huang
e-mail: tao.huang.new@gmail.com

U. Häger et al. (eds.), *Monitoring, Control and Protection of Interconnected Power Systems*, Power Systems, DOI: 10.1007/978-3-642-53848-3_4,

technologies. Since dynamic events have not been detected as necessary for sufficient control actions, the application of new technologies such as wide area monitoring systems (WAMS) based on time synchronized measurements in phasor measurement units (PMU) would be an improvement for the detection of dynamic changes.

In general it is expected that with the rise of renewable energies and new control technologies power flows will change more and more dynamic then in the past. Similarly to these changes also computer and communication technologies with improved performance enable new applications. On the one hand just the improvement of current communication technologies may result into an improvement of power system state monitoring. On the other hand the impact of new technologies such as global satellite systems is essential.

In the range of Monitoring Technology the following aspects must be taken into account.

- Monitoring of the power system state is based on a reliable communication system across the power system. Beside data exchange from the acquisition units in substations up to control centers communication technology must also enable the information exchange between control centers. The following subchapters contain a description of ***communication principles*** in power systems and a comparison of equipment for protection, monitoring and control with regard to requirements on communication technology.
- The benefits of ***wide area monitoring systems*** (***WAMS***) based on phasor measurement units (PMUs) and phasor data concentrators (PDCs) are given in Chap. 5. This chapter deals with the concept of a Large Scale WAMS and gives an example of an installed Large Scale WAMS.
- ***State estimation*** can be enhanced by the application of PMUs. The algorithms which apply this new measurement technique must be evaluated in terms of their benefit for the monitoring of large scale power systems. In Chaps. 6 and 7 Distributed and Dynamic State Estimation are introduced. Distributed State Estimation can be applied for the coordination of data exchange between system operators which must collaborate in an interconnected power system. Dynamic State Estimation enables monitoring of dynamic changes. Finally, an evaluation of the efficiency of the State Estimation algorithms is provided in Chap. 8.

4.2 Communication Exchange Over TSO Interconnections

The communication of a TSO can be divided into internal and external communication. Internal communication refers on internal processes within the TSO. External communication is necessary for operation of interconnections with neighboring power systems. The operational handbook for system operators in continental Europe defines [2]: "TSO … participating in creation of Common Information … related to a specific Operational TSO Business has to share this Common Information or the data derived from it with other contributing … TSOs."

Because the capacities of interconnections are limited market based auctions have been established. Additionally system operators must make available their network for third party access. In order to coordinate power flows in their area system operators have to purchase energy reserves on markets.

In accordance to local and European laws, e.g. the needs for transparent pricing, rules have been established for energy markets within the European Union. In order to develop a common European market an exchange of information between European TSOs is necessary.

In the power system within the Russian Federation (Unified Power System—UPS) a centralized control is applied. According to the current practice in UPS and in other large power interconnections, system operators of each power system within the interconnection are charged with the task of planning and dispatching their power systems. Values of power flows between power systems depend on commercial contracts between market participants, located at different power systems. In power interconnections, consisting of power systems of independent countries, values of power flows between power systems can be determined by interstate agreements on energy exchanges under consideration of power systems security limitations.

The Baltic region is connected to IPS/UPS but is situated on the border between ENTSOE RG Nordic (formerly known as Nordel) and UPS. System operators from Baltic countries are furthermore members in ENTSO-E. The communication structure is similar to the rest of IPS/UPS, but also has to meet requirements on communication within ENTSO-E.

4.3 Hierarchy of Control Centers in Power Systems

Control centers are organized on a hierarchical way [3, 4] as displayed in Fig. 4.1 [5]. They can be divided in three categories, depending on the area they control. The control area is linked with the voltage level. On the first level of the hierarchy there are local control centers. Each of these centers directly manages many substations for distribution in low voltage and medium voltage grids. Then data is transferred to regional control centers, which have the resources to implement a lot of managing functions that give them full control on system elements depending on them. At the end of the hierarchy, there is a national control center, which is responsible for all the operations performed on the grid. However, the complexity of the control network led to the definition of a distributed approach, where regional control centers have almost the same functions as national control centers. The latter mainly has an overview function on the other centers and it is responsible for the communication with the foreign control centers of the neighbouring interconnected grids.

Control centers of power systems are today equipped with good data exchange systems. These systems work sufficiently well in normal conditions. In the emergency and post-emergency operation conditions the control centers need more

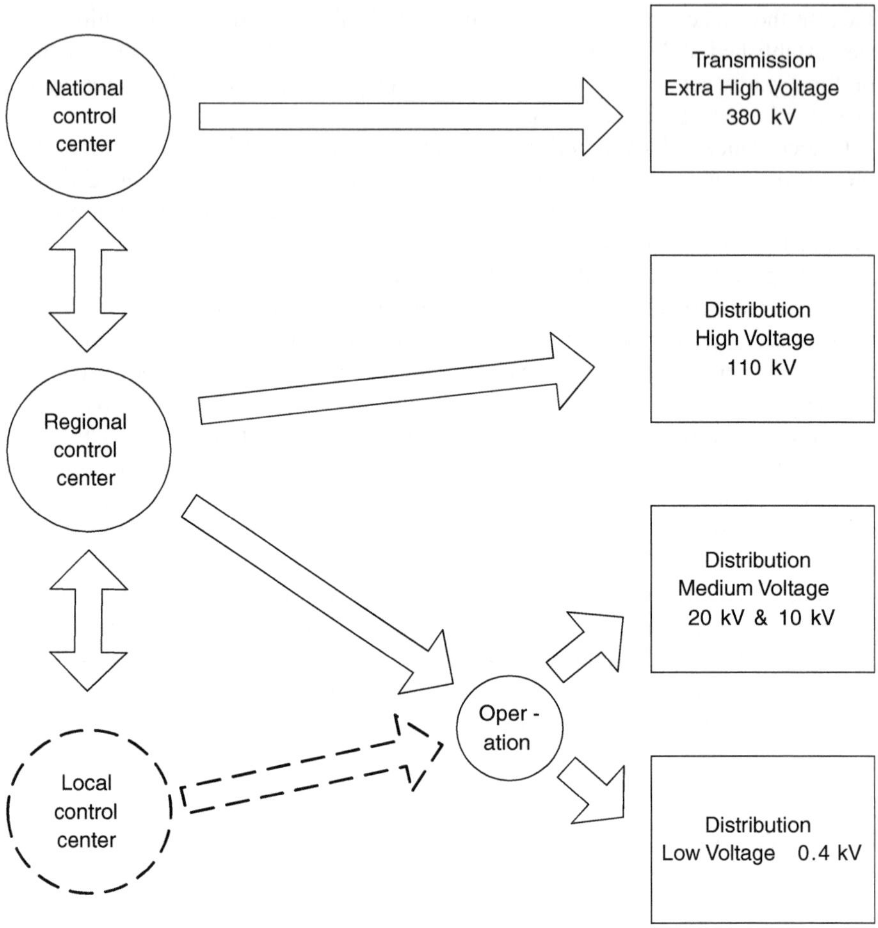

Fig. 4.1 Example for hierarchy of control centers allocated to voltage level as applied in continental Europe

information from other control centers. As a consequence the number of control centers, which interchange information, must increase.

Instead of having local control centers, low voltage grids are very often operated manually by local staff [6]. Such an approach is reasonable as long as the operated grid can be regarded as static because load curves and power feed in are known in advance or networks are planned with big reserves so that the system can be operated within its operational limits. However, because of the rise of photovoltaic systems parameters such as loading of network elements must be monitored more and more online. The approach is to enhance the functionality of transformer stations in low and medium voltage grids by adding more measurements in order to detect the state of the system [7] or of specific network elements such as transformers [8].

4.4 Data Communication Network

4.4.1 Communication Network in Control Centers

The data communication network that can be found in a control center is usually divided into different sub-networks. One sub-network in control centers and substations, each consisting of several servers and clients, is dedicated to the communication for grid control. Another sub-network is dedicated to typical operations that are performed on an office network which is part of the corporate Intranet. This second network is more vulnerable from a cyber-security point of view, as it can access to Internet, e.g. for participating in market auctions. There are different possible configurations in order to make possible that both networks work properly.

In some cases, these two networks are physically separated and it is not possible to exchange data between the networks without using external media, like USB drives. In other cases, these two networks are interconnected through a firewall, but there is a big security problem as a machine in the Intranet section could have access to all the data in the control system and perform some operations on them.

However, the most common configuration is a mix of the two configurations described above. In the following example a third sub-network is used to exchange data between the Intranet and the control network. A demilitarized zone (DMZ) is a computer network with controlled access to the connected server by the application of safety rules. The network in the DMZ is protected by firewalls against one or more other external networks. This separation can limit access to public available services and at the same time it protects the internal network from unauthorized access.

These three sub-networks are connected through a firewall, configured in order to allow that data originating from the control network can flow to the DMZ zone, while the computer in the Intranet can communicate only with the server located in the DMZ area. With this configuration, an operator cannot issue commands to the control system, unless he directly logs on a machine in the control area, but he is able to check the measured values, elaborate them and make decisions, increasing the security level of the whole system.

Figure 4.2 shows an example of a possible network configuration that can be implemented in a control center, with the use of a DMZ network between the corporate Intranet and the Process network, which is the network used to control the grid. The example network is Microsoft Windows based.

The ***corporate Intranet*** is composed by a number of clients, some of them configured with the software allowing the creation of ***Virtual Private Network*** (VPN) channels towards the DMZ zone. A VPN channel is a commonly used security measure in order to create secure communication channels between two machines even if the data must flow through various devices. In the Intranet a Windows Domain Server is employed to authenticate the user on the network.

The ***DMZ network*** contains the PI server, devoted to data collection coming from the field, and that it can be accessed by the clients in the Intranet to elaborate them and make decisions upon the results.

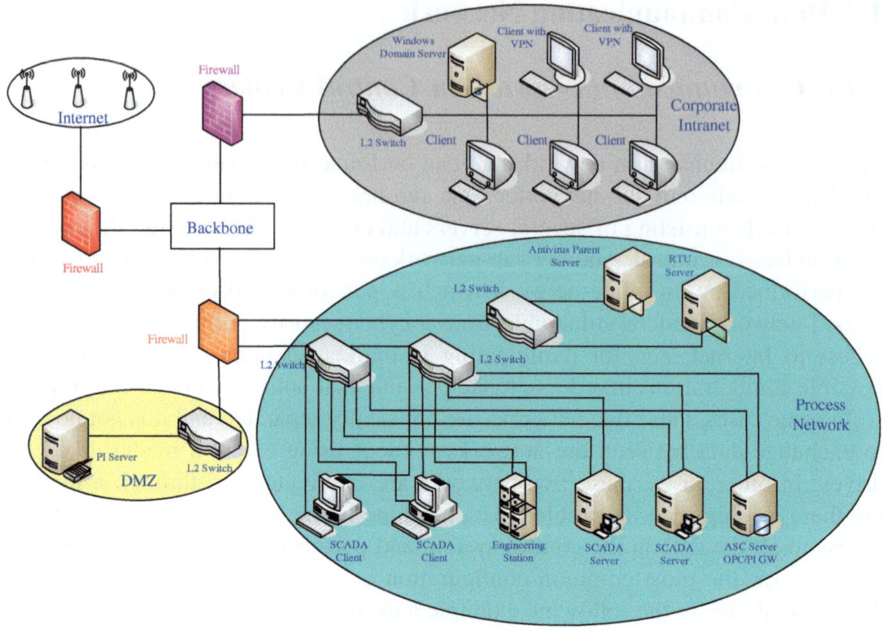

Fig. 4.2 Example of network architecture implemented in a control center in Italy

The ***Process Network*** is the network responsible of the grid management and it contains all the servers and clients needed to perform the various operations. In the figure the use of SCADA is indicated, that is the most common control system in the field of industrial automation worldwide. In the example, an antivirus parent server has been included in the process network. Due to the fact that the system is Windows based, viruses can be a serious issue for the system integrity. Therefore the antivirus is the only component that is continuously updated. To simplify the updating procedures and reducing the potential security risks, only the antivirus parent server can be connected to an external server to receive the updates, then it will propagate these updates to the other machines in the network.

4.4.2 Communication Between Data Acquisition Units in Substations and Control Centers

Each substation contains a kind of data acquisition unit which collects data from the bay level devices in the substation. The structure of the communication in a substation and to the control center is displayed in Fig. 4.3 [9]. Data acquisition units communicate with the assigned control center. They can be a server system implemented on a station computer or dedicated remote terminal units (RTU). The data from the bay level includes process data such as voltage and current

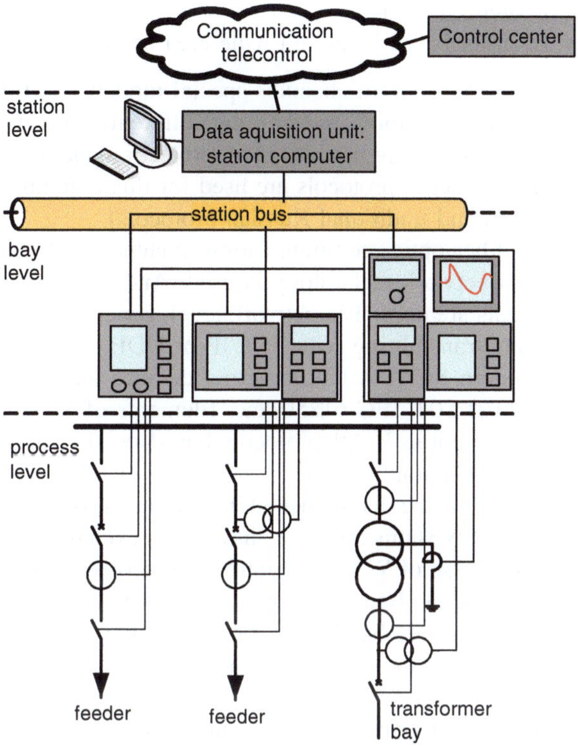

Fig. 4.3 Communication in a substation

measurements and statuses. The control center is able to transmit switching commands. The data acquisition unit forwards the commands to devices in the bay level.

Close to the process level, the switchgear, bay devices acquire the currents and voltages through instrument transformers. Although this link can be implemented using IEC61850-compliant process busses, conventional hard-wired connections are still preferred. Deterministic messages between bay-devices can also be transmitted via the station bus as so called GOOSE-messages. However, in older substations hard-wired connections are still applied.

When no RTU or server is applied as data acquisition unit then data requested from the control center must be transferred through a switch and a star coupler to the bay level device directly. In this case, a dedicated protocol is necessary. Server system and RTUs enable standardized protocols.

4.4.3 Protocols

Communication channels are organized via:

- Currier-current communications by HV transmission lines;
- Fiber-optic communication channels on HV transmission lines;

- Rented channels;
- Ethernet based local and wide area network LAN/WAN.

In curried-current channels special modems and multiplexing devices are used. These devices allow transmitting different types of information by one channel. These channels are used for relay protection and low speed SCADA measurements. As a rule, serial protocols are used for this communication (IEC 60870-5-101 and outdated and traditional Russian protocols).

In fiber-optic communication channels an Asynchronous Transfer Mode (ATM)-network technology is used. And on communication level corresponding protocols are used. The network level, generally, is presented by Ethernet protocol for data transmitting with TCP/IP or UDP on a transport level.

Modern SCADA and metering devices maintain standard protocols like: IEC 870-5-101/104, IEC 61850, etc. Elements of WAMS (PMU and PDC) use IEEE C37.118 protocol. All protocols based on the internet protocol enable TCP and UDP connection.

Data must be distinguished into status messages and streaming data. In the past usually only status messages have been exchanged between control centers. However the arise of WAMS requires the incorporation of instantaneous synchronized measurements. Bandwidth requirements depend on the number of streaming data plus the required bandwidth for all possible status messages in case of endangered system states as well as during and after emergency states. Dedicated data handling is necessary because status messages must be exchanged by secure data exchange via TCP. But for online operation instantaneous measurement values are necessary. These values can be mapped in streaming data which only provides information as long as it represents the actual values. Older values should be neglected. TCP can be applied for streaming data but it must be taken into account that UDP must be applied for efficient data exchange via multicast [10].

In contrast to dedicated wide area communication for WAMS the standard IEC 61850 covers the communication in substations and control centers as well as for the communication between them. The main objective is the interoperability of equipment control in centers and substations, especially for the Intelligent Electronic devices (IED). IEDs are protection devices fault recorders, metering devices, etc. Also PMUs can be deployed as IEDs but also protection devices or fault recorders can be upgraded with PMU functionality. That means that PMU data will be wrapped into IEC 61850 or IEC 60870-5 messages. The standard IEC61850 replaces the standard for substation communication IEC60870-5. In contrast to IEC 60870-5 the new standard for communication between control centers IEC 61850 may not be used yet, as necessary extensions like (618560-90-10 and -90-20) still need to be approved.

4.4.4 Peer to Peer Communications for Data Transmission Optimization

Distributed architecture of data communication for improved availability, performance and flexibility of applications for power distribution can be based on a peer to peer principle, commonly abbreviated to P2P [11]. It is any distributed network architecture composed of participants who make a portion of their resources (such as processing power, disk storage or network bandwidth) directly available to other network participants, without the need for central coordination instances (such as servers or stable hosts).

In comparison with client–server architectures P2P offers advantages, such as self-organization, fault tolerance to loosing connection with a single node, the possibility of sharing resources, increasing communication speed thanks to simultaneous use of several data sources, wide bandwidth and flexible communication load balancing.

The main difference compared to traditional data sharing is that data for routing only needs to be searched before establishing communication with data source. Further on the node is receiving data cyclically. Protocols such as IEC 60870-5-104 or IEEE C37.118 are sufficient for cyclic data transmitting between nodes.

It is suggested to implement Intelligent Search Mechanisms (ISM) for searching needed information [12]. ISM is a new method of searching in P2P networks. It allows decreasing time of search due to minimization of the number of queries. ISM makes a query only to nodes which correspond to intellectual criteria:

- By neighborhood profiles. This mechanism is used to store last answers from each known node with time and consistency characteristics.
- By relevancy. All known nodes are sorted by a complex characteristic combined from communication speed with this node and amount of available information.

To prevent circularity of the search process additional random subsets of nodes should be added to the set of relevant nodes. As a result ISM can reach most of nodes in P2P networks.

This method is effective for P2P networks with special information which can be structured, as example information for emergency control.

4.5 Comparison of Local Protection Devices, Wide Area Monitoring System and SCADA/EMS for Detection of Power System States

Situational aspects (i.e. in the normal and abnormal conditions, in particular, in the emergency and post-emergency conditions of operation) have an impact on methods for the operation and control of large scale power systems. Sudden changes in the power system such as short circuits, disconnection or connection of

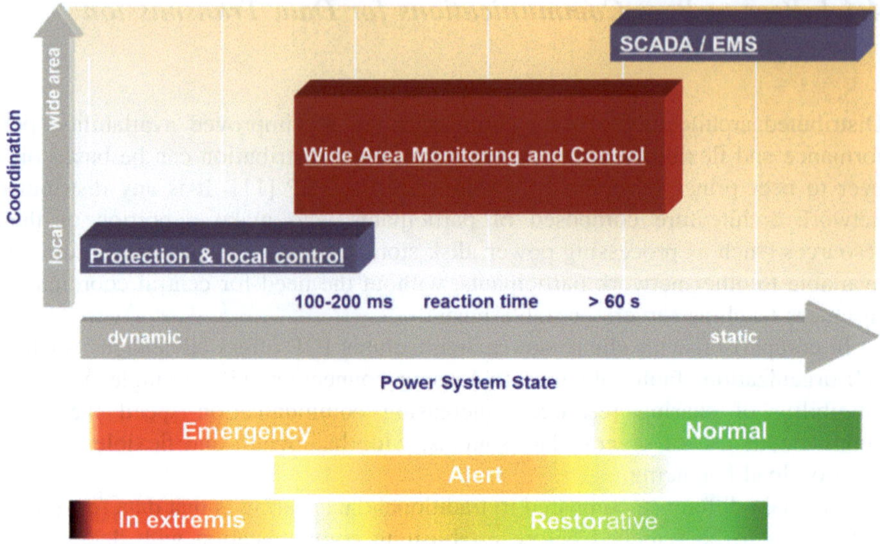

Fig. 4.4 Protection, monitoring and control equipment reaction time in comparison to scale of coordination for different power system states

a large load may lead to long term distortion in power balance. In this case different equipment must be coordinated in accordance to the geographical extent and the requested reaction time. This means that status messages containing information about the switching and subsequently topology information must be sent to state estimation. Data can be evaluated correctly when state estimation is based on consistent topology information. Otherwise, if topology information is wrong then the state estimation solution is not feasible. However state estimation can be improved with the use of time-synchronized phasor measurements by PMUs. This enables faster detection of dangerous power system states. Thus, information exchange for system control between control centers and substations as well as WAMS must be compared in terms of reaction time of the equipment, scale and coordination of data processing.

Thus, monitoring for power systems must contain.

1. data acquisition,
2. detection of system state information,
3. exchange of information between system operators.

A comparison of different equipment in substations and control centers is displayed in Fig. 4.4. In case of fault or overload which can lead to emergency or extremis state, protection devices must react very fast in order to protect power system elements from severe damage. This should be done within the time period of tens to one hundred milliseconds. Data exchange with devices from neighboring nodes may require too much time. However status messages to control centers are

Table 4.1 Comparison of telecontrol and protection equipment corresponding to requirements on communication

	Protection devices	WAMS	SCADA/EMS
Reaction time	200 ms max	Few hundred ms	Few seconds
Scale/Area	Local for variables Status messages to control center	Regional/global	Regional/global
Reporting rate	Only in case of event	Continuous up to 20 ms	Continuous up to few seconds
Data amount	Few status messages, high for fault records	High	High

sent via SCADA/EMS by protection devices. Fault recorders transmit their recordings either on request or automatically.

For the protection of the whole power system itself, information from several nodes is required. WAMS enables time synchronized phasor measurement data with a reporting rate of 20 ms from several locations across the power system. However, dedicated signal processing and data transfer can require additional time in ranges of a few hundred Milliseconds. Because of synchrophasor time synchronization and a resolution of 20 ms, events can be detected much faster than by SCADA/EMS systems with resolution of a few seconds. The same applies to corrective actions after disturbances during a restorative state. SCADA/EMS finally is used to manage power flows in normal state of power system. Events are displayed in control centers and corrective actions are usually executed manually. Here information exchange by phone is still sufficient. Table 4.1 compares types of equipment for protection, monitoring and control corresponding to communication requirements.

Comparison of protection devices, WAMS and SCADA/EMS revealed that all power system states can be only detected when the full equipment in substations as well as in control centers is taken into account. WAMS is able to fill the gap in the reaction time between conventional SCADA/EMS and protection. However the communication network must provide sufficient bandwidth for the data stream.

4.6 Conclusions

Reliable communication technology is a base for monitoring power systems. Communication principle is the organization of data exchange by distinguishing data types and the application of required architecture and protocols. The architecture consists of dedicated networks for internal and external processes. These networks must be coupled in order to acquire data from the process network and to make them available for external data exchange for example for the exchange with other system operators or for market based auctions. Dedicated protocols for monitoring power systems are standardized and usually wrapped in Ethernet protocols. In terms of reaction time communication can be distinguished on the one hand in fast local communication within a substation suitable for protection applications. In case of cascading outages their operations can lead to an emergency state. On the other hand in steady state the communication intervals can be much higher. In order to monitor dynamic changes and if necessary to alert the system operator wide area monitoring systems are applied which provide information of the whole system with short communication interval.

References

1. V Madani, M Begovic, Global industry experiences with system integrity protection schemes (SIPS). IEEE PSRC Working Group C **4** (2009)
2. ENTSO-E, *Continental Europe Operation Handbook*, 2nd release - Policy 7, (2013), Download Jan 14th 2014
3. K.-P. Brand, V. Lohmann, W. Wimmer, *Substation Automation Handbook* (Utility Automation Consulting Lohmann, 2003), (http://www.uac.ch), ISBN 3-85758-951-5, p. 24
4. D. Rumpel, J.R. Sun, *Netzleittechnik* (Springer, Berlin, 1989)
5. E.-G. Tietze, *Netzleittechnik, Teil 1: Grundlagen Anlagentechnik für elektrische Verteilungsnetze—Band 20* (VDE Verlag, 2006)
6. E. Handschin, W. Kaufmann, G. Schaffer, *Netzleittechnik für Verteilnetze* (VWEW-Verlag, Frankfurt a. M., 1992)
7. N. Neuel-Lange, C. Oerter, M. Zdrallek, in *State Identification and Automatic Control of Smart Low Voltage Grids*. IEEE PES innovative smart grid technologies (ISGT) Europe conference, Berlin 2012
8. K. von Sengbusch, Intelligente nachrüstbare Überwachung von Ortsnetzstationen als Teil von Smart Grids. FNN/ETG-Tutorial Schutz- und Leittechnik (2010)
9. M. Kleemann, C. Rehtanz, in *Centralized Substation Automation Systems Based on off-the-Shelf Devices*. 4th European conference on HV & MV substation equipment, Lyon, 23–24 Nov 2011
10. Cisco Systems, *PMU Networking with IP Multicast*—White paper, http://www.cisco.com/en/US/prod/collateral/routers/ps10967/ps10977/whitepaper_c11-697665.html. Download, 1st Oct 2012
11. O.J. Vahamaki, A.J. Allen, J.T. Gaff, in *High Speed Peer-to-Peer Communication System for Integrated Protection and Control in Distribution Networks*. Developments in power system protection, sixth international conference on (Conf. Publ. No. 434), pp. 243–246, 25–27 Mar 1997
12. V. Kalogeraki, D. Gunopulos, D. Zeinalipour-Yazti, in *A Local Search Mechanism for Peer-to-Peer Networks*. Proceedings of the eleventh international conference on information and knowledge management, McLean, Virginia, USA (2002)

Chapter 5
Wide Area Monitoring System

Tadeja Babnik, Kay Görner and Bojan Mahkovec

5.1 Introduction

As presented in Chap. 1 recent power system blackouts showed that operation of interconnected power systems with high level on operation security and stability is not achievable with limited system knowledge and monitoring only of separate interconnection patches. Secure operation of the power system can be achieved if transmission system operators (TSOs) have reliable information of the total system state.

Conventional Supervisory Control and Data Acquisition (SCADA) systems provide steady, non-synchronous information of the power system with time resolution between 1 and 10 s. Conventional SCADA systems are limited to steady-state measurements and cannot be used for observing the system dynamics. The use of Wide Area Monitoring System (WAMS) enables power system dynamic behavior to be permanently monitored. WAMS provides time-synchronized information every 20 ms (in 50 Hz systems); each data sample is equipped with a Coordinated Universal Time (UTC) stamp. Moreover, data is synchronized with a minimal accuracy of 1 µs.

For big interconnected systems like ENTSO-E RG Continental Europe and IPS/UPS reliable operation of the interconnections is important. Each interconnected system consists of several TSOs. To observe the current state of the interconnected system the real-time information from all TSO is required. Hence, interconnected power systems require large scale LS-WAMS.

T. Babnik (✉) · B. Mahkovec
ELPROS, Ljubljana, Slovenia
e-mail: tadeja.babnik@elpros.si

B. Mahkovec
e-mail: bojan.mahkovec@elpros.si

K. Görner
TU Dortmund University, Dortmund, Germany
e-mail: kay.goerner@tu-dortmund.de

U. Häger et al. (eds.), *Monitoring, Control and Protection of Interconnected Power Systems*, Power Systems, DOI: 10.1007/978-3-642-53848-3_5,

The first part of this chapter presents main components and the concept of LS-WAMS. The second part of this chapter concentrates on the LS-WAMS experiences based on the low voltage (230 V) WAMS. Additionally some detected events captured by the WAMS are given.

5.2 WAMS Overview

A typical WAMS architecture based on PMUs is shown in Fig. 5.1. The main components are: PMUs, Phasor Data Concentrators (PDCs), communication networks, data storage and application software. The power system requirements dictate the number of substation PDCs.

PMUs installed at substations measure voltage, current and frequency. These measurements are sent directly to the central PDC or to a substation PDC.

The substation PDC has the following functionalities:

- acquires data from PMUs
- data time synchronization
- evaluates received data
- sends data to central PDC
- exchanges data with the local SCADA
- archives data locally
- performs local data analysis and protection actions.

The use of a substation PDC is recommended in case of bad communication links between the substation and the central system. In such situations the substation PDC serves as local data recorder in case of communication outage. Additional if many PMUs are installed within the substation it can perform data reduction (sends only selected data) for the central PDC.

A regional PDC can be applied between the substations and the central PDC in case of bigger power systems.

A central PDC acquires data from substation PDCs and PMUs and synchronizes data by the time tag and performs data evaluation. The synchronised data is used for different WAMS application, data exchange with other systems like SCADA/EMS or data exchange between companies. Additional, data is used for real-time visualization and data archiving.

The communication links between PMUs, substation PDC and central PDC most commonly use IP over the network communications. The IEEE standard C37.118 [1, 2] is used for synchrophasor data transfer. Based on this standard, synchrophasor data can be carried over any communication system that has sufficient bandwidth. The required bandwidth is dedicated by the reporting rate and the message size. Assuming that the message size is 74 bytes (containing 6 phasors and 2 analog values) and the reporting rate is 50 times per second a bandwidth of

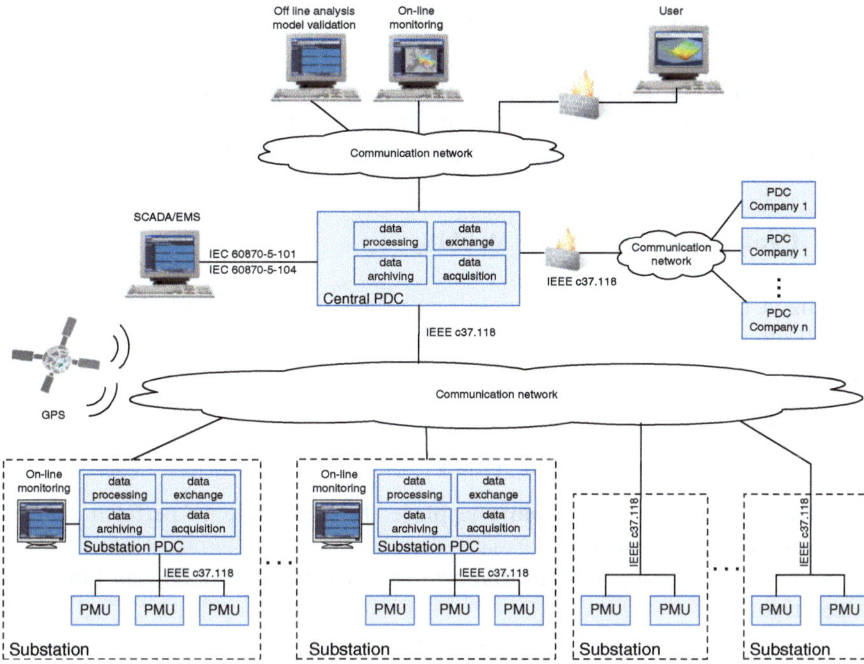

Fig. 5.1 Generic architecture of the WAMS

30 kilobits (kb) per second is required. If for example a 1 Mb communication link is available, then 30 kb presents 3 % of the total communication bandwidth.

The common methods for communication are [2]:

- Client-server. The device providing data is the server and the device receiving data is the client. The device providing data can be a PMU, a PDC, or any other device that will output synchrophasor data. The device receiving data can be a PDC or any other device that receives synchrophasor data. In cases where data transmission is initiated by command, the client initiates contact and controls data flow with commands.
- Basic modes of operation: spontaneous and commanded. With spontaneous mode, the server sends data by User Datagram Protocol (UDP) to a designated destination without stopping, whether a receiving device is present or not. The stream is initiated by a function in the device accessed separately from data operations. With commanded operation, the server only sends data when a client requests it using the standard Start and Stop commands. Both modes may support commands to retrieve configuration and header data.
- Transmission Control Protocol (TCP), UDP, and multicast communication. All modes are supported in various appropriate configurations [2].

5.2.1 Large Scale WAMS Concept

Large scale WAMS is an enlargement of the common concept of WAMS since it incorporates synchrophasor measurements form several TSO-WAMS. The concept of large scale WAMS can be centralized or distributed as shown in Fig. 5.2.

5.2.1.1 Centralized Large Scale WAMS

Centralized LS-WAMS contains only one central PDC. This PDC collects data from all other sources like:

- PMU devices
- Local PDC WAMS from individual TSOs.

Advantages of the centralized LS-WAMS

- All data can be accessed from one place. It is possible to perform stability analysis in a great extent.
- Coordinated alarming and remedial actions in case of detected event (abnormal operation).
- Easier administration of coordination for data exchange since data exchange between TSOs and the centralized LS-WAMS should be formalized only once.

Disadvantages of the centralized LS-WAMS

- In case of a PDC fault, the whole system is out of operation. This can be improved by redundancy.

5.2.1.2 Decentralized Large Scale WAMS

Distributed LS-WAMS contains several PDCs which can operate on the following levels:

- TSO
- Region of several countries.

Data sources can be:

- PMU devices
- Substation PDCs
- Local PDC WAMS from individual TSO.

Advantages of the decentralized LS-WAMS

- More reliable in terms of a failure of individual units since all other units can still operate with limited amount of data.
- PDCs of LS-WAMS may contain dedicated functions for the region where the PDC is installed.

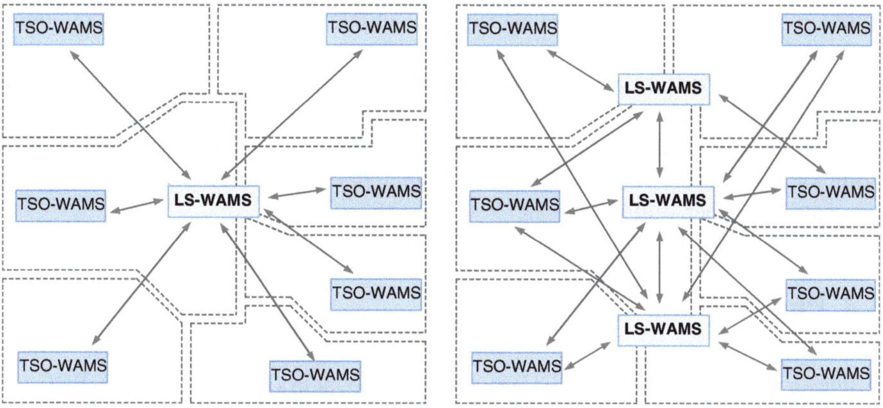

Fig. 5.2 Centralized LS-WAMS (*left*), decentralized LS-WAMS (*right*)

Disadvantages of the decentralized LS-WAMS

- Smaller extend of data enables a limited extend of stability analysis.
- Higher communication bandwidth for data transfer since the same data is sent to several distributed LS-WAMS.
- Coordination in case of events is more complex.
- Higher implementation costs.

5.3 Phasor Measurement Units

A phasor measurement unit (PMU) is a device which measures the voltages and currents on an electric power system, using a common time source for synchronization. Time stamping of each measurement to a common time reference allows synchronized real-time measurements from different locations in the power system to be synchronised [3].

The PMU is a function or logical device that provides synchrophasor and system frequency, as well as other optional information such as calculated active power (MW) and reactive power (MVAR), sampled measurements, and Boolean status words. The PMU may provide synchrophasors from one or more voltage or current waveforms. The PMU can be realized as a stand-alone physical device or as a part of a multifunction device such as a protective relay, fault recorder, or meter [2]. This information may be recorded locally or transmitted in real time to a substation or central location as illustrated in Fig. 5.1.

Phasors measured throughout an interconnected grid require a common timing reference provided by a synchronizing source. The signal provided by the synchronizing source shall be referenced to Coordinated Universal Time (UTC). The signal shall be accurate enough to allow the phasor measurement equipment to maintain synchronism within 1 μs of UTC including both synchronizing source and local receiving equipment error.

The Global Positioning System (GPS), which is a satellite based system, is the main synchronizing source that is used to provide a time reference on the communication networks, and its widespread availability makes it possible at each point of the system to obtain a clock signal that is synchronized with the one generated in other remote places.

5.4 Phasor Data Concentrator

A phasor data concentrator (PDC) is a core component of the WAMS. A PDC can be implemented as a stand-alone unit that collects data and re-distributes it to other applications. PDC functions can also be integrated into other systems, for example monitoring/control platforms.

Since this book deals with large scale power systems, the PDC requirements will as well be related to large scale PDCs hereinafter referred to as PDC.

In the following the PDC is described from the following aspects:

- Real-time data exchange
- Real-time data processing—detection functions
- Data storage
- Data visualization.

5.4.1 Real-Time Data Exchange Requirements

This section presents requirements for real-time data exchange between the PDC and PMUs, between PDCs, between the PDC and SCADA/EMS systems and between the PDC and visualization.

5.4.1.1 Data Acquisition from PMU Devices

A PDC must support the following standard communication protocols for connection with PMU devices:

- IEEE C37.118.2 (2011) client driver
- IEEE C37.118 (2005) client driver
- IEEE 1344 (1995) client driver.

The PDC must be able to simultaneously acquire data from at least 100 PMUs with full resolution of 50 samples per second for 50 Hz power system.

5.4.1.2 Data Exchange with Other PDCs

A PDC must support the following standard communication protocols for data exchange with other PDCs:

- IEEE c37.118.2 (2011) server driver
- IEEE c37.118 (2005) server driver
- IEEE 1344 (1995) server driver.

The PDC as a server driver must fulfill the following requirements:

- Simultaneously send data to several (at least 10) clients with full resolution of 50 samples per second for 50 Hz power systems.
- Optionally the data transfer resolution can be lower e.g. 10 samples per second depending on the client and transfer capacities.
- Filtering functionality for restricted signals (selection of which data will be exchanged).
- Protecting the PDC from unauthorized access.

5.4.1.3 Data Exchange with SCADA/EMS Systems

A PDC must have the ability to exchange data by standard protocols like IEC 60870-5-101/104, DNP3, etc. The purpose of this connectivity is to send data to a control center (SCADA/EMS systems) and to get commands from a control center.

5.4.1.4 Data Exchange for Control and Protection

A PDC must have the ability to send commands like trig and block signals by standard protocols like IEC 60870-5-101/104, DNP3, etc. The purpose of this connectivity is to send data to devices in substations, e.g. protection and control devices for the purpose of control and automatic remedial actions.

5.4.1.5 Data Exchange with Real-Time Visualization

A PDC must have the ability to send data to a system (software) for visualization. The visualization system must acquire data with a maximal delay of 3 s. This way monitoring of the trend through visualization is assured.

5.4.2 Detection and Protection Functions

Based on voltage and current synchrophasors different applications for stable power system operation can be implemented in wide extend. Some of them are summarized below.

5.4.2.1 Angle Difference Detection

By synchrophasor technology it is first time possible to compare voltage phase angles from different locations in the power system. Phase angle difference is an important indicator of the stress in the power system as a result of heavily loaded lines. It is also of great value during system reconnection maneuvers.

Based on the phase angle difference detection the operator obtains a warning signal when the set value is exceeded and has an opportunity to take remedial actions.

5.4.2.2 Low Frequency Oscillation Detection

The interconnection of power grids enables the reduction of capacity for reserve energy for the operation of the power system, while the whole dynamic process becomes more complex. In large interconnected electric power systems, inter-area electromechanical oscillations may occur because of weak links between power systems or local problems related to system operation. These oscillations involve power exchanges and also voltage and current oscillations that may cause tripping of lines or generators and thus lead to severe disturbances [4].

The purpose of the oscillation detection is detection of oscillation modes which may be inter-area when they involve many generators belonging to different areas, intra-area if they involve generators of the same area, or local if they involve only one generator (or generators inside one power plant).

Oscillations present a normal behavior following perturbations in the power system. The power system is in a stable operation mode when oscillations are well damped based on the stability controllers. If an adequate damping is not guaranteed oscillations may become dangerous for the integrity of the interconnected power system.

Detection of oscillations is based on amplitude, frequency and damping of the detected mode. Usually the low frequency oscillation range is between 0.01 and 4 Hz.

5.4.2.3 Oscillation Source Location

When oscillations occur in the power systems it is desirable to know the location i.e. the source of oscillations. The frequency oscillation mode locator is used for locating generators (if PMUs are installed at generators) or the direction towards the generators causing the inter-area oscillations as well as local oscillations.

Analysis and localization of dominant frequency modes can be performed on available measurements such as active power, reactive power, power system frequency or calculated values like the voltage angle difference.

5.4.2.4 Islanding Detection

Islanding of the power system can be related to:

- The separation of a large interconnected system
- The splitting of a power system into several islands
- The separation of a significant portion of the power system
- The separation of a single generator.

WAMS enables the detection of islands in the power system. Methods for islanding detection can be based on frequency or/and voltage phase angle differences.

Criteria for frequency based detection are frequency differences and the speed of the frequency difference. Criteria for voltage angle difference based detection are speed and acceleration of the angle difference.

5.4.2.5 Voltage Stability Detection

The electricity market forces the system operators to utilize the maximum available capacity of the power system infrastructure. This requires a close look at security and stability issues. One of such issues is related to voltage stability.

Voltage stability monitoring provides real-time information about the power margin with respect to the voltage stability. The power margin is the amount of additional active power that can be transported on a transmission corridor or line without endangering the voltage stability of the grid.

5.4.3 Data Storage

A data storage contains a real-time circular database, an event database and a disturbance recordings (oscillography) database. Important parameters of data storages are data nodes (phasors, analog values, digital statuses) storage capacity and the access response time.

Circular Real-time database

The real-time circular database enables data storage of measured and calculated values (i.e. currents, voltages, power, different stability indices, etc.) with user selectable sample rate.

The concept of the circular database is a data array, which works by the FIFO (first in, first out) method. The storage time period depends on the available hard disk storage and the sample rate.

Events database

The event database enables archiving of events and alarms. The events database is a permanent database.

Disturbance recording database

The main purpose of the disturbance database is recording of important user-selected data when some events occur. Recordings are used for post-mortem analysis.

The disturbance recording database is separated permanent data storage.

5.4.4 Data Visualization

User-friendly designed graphical interfaces are the most important tool for power system operators and analysts. The visualization system must enable the presentation of:

- real-time measurements and calculated values
- data stored in the circular data base
- event disturbance recordings
- stored data on the local user computer
- events and alarms.

It is important that different charts like two and three-dimension time-base charts, polar charts, etc. have user-friendly flexibility which can be reflected in:

- adjustable trend time interval
- adjustable selection of presented data
- adjustable graphical design enabling design of data views according to the user requirements.

An example of data visualization is given in Fig 5.4.

5.5 WAMS Experiences

Since the communication network of TSOs is closed for external users and therefore the access to installed PMUs in the transmission network is not possible, a low voltage (230 V) centralized LV-WAMS was built for testing purposes. This LV-WAMS is explained in the present sub-section. PMUs are connected to an electrical socket (230 V) at universities and institutions. The concept of LV-WAMS is illustrated in Fig. 5.3:

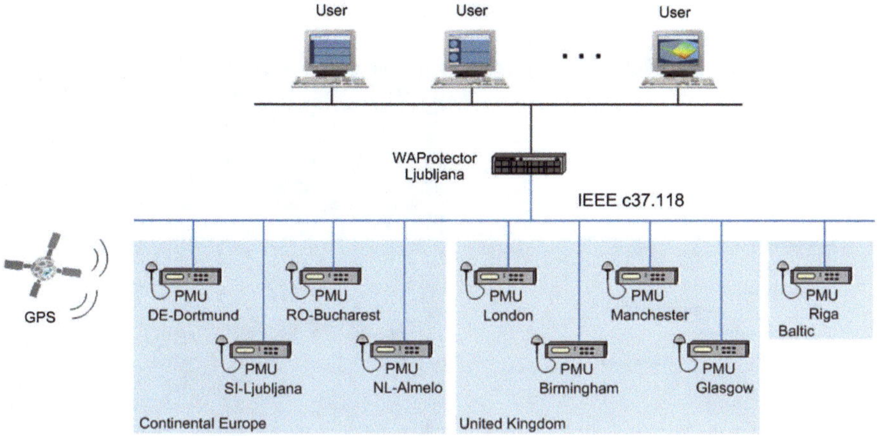

Fig. 5.3 Low-voltage WAMS configuration

- Acquisition units-PMUs are installed at different locations across Europe. The PMUs belong to different regional groups (RG) according to Interconnection networks i.e. RG Continental Europe, RG Baltic and RG UK.
  - Continental Europe:

    Dortmund (Germany)
    Almelo (The Netherlands)
    Ljubljana (Slovenia)
    Bucharest (Romania)
  - Baltic:

    Riga (Latvia)
  - United Kingdom:

    Glasgow
    Manchester
    London
    Birmingham
- The WAProtector (a product from the company ELPROS) was installed in Ljubljana as a phasor data concentrator. It acquires data from PMUs by the standard Internet communication network. For the communication link between PMUs and PDC the public Internet is used.
- Data visualization is performed by web SCADA which is the integrated part of the WAProtector.

Beside the PDC functionality the WAProtector system performs:

- Data acquisition from PMUs.
- Phase angle difference monitoring.

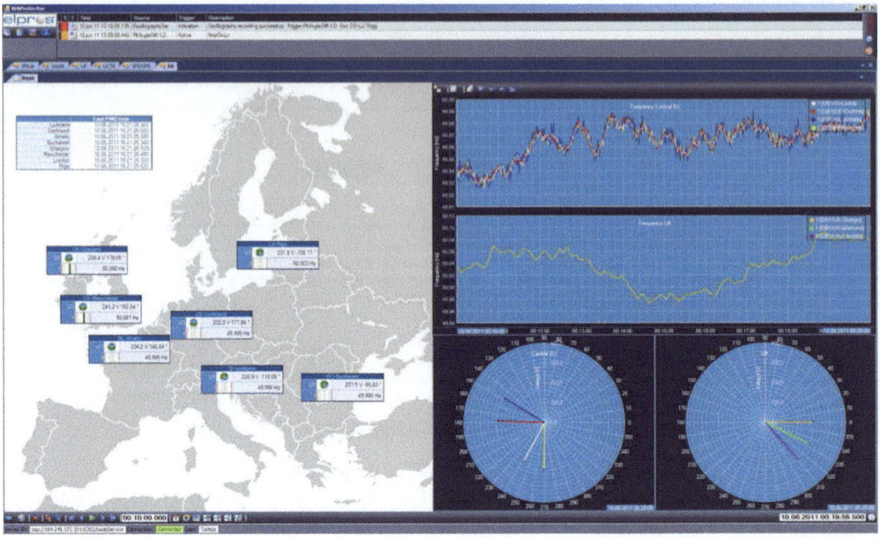

Fig. 5.4 Example of WAProtector visualization

- Low-frequency oscillation detection based on phase angle difference and frequency.
- Under/over frequency detection.
- Under/over voltage detection.
- Archiving of data into a cycling data base. The storage is large enough for storing the last 14 days with a sampling rate of 20 ms.
- Visualization for simultaneous users with data exporting capabilities. Example of visualization is given in Fig. 5.4. The left part of the figure presents current values of frequency and voltage phasors at different locations around Europe. The upper right part of the figure presents the frequency trend chart for Continental Europe and United Kingdom. The lower right part of the figure presents voltage phasors in Europe and United Kingdom.

The project of LV-WAMS started in March 2009. Since that time some events have been detected. In the following three selected events are presented:

- Un-damped low-frequency oscillation.
- Low-frequency oscillation caused by synchronization of the Turkish power system with the interconnected power systems of Continental Europe.
- Hydro power plant outage in Turkey.

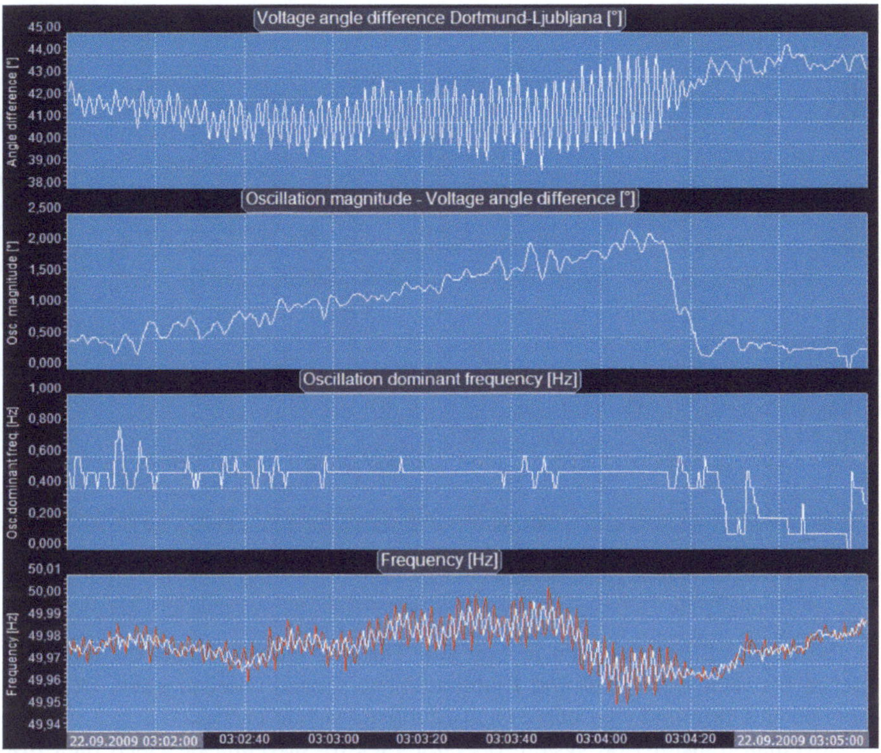

Fig. 5.5 Low frequency oscillation recording

5.5.1 Example 1: Un-Damped Low-Frequency Oscillation

On September 22nd, 2009 at 03.02:10 a.m. un-damped oscillations were detected on the voltage phase angle difference between Dortmund and Ljubljana. The oscillation phenomenon was also present on the power system frequency.

The dominant frequency of oscillation was around 0.5 Hz. The duration of the un-damped oscillation was longer than 2 min. Recording of the event is given in Fig. 5.5. The first chart presents the voltage angle difference between Dortmund and Ljubljana. The second chart presents the oscillation magnitude on the voltage angle difference between Dortmund and Ljubljana. The third chart presents the dominant frequency of the oscillation on the voltage angle difference between Dortmund and Ljubljana. The fourth chart presents the frequency captured in Dortmund (red curve) and Ljubljana (white curve). Time window of the recording is 3 min.

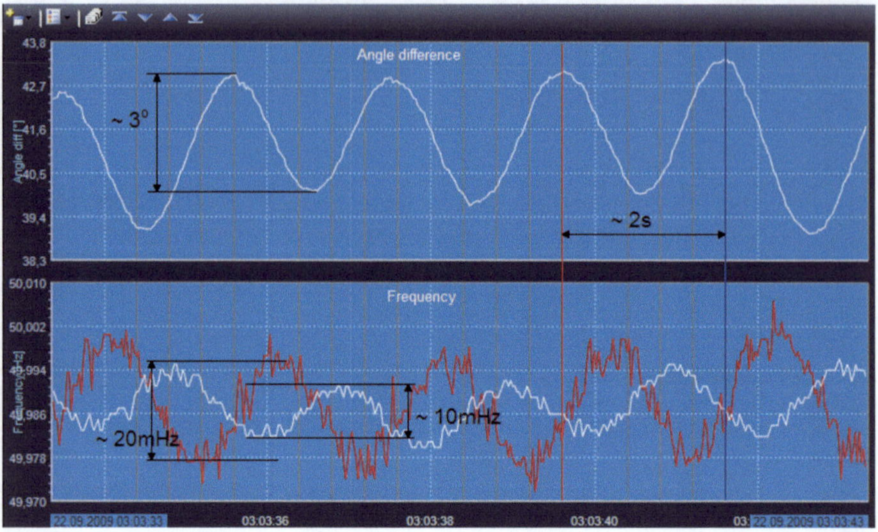

Fig. 5.6 Detail of frequency oscillation from Fig. 5.5

The detail of 10 s is given in Fig. 5.6. The first chart presents the angle difference between Dortmund and Ljubljana. The second chart presents frequency captured in Dortmund (red curve) and Ljubljana (white curve).

Slovenian transmission system operator Elektro-Slovenija provided data from high voltage (HV) 400 kV WAMS. This enables comparison of data obtained by LV WAMS with data obtained by HV WAMS. Figure 5.7 presents a comparison of data obtained by LV WAMS and HV WAMS. The first chart presents the frequency in Dortmund and Ljubljana captured by the 230 V WAMS and the second chart presents the frequency at the 400 kV substation Divača (Slovenia) captured by the 400 kV WAMS. We can see that the oscillation captured on 400 kV and 230 V in Slovenia are in strong correlation regarding oscillation magnitude and frequency. The time window is 1 min and 10 s.

According to some past experiences with un-dumped low-frequency oscillations [5] the reason for oscillation can be non-normal operation of generation units. From the figures above we can see that the magnitude of the dominant frequency oscillation of the power system frequency is higher in Dortmund (oscillation magnitude is approximately 10 mHz) than in Ljubljana (oscillation magnitude is approximately 5 mHz). Therefore we can conclude that the source of disturbance originated from the part of Europe which is closer to Dortmund.

Regarding on the recorded event we can conclude that the event (low frequency oscillation) originated in one part of the transmission network affect to the whole power system and can be well detected on the low voltage side.

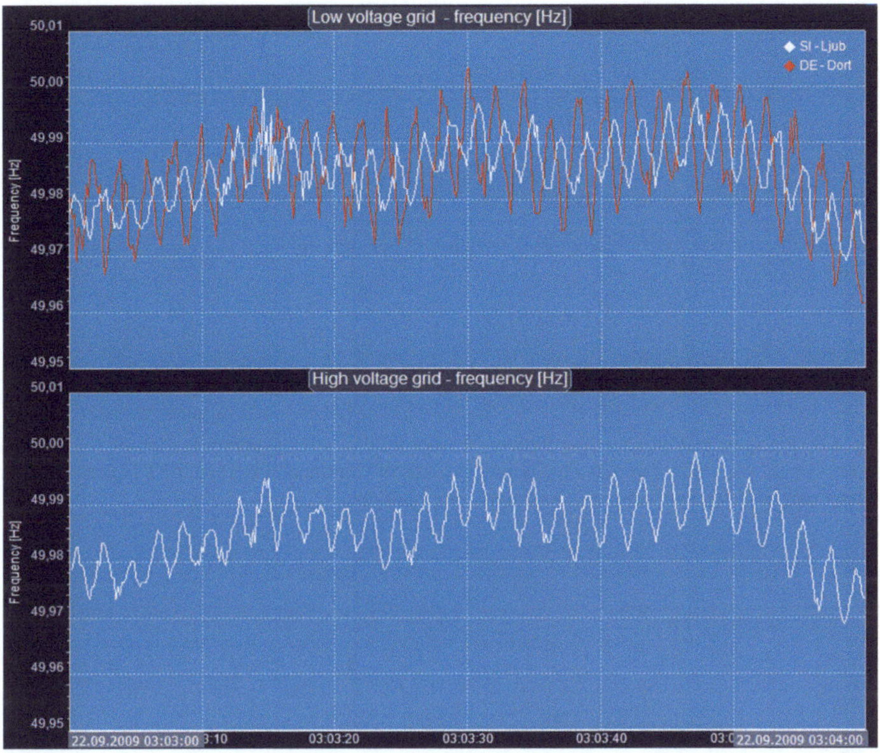

Fig. 5.7 Correlation of low voltage data with high voltage grid for the case of un-damped oscillation

5.5.2 Example 2: Recording of Synchronization of Turkish Power System with the Interconnected Power Systems of Continental Europe

Low frequency oscillations in the interconnected system of continental Europe can be divided into four dominant inter-area modes (global modes) in the low frequency range between 0.2 and 0.5 Hz [4]. After the scheduled interconnection with Turkey a new Global mode (Turkish Mode) is expected with frequency around 0.15 Hz.

On September 18th, 2010 at 9:25 a.m. the Turkish power system was synchronized with the interconnected power systems of Continental Europe [6]. After the interconnection of ENTSO-E with Turkey the new low-frequency oscillations occurred as shown in Fig. 5.8 which presents a FFT analysis performed on the voltage angle difference between Dortmund and Ljubljana. The time window is 2 days. Before the synchronization of Turkey we can see permanent dominant frequencies in the range between 0.2 and 0.3 Hz and in the range between 0.3 and 0.4 Hz. After the synchronization new permanent oscillations in the range between 0.1 and 0.2 Hz occurred.

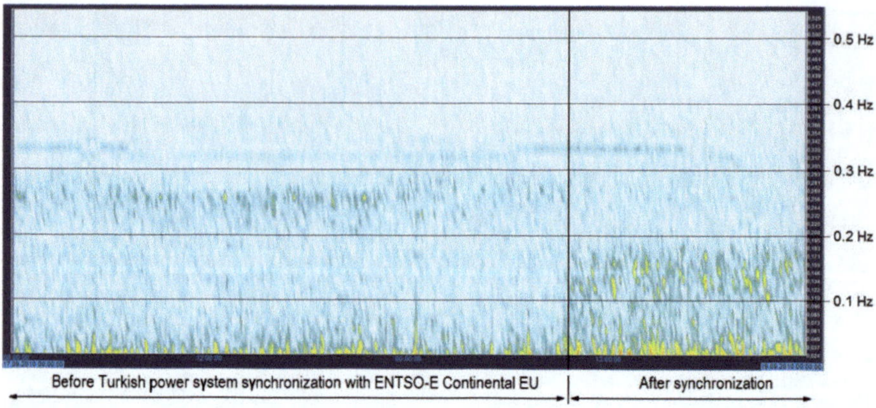

Fig. 5.8 Permanent dominant low frequencies before and after synchronization of the Turkish power system with the interconnected power systems of Continental Europe. FFT is performed on voltage angle difference between Dortmund and Ljubljana

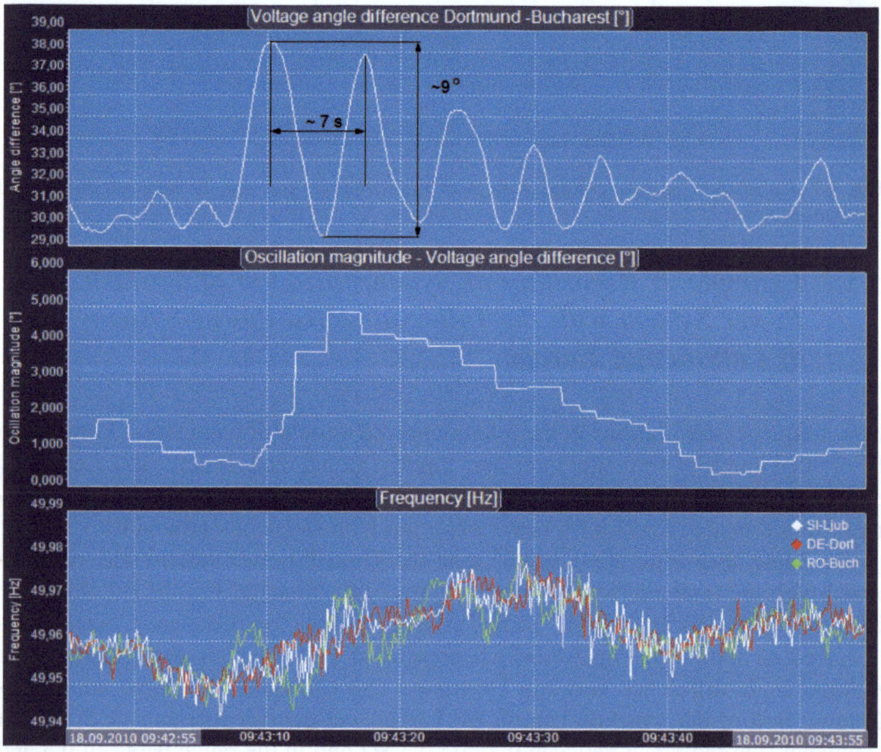

Fig. 5.9 Low frequency oscillation ($\sim$0.15 Hz) detected on voltage angle difference between Dortmund and Bucharest

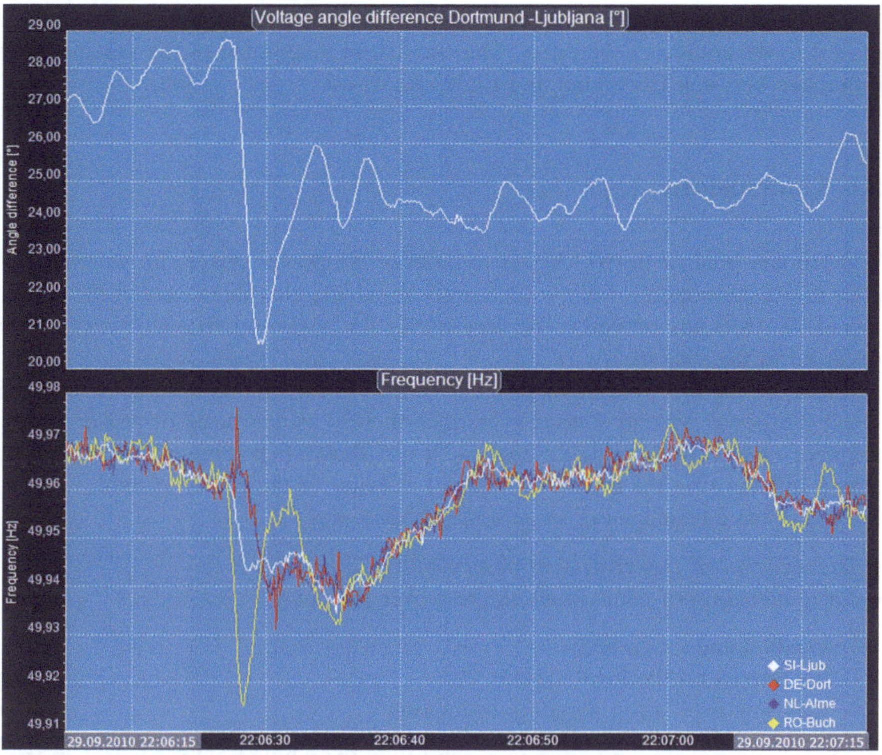

Fig. 5.10 Outrage of hydro power plant in Turkey recording

One of the examples of low-frequency (0.15 Hz) oscillations is given in Fig. 5.9. It presents the voltage angle difference between Dortmund and Bucharest, oscillation magnitude and frequency at Dortmund (red curve), Ljubljana (white curve) and Bucharest (green curve). The time window of recording is 1 min.

5.5.3 Example 3: Hydro Power Plant Outage in Turkey

On September 29th, 2010 the LV-WAMS captured an event which caused a frequency drop and a following damped oscillation. The event also caused a change in voltage angle difference between Dortmund and Ljubljana.

The event was caused by the outage of Atatürk hydro power plant in Turkey [7]. The power system behaviour after the outage of 776 MW is given in Fig. 5.10. The first chart presents the voltage angle difference between Dortmund and Ljubljana. The second chart presents the frequency captured in Ljubljana (white curve), Dortmund (red curve), Almelo (blue curve) and Bucharest (yellow curve). The time window of the recording is 1 min.

The highest frequency drop of 46 mHz was recorded in Bucharest which is closest to the source of the event. The oscillation magnitude of the voltage angle difference between Dortmund and Ljubljana was 4°.

5.6 Conclusions

The primary goal of the WAMS is to monitor the power system in real-time and alert the system operator if the measured or calculated values exceed the setting values. WAMS has proved to be very useful for post-mortem analysis of events occurred in the system.

Based on the low-voltage large scale WAMS is shown that several conclusions about the power system state can be done merely on the basis of power system frequency and voltage angle differences. It was shown that some events can have a strong influence on the global or regional stability of the power system.

Additional advantages of large scale WAMS are:

- Higher level of power system observability
- Early detection of oscillations in the power system
  - location and magnitude
  - the impact on the local system
  - real-time oscillation damping estimation
- TSOs have real-time monitoring capability of the whole ENTSO-E system from the viewpoint of voltage angles and frequency.
- Recording and archiving of events.
- Help based on real-time synchronized data for TSO operators when corrective actions should be taken.

References

1. IEEE C37.118-2005, *IEEE standard for synchrophasors for power systems* (Revision of IEEE Std 1344-1995), pp.1–57 (2006)
2. IEEE C37.118.2-2001, *IEEE standard for synchrophasor data transfer for power systems* (Revision of IEEE Std C37.118-2005), pp.1–53, (2011)
3. A.G. Phadke, J.S. Thorp, *Synchronized Phasor Measurements and Their Applications* (Springer, New York, 2008)
4. E. Grebe, J. Kabouris, S. Lopez Barba, W. Sattinger, W. Winter, Low Frequency Oscillations in the Interconnected System of Continental Europe, in *IEEE Power and Energy Society General Meeting*, (*IEEE*, 2010), pp. 1–7
5. T. Babnik, U. Gabrijel, B. Mahkovec, M. Perko, G. Sitar, The road from WAMS to WAPCS, in *Paper PS3-304, CIGRE B5 Colloquium October* (Jeju, Korea, 2009), pp. 19–24
6. ENTSO (European Network of Transmission System Operators for Electricity), https://www.entsoe.eu/
7. E. Grebe, H. Weber, W. Sattinger, Systemstudien und Messungen zum Anschluss der Türkei an das europäische Verbundsystem, ETG-Mitgliederinformation, Juli 2011

Chapter 6
Distributed State Estimation

Irina Kolosok, Elena Korkina and Oleg Soukhanov

6.1 Introduction

The previous chapter focused on Wide Area Measurement Systems, that are typically used to oscillations within the transmission systems by comparing time synchronized measurements of different locations. By contrast, this and the following two chapters focus on the use of modern monitoring technologies (compare Chap. 4) for static observation of large scale transmission systems with State Estimation (SE) tools. SE was firstly introduced in the 1970s to improve monitoring of Electric Power Systems (EPS). SE is an important procedure that allows for on-line calculation of state variables for a current situation of the EPS on the basis of measured state variables and signals about the switching state of power system equipment. The obtained SE model of the power system is then used in control centers for observation and control of the EPS.

As introduced in Chap. 1, now the Unified Power System (UPS) of Russia operates in parallel with the interconnected power systems of the Newly Independent States and the Baltic States that constitute the Eastern synchronous zone (IPS/UPS). The UPS of Russia, in turn, includes 6 simultaneously operating interconnected power systems [1]. Currently the computational model of the Russian IPS/UPS includes about 7,000 nodes, 10,000 branches and 800 generators. A similar situation is observed in the dispatching practices in other countries. As an example the European Network of Transmission System Operators

I. Kolosok (✉) · E. Korkina
Energy Systems Institute, Siberian Branch of Russian Academy of Sciences (ESI SB RAS), Irkutsk, Russia
e-mail: kolosok@isem.sei.irk.ru

E. Korkina
e-mail: korkina@isem.sei.irk.ru

O. Soukhanov
VEI, Moscow, Russia
e-mail: soukh@enersys.ru

U. Häger et al. (eds.), *Monitoring, Control and Protection of Interconnected Power Systems*, Power Systems, DOI: 10.1007/978-3-642-53848-3_6,

(ENTSO-E) or the North American Electric Reliability Corporation (NERC), that embraces most of the North-American power systems, has led to the necessity to make calculations for very large and sophisticated systems.

The schemes of such power systems are as a rule incompletely observable. There can be data distortion and the synchronization can be very poor. Erroneous measurements (bad data), deviation between calculated and real states of the EPS as well as loss of observability due to failures in the system of data acquisition and transmission can distort the results of SE, deteriorate the convergence of the computational process down to divergence and, as a result, lead to wrong solutions derived on the basis of the obtained calculation model.

In this book a further increase of the size of the synchronous area is considered, as for example at a joint synchronous operation of the power systems between the Eastern synchronous zone and Western Europe. Such a situation would increase the dimension of the problem and exacerbates the above enumerated problems.

Distributed data processing and decomposition of the SE problem is an effective method to solve these problems, to improve the quality of the results and to increase reliability of the computational procedure of the EPS SE. Furthermore, such techniques decrease the computation time of the calculation. With parallel computing the computation time might be reduced even further, but this requires the full system to be separated into subsystems.

6.2 State Estimation with Phasor Measurement Units

Until recently the measurements employed in the EPS SE were mainly the measurements received from the Supervisory Control and Data Acquisition (SCADA) system. The disadvantages of SCADA systems in the field of power systems are insufficient volume and low accuracy of the measurements. Furthermore, SCADA systems do not provide any absolute synchronization of the data because measurements are obtained successively. Non-simultaneous arrival of measurements is particularly noticeable when subsystems that operate in parallel have their own SCADA which is included in the SE of the global system. The state variables calculated on the basis of such data lag behind the current state of the EPS.

An essentially higher level of observability and controllability of the EPS can be achieved by use of the Wide-Area Measurement Systems (WAMS) technology (compare Sect. 6.5). The basic measurement equipment in WAMS are Phasor Measurement Units (PMUs) [2] (compare Sect. 6.4). These devices allow for a synchronously and accurate control of the EPS state. Compared to conventional measurements received from the SCADA system, PMUs provide accurate measurements of voltage magnitudes (the error is $\pm 0, 15\ \%$ [3]) and phase angles (the error is $\pm 0{,}01^{\circ}$) at the node where they are installed, as well as the magnitudes and phase angles of currents in the branches adjacent to this node.

PMU measurements used in combination with SCADA measurements reflect more fully the operation of the EPS scheme and improve considerably the results of the SE problem. The use of PMU measurements makes it possible to solve some problems encountered in the decomposition of the SE problem, such as:

1. The need to make iterative calculations on the subsystem level to solve a coordination problem.
2. The need for recalculation of the entire system after the calculations on the subsystem level.
3. The occurrence of bad data interacting at the boundaries between subsystems due to inaccurate synchronization of the measurements.

In this chapter algorithms for the SE problem decomposition including PMU measurements are presented, which are intended for the calculation of large interconnected power systems on the basis of the Test Equations (TEs) method [4], developed at ESI SB RAS as well as the Functional Modeling (FM) method [5], developed at VEI.

6.3 The EPS State Estimation Problem and Solution Methods Based on Test Equations

The SE problem calculates the EPS steady-state conditions based on measurement values. For this purpose we introduce the notion of the state vector x that comprises the magnitude U_i and the phase angle δ_i of the bus voltages, i.e. $x = (\delta_i, U_i)$. This vector uniquely determines all the state variables, including the measurements y and all not measured quantities z.

The measurements used for state estimation include mainly the measurements received from SCADA, magnitudes of bus voltages U_i, power flows P_{ij}, Q_{ij} on lines and less often line currents I_{ij} and nodal currents I_i. Additionally pseudo measurements of bus injections P_i, Q_i are used:

$$\bar{y} = \{U_i, P_{ij}, Q_{ij}, P_i, Q_i, I_i, I_{ij}\} \tag{6.1}$$

The traditional mathematical statement of the SE consists of a minimization of the objective function:

$$\varphi = (\bar{y} - \hat{y})^T R^{-1} (\bar{y} - \hat{y}) \tag{6.2}$$

subject to constraints in the form of steady state equations:

$$w(y, z) = 0 \tag{6.3}$$

where R is a diagonal matrix, in which the elements are equal to measurement variances.

By using dependencies $y = y(x)$ in Eq. (6.2) the SE problem is reduced to the search for the values (estimates) of the state vector $\hat{x}$, such that the values $y\,(\hat{x})$ are closest to the measurements in a sense of the criterion:

$$\varphi = \left(\bar{y} - y(\hat{x})\right)^T R^{-1} \left(\bar{y} - y(\hat{x})\right) \tag{6.4}$$

This method is presented in detail in [6, 7].

To solve the SE problem the so called Test Equations (TEs) can be used:

$$w_k(y) = 0 \tag{6.5}$$

which can be obtained from the system of steady state Eq. (6.3) by elimination of unmeasured variables.

After formation of the TEs the SE problem can be reduced to a minimization of the objective function (6.2) with constraints in the form of a system of TEs (6.5). In this statement the SE problem can be solved directly in coordinates of the measured parameters y. It allows for the distortion of SE results that is caused by the effect of error “smearing” [8] to be avoided. Moreover, the SE algorithms using the TEs are less laborious and have a high speed, since the order of the system of TEs is as a rule substantially lower than that of the initial system (6.3). Besides that, the obtained TEs make it possible to detect bad data before implementation of the SE procedure. The substitution of the obtained measurements into the TEs causes discrepancies due to measurement errors. If the magnitude of the TEs discrepancy is higher than a threshold value d_i, which means that the condition

$$\mid w_{ki}\left(\bar{y}_i\right) \mid \; > \; d_i \tag{6.6}$$

is met, then the measurements that belong to it are considered to have bad data. The value of d_i is determined by statistic properties of normal measurement errors that enter in this i-th TE. Algorithms for logical analysis have been developed to detect bad data among the measurements entering into the TEs with large discrepancies [4].

All the identified erroneous measurements are revealed simultaneously with no repetition of the SE procedure. The possibility to retain absolutely precise measurements, for example, zero injections at transit buses is an important advantage of the SE method using the TEs.

All unmeasured state variables are calculated by the obtained estimates of the measurements. For this purpose so called *basic* set of measurements, which is at least necessary for the unique determination of the system state, is selected.

PMU measurements used in combination to the SCADA measurements reflect more fully the operation of the EPS scheme and improve considerably the results of the SE problem (to increase the accuracy of measurements, to provide observability of a more complete calculated network, to enhance the efficiency of methods for bad data detection (BDD) in measurements and to increase the accuracy of the obtained estimates).

6.3.1 Decomposition Methods for State Estimation of Large-Scale Power Systems

A distributed SE approach implies the procedures of ***decomposition and aggregation***. The SE procedure thereby consists of the following stages:

1. Division of the calculated system into subsystems at the decomposition stage.
2. SE for each subsystem.
3. Solution of *the coordination problem* consisting of the calculation of the boundary variables and checking *the boundary conditions*. If the conditions are not met, the subsystems are recalculated with new values of the boundary variables.
4. Formation of the general solution for the complete system by combination of the solutions of the individual subsystems with the solution of the coordination problem at the aggregation stage.

So far a large number of methods have been suggested for decomposition of the SE problem [5, 9–15].

A review of these methods shows that algorithms for decomposition of the SE problem in large scale systems allow:

- to reduce the burden on the computational resources in the control center of the interconnected power system
- to minimize the volumes of information to be transferred between the control centers of the subsystems operating in parallel and the coordination center
- to reduce the computation time of the state estimation procedure for the entire scheme
- to enhance the efficiency of the algorithms for bad data detection by reducing the dimension of the problem.

But there are also problems encountered in the decomposition:

- There is a necessity for iterative calculations between the subsystems to solve the coordination problem.
- There is bad data interaction at the boundaries between subsystems due to inaccurate synchronization of measurements and decreasing efficiency of BDD methods at boundary regions.

These problems can be solved by placing PMUs at boundary regions of subsystems.

6.3.2 Decomposition of the SE Problem with the Test Equations Method and PMU Data

A method for distributed SE on the basis of the TEs method with PMU measurements was developed at ESI SB RAS. In accordance with this method the minimum of the objective function for each of the subsystems [13]

$$\varphi_v = (\bar{y}_v - \hat{y}_v)^T R_{y_v}^{-1} (\bar{y}_v - \hat{y}_v) \tag{6.7}$$

is determined under constraints in the form of a set of TEs

$$w_k(y_v, y_b) = 0 \tag{6.8}$$

in which v is the number of subsystem and y_b are measured boundary variables. The obtained estimates of the boundary variables are transferred to the upper level where the coordination problem is solved.

The coordination problem consists of the minimization of

$$\varphi_b = (\bar{y}_b - \hat{y}_b)^T R_{y_b}^{-1} (\bar{y}_b - \hat{y}_b) \tag{6.9}$$

under constraints in the form of TEs that include only boundary variables:

$$w_k(y_b) = 0 \tag{6.10}$$

The problems (6.7), (6.8) (6.9), and (6.10) are solved iteratively until the required accuracy of estimates is achieved. This technique does not require any modification of the SE algorithms and can be implemented in the available software, but several iterations are necessary to coordinate the solution for several subsystems.

The algorithms of the SE decomposition divide the calculated scheme into subsystems whose boundaries are either nodes (a) or branches (b). It is so called structural decomposition. Decomposition with boundary branches is used more often. In this case the SE problem is solved iteratively unless *the boundary conditions* are met [16]:

- if decomposition (a) is used then the equality of voltage magnitudes and phase angles at the boundary nodes should be met. The boundary relationship of power balance should also be met;
- if decomposition (b) is applied then the difference of the power flow (active and reactive) at both ends of a boundary branch must be equal to the branch losses. Also voltage magnitudes and phase angles at the boundary nodes must be subordinated to the equations of the electrical circuit.

The placement of PMUs at boundary and slack nodes of the subsystems makes it possible to considerably simplify the solution of the coordination problem and does not require iterative calculations between subsystems.

Advantages of the application of PMU data at the SE problem decomposition are discussed further.

The idea of SE problem decomposition with PMU placement at boundary nodes is rather attractive. Nowadays in Russia they can only be used when the system is decomposed into rather large subsystems with a small number of boundary nodes, which does not solve the problem of inhomogeneity of the systems.

To calculate large inhomogeneous networks the authors propose a method of dividing the calculated scheme with respect to voltage levels [16] in Sect. 6.3.2.1. This method decreases essentially the negative impact of inhomogeneity of the calculated scheme and telemetric information in calculation of the subsystems of one voltage class. However, for the complex scheme it inevitably leads to a large number of boundary nodes. Therefore, a two-stage algorithm is proposed in [17] to decompose the calculated scheme into subsystems that combine the positive features of both approaches. The functional decomposition is suggested in Sect. 6.3.2.2. Section 6.3.2.3 describes the use of PMU measurements for decomposition of the SE problem.

6.3.2.1 A Two-Stage Algorithm for Subsystem Allocation with the Test Equations Method

In the first stage the system is divided into rather large subsystems with a small number of tie lines and boundary nodes. This decomposition can be made according to the administrative division. For example, the entire scheme of Russia's UPS is divided into interconnected power systems of large regions inside the country that operate in parallel (Fig. 6.1). As an alternative it can be decomposed artificially into separate areas by special algorithms.

The PMUs are placed at the boundary nodes. Highly accurate measurements from these devices allow to fix the values of voltage magnitudes and phases at the boundary nodes and to estimate the state of all subsystems simultaneously. The algorithm for PMU placement at boundary nodes is based on the simulated annealing method [18]. According to this method a small number of PMU installations make it possible to obtain the calculated values of U, δ (using the so called "calculated" PMUs) at the other boundary nodes. The accuracy of the calculated values, as it is shown below, is comparable with the accuracy of measurements provided by real PMUs.

In the second stage the calculated scheme of each subsystem in turn is divided into areas that correspond to voltage levels.

The calculation starts with the subsystem of the highest voltage level (750–500 kV in Russia). Normally this part of the scheme is well provided with highly accurate measurements and contains the slack node. Then the calculations are made successively for the rest of the subsystems. The subsystems are ranked by the voltage level (220, 110 kV, etc.). In each step the bordering node of the subsystem belonging to the higher voltage level is chosen to be the slack node.

The TEs method makes it possible to fix the estimates of boundary measurements.

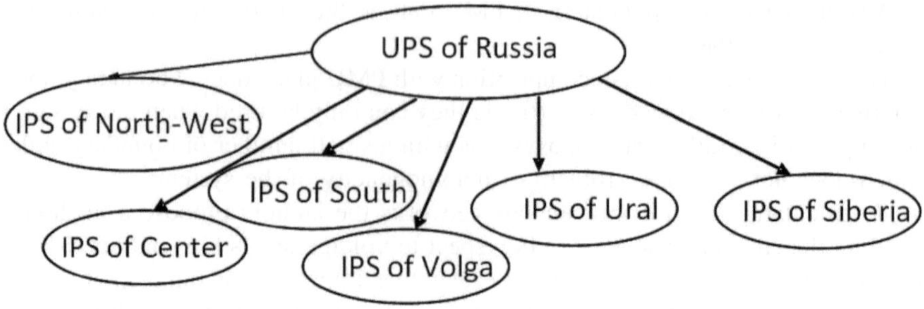

Fig. 6.1 Administrative division of Russia's UPS

As soon as the lower level areas of all subsystems are calculated, the coordination problem is solved. In this case the calculation of nodal injections at the boundary nodes is implied by the power flow estimates obtained from each subsystem.

6.3.2.2 Functional Decomposition of the SE Problem

With functional decomposition the SE problem is decomposed into sub-problems. The main items are: analysis of the network topology; analysis of the observability; analysis of bad data; calculation of estimates as well as the steady state estimates.

This section deals with the problems of bad data detection as well as SE with quadratic and non-quadratic criteria. Methods of a priori bad data detection (that will be addressed to later) do not always reveal bad data entirely. The efficiency of these methods enhances with decreasing size of the power system. The functional decomposition combined with decomposition of voltage levels allows for varying the selection of a method for the solution of the SE problem for each subsystem. In regions of higher voltage levels redundancy of the measurements is higher, hence the weighted least squares method is applicable for these calculations. In regions with lower voltage levels it may be impossible to detect all bad data due to the discrepancies of the TEs caused by low redundancy of the measurements. Therefore, a combination of the weighted least squares method (WLSM) and the robustness criteria of SE are required. These methods will be explained in the following.

A priori bad data detection with the TEs. Bad data detection or validation of measurements is a very important problem in EPS SE. In this section bad data is detected by the TEs. This method allows for validation of data before the SE problem is solved (a priori).

Regardless of the method for measurement validation of the SE solution, a low redundancy of measurements (which is typical for EPSs in Russia) leads to critical measurements. It is impossible to explicitly determine bad data in these measurements. Critical groups are the groups of measurements in which a drop-out of any measurement makes all the remaining measurements in the set critical. One of the possible approaches to improve this is to apply measurements from PMUs. With optimal placement they increase the redundancy of SCADA measurements and eliminates critical measurements, i.e. bad data in measurements is detected completely [18].

SE with the weighted least squares method. The weighted least squares method minimizes the sum of weighted squares of the deviations between estimates and measurements (6.2). This method is very frequently applied to solve the SE problem [6, 7]. The obtained estimates are called the estimates by WLSM. The TEs method is also based on the WLSM.

In estimation theory [19, 20] of the WLSM every outlier can considerably distort the SE results, such as erroneous measurement, errors in the setting of scheme parameters or topological errors. Therefore, different robustness criteria are suggested for obtaining a stable solution when outliers occur [21, 22].

SE with the TEs method based on the robustness criterion. Robustness methods of SE allow for detection of bad data and simultaneously suppression of their impact on the SE results. At present the robustness methods of SE that are based on the search of variants of basic measurements are the most effective. They provide a stable solution even though half of the redundant measurements are bad data. These methods are called highly robust methods of SE and are, as a rule, indirect methods that require the search for a large number of variations and the selection of the best one satisfying the estimation criterion to determine an optimal solution.

Despite their high robustness properties, the application of this group of methods involves certain difficulties, particularly for the solution of the SE problem as classical statement, i.e. with the state vector x. In this case the estimates are calculated by solving the load flow problem for each variation of the basis. Provided that the number of variations is large, then the time required for the solution becomes unacceptable. The publications suggest a series of approaches aiming to overcome these difficulties [22]. The implementation of SE algorithms searching for bases is much easier by using TEs, because of the following factors: basic and redundant measurements are included in an equal way in the set of TEs. Therefore, there is no need to generate new TEs for each change of the basis. Residuals can be calculated without solving the load flow problem, since the SE problem is solved directly in coordinates of the measurement vector y, and the estimates of the measured variables are calculated during the solution process.

Note that decreasing the problem dimension of the subsystems sizably enhances efficiency and speed of the robustness methods of SE. Besides this, when highly accurate measurements from PMUs are applied in the SE algorithms with the TEs method, these measurements as well as the absolutely accurate pseudo measurements of zero injections at the transit nodes are always placed into the basis. As a result the conditionality of the basic Jacobi matrix is improved and the number of variations of basic measurements decreases.

Interaction between the problems. The criterion of EPS SE (weighted least squares or robustness) is selected depending on the solution to the a priori bad data detection problem of the TEs. If bad data are detected or there are no bad data at all, which is indicated by a low value of discrepancies when checking the condition (6.6) for all TEs, the criterion of weighted least squares is selected. If bad data is not identified, then the robustness criterion is chosen with a search for the bases of their suppression.

6.3.2.3 PMU Measurements for Decomposition of the SE Problem

The use of synchronized highly accurate phasor measurements provided by PMUs makes it possible to solve certain problems arising in the SE problem decomposition:

Placement of PMUs at boundary nodes makes it possible to fix the boundary variables U and δ at their most accurately measured values (i.e. the variance of the measurement is close to zero). In this case the boundary conditions can be met automatically and the solution of the coordination problem consists of calculating nodal injections at boundary nodes using the estimates of the power flows in lines that were obtained from calculations of individual subsystems. In this case the states of the individual subsystems can be calculated independently of each other and iterative calculation of the subsystems is not required.

If the boundaries of subsystems are branches, then a real PMU is placed at one of the nodes of the boundary branch. On the other end we can obtain a "calculated" PMU, i.e. the value of voltage magnitude and phase angle can be calculated through the measurements of a real PMU placed at the adjacent node. The accuracy of measurements provided by the "calculated" PMU is almost the same as the accuracy of measurements provided by a real PMU [16].

The placement of a PMU at one of the nodes of a boundary branch in combination with measurements received from the "calculated" PMU at the neighboring node provides the fulfillment of the boundary conditions. In this case the states of the subsystems can also be calculated independently of each other and iterative calculations of the subsystems are not required.

To coordinate voltage phase angles obtained by the solution of the SE problem with subsystems we use the phase angle measurements provided by PMUs which are placed in each subsystem. These nodes are assumed to be slack nodes of the subsystems. The PMU measurements coordinate the SE results of the individual subsystems.

For the decomposition of boundary nodes it is appropriate to place PMUs at boundary nodes that are common for several subsystems. This PMU placement along with the coordination of voltage phase angles of the subsystems simplifies the solution of the coordination problem.

For the decomposition of boundary branches the slack can be chosen to be the node of the boundary branch at which the PMU is placed. Despite the fact that the "calculated" PMU with sufficiently accurate measurements of voltage magnitude and phase angle can be obtained at the second node of the boundary branch, it is not advisable to use this node as a slack node of the second subsystem because there is a probability that a gross error will occur in the phase angle measurement.

This error can distort the value of the phase angle in the slack nodes of both subsystems. A more appropriate approach is to place real PMUs at the slack node of each subsystem.

Measurements from PMUs placed in the vicinity of boundary nodes enhance essentially the efficiency of *bad data detection in boundary regions* and improves the accuracy of the obtained estimates. When calculating simultaneously operating subsystems that use SCADA systems for the collection and processing of measurements, inaccurate synchronization of data in the boundary regions of subsystems can cause bad data, which considerably complicates the procedure of data processing and could affect the convergence of SE.

Accurately synchronized measurements provided by PMUs placed near the boundary nodes essentially increase the redundancy of measurements and the efficiency of methods for detection of bad data in boundary regions.

6.3.3 Calculation Example

The capability of the algorithm to identify bad data is demonstrated at the IEEE 14-node system (Fig. 6.2) with active power measurements.

Gaussian errors are added to the measurement values obtained via load flow. The standard deviation of the measurement errors are: $\sigma_{P_i} = 5$ MW, $\sigma_{Pij} = 2,5$ MW. A PMU installed at a node can guarantee an accurate measurement of the voltage phase angle and magnitude at this node (the error is below 0.01° and 0.15 %). The calculated values of U, δ at the adjacent nodes (by the so called "calculated" PMUs) and also current magnitudes I_{ij} and angles δ_{ij} or active P_{ij} and reactive Q_{ij} power flow measurements in the branches match with the node that has a PMU.

The standard deviations of real and calculated PMU measurement errors are: $\sigma_U = 0.1$ kV, $\sigma_\delta = 0.1°$, $\sigma_{Pij} = 0,5$ MW. The entire system was divided into two subsystems corresponding to the voltage levels of 230 and 115 kV. The first subsystem includes the nodes 1, 2, 3, 4, 5 and the second subsystem includes the rest of the nodes. Nodes 4 and 5 are boundary nodes. The redundancy of measurements in the first subsystem is equal to 2 and in the second subsystem to 1.3. The available measurements were used to generate TEs for both subsystems.

The first subsystem does not contain critical measurements and critical sets. In the second subsystems the measurement P_{5-4} does not belong to the test equation. Hence, it is critical. The measurements $(P_{7-8}, P_{8-7}), (P_{10}, P_{13}, P_{6-11})$, (P_9, P_{4-7}, P_{4-9}) form 3 critical sets.

For an iteration-free simultaneous calculation of the subsystems it is necessary to place a PMU at the boundary nodes 4 and 5. For elimination of critical measurements and critical sets, the simulation algorithm identified the nodes 7 and 13 to be used for the placement of PMUs. Since a PMU at node 7 makes it possible to obtain a "calculated" PMU in the adjacent node 4, the final list of nodes for PMU placement includes the nodes 5, 7 and 13.

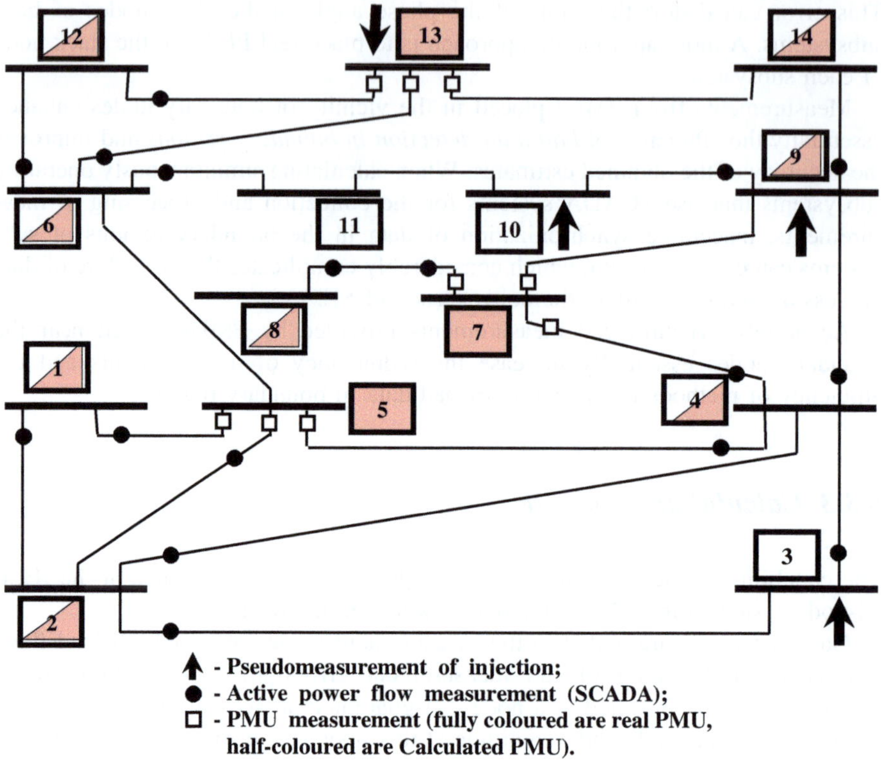

Fig. 6.2 IEEE 14-node system and measurement configuration

Thus, we obtain accurate measurements of U_i and δ_i at the nodes with PMU installations and calculated values of U_i and δ_i at the nodes 1, 2, 4, 6, 8, 9, 12, 14, as well as accurate measurements of P_{ij} from all branches with connection to the nodes that have PMUs. Measurements of voltage magnitudes U_i and phase angles δ_i will be used as reliable initial approximations of the state vector components for SE according to the WLS criterion, and as basic measurements for SE according to the robustness criterion. Measurements P_{ij} received from PMUs will be used to form additional TEs. It should be noted that the gross error that occurs in the measurements of δ_i received from PMUs does not lead to distortion of the measurements of P_{ij} from PMUs. By adding PMU measurements to SCADA measurements we obtain additional TEs: 3 in the first subsystem and 6 in the second subsystem.

Based on the assumption that the amount of bad data does not exceed 20 % of the total amount of measurements the gross error (10–20) σ was simulated under different combinations of erroneous measurements. The calculations have shown the following:

For the first subsystem: despite the fact that the redundancy of measurements in this subsystem is sufficiently high (equal to 2) the TEs method does not allow for correct identification of all combinations of erroneous measurements. For example bad data in measurements of P_{1-2} and P_{1-5} are identified incorrectly. This leads to distortion of the SE results obtained by the high breakdown point estimator. The value of breakdown point is a numerical characteristic of the SE method capability to resist bad data or a measure of its robustness and it is determined as a relationship between the amount of identified bad data m_{bd} and the total amount of measurements $m : \varepsilon^* = m_{bd}/m$

With three additional measurements of power flows P_{ij} obtained from PMU this problem is solved completely.

In the second subsystem the SCADA measurements contain critical measurements and critical sets in which errors cannot be detected by the TEs method. The robustness criterion makes it possible to choose a correct measurement from the critical sets and include it in the basis. However, it does not solve the problem of the gross error in the critical measurement which always belongs to the basis. Besides that, the search for all possible basis variants requires a lot of time (6.6 s, Intel(R) Core(TM) i5-2400 CPU@3.10 GHz, programming language—Fortran Power Station 4.0) at the present combination of measurements.

Accurately synchronized measurements provided by PMU placed near boundary nodes essentially increase the redundancy of measurements and efficiency of the methods for bad data detection in subsystems and in boundary regions.

After placement of PMUs at the nodes 7 and 13 and the use of additional measurements of the flows P_{ij}, that are received from PMUs, the efficiency of the TEs method is noticeably improved. However, another problem arises here: the SCADA measurement of P_{4-5} stops to be critical but forms a new critical set along with the PMU measurement of P_{5-4}. Therefore, if a gross error occurs in one of these measurements it will be necessary to apply the robustness criterion of SE. But since now all nodes of the system except the nodes 10 and 11 have measured or calculated values of δ_i, that are highly accurate and always included in the basis, it takes much less time (1.6 s, Intel(R) Core(TM) i5-2400 CPU@3.10 GHz, programming language—Fortran Power Station 4.0.) to search for the basis variants. In the cases where measurements of δ_i obtained from PMUs contain a gross error, this measurement is replaced by one of the measurements of the flow P_{ij} in the branches adjacent to the node that has a PMU.

6.4 Distributed Hierarchical SE of Large Power Systems on Basis of the Functional Modeling Method

For the calculation of large-scale power systems the functional modeling (FM) method is applied as a basic method of distributed hierarchical SE [23]. A wide survey of literature on hierarchical SE is given in [24].

The main principles of FM methods and its application for solving the SE problem are presented in [25, 26].

The main principle of the FM method lies in the construction and solution of a hierarchical model of the EPS in the form of a system of sets of equations of different levels. The FM methods belong to the technology of state calculations in the distributed systems with operational organization.

The sets of equations of the lower level in the hierarchical model of the EPS are the sets of subsystem equations that consist of equations related to internal and boundary nodes of the subsystem. These equations include internal and boundary variables of the subsystems, which both need to be determined in the solution process.

The set of the upper-level equations, called the set of connection equations (SCE), includes the equations related to all boundary nodes involved in the system model. In this set of equations the boundary variables are the only variables to be determined.

One of the main mathematical formulations introduced in the FM method is the introduction of functional characteristics of the subsystems. These are used to make a transition from a set of subsystem equations to the set of upper-level equations. This characteristic represents a relationship between boundary variables of the subsystems in which all internal equations as well as equality and inequality constraints are satisfied.

In the system of hierarchical distributed SE of the EPS the actions related to the sets of subsystem equations are performed by lower-level computers that are situated within the respective subsystems. The actions that concern the set of upper-level equations are performed by the upper-level computer (server).

6.4.1 Description of Distributed Hierarchical FM Algorithm

The problem of SE in power systems is usually formulated as an optimization problem in which the objective is to minimize the sum of the squares of weighted deviations of the estimated variables from the actual measurements. In this problem the state variables are the voltage magnitudes and phase angles at the system nodes.

As a basic algorithm for the elaboration of the hierarchical FM algorithm the second order Newton method has been chosen. In this method the vector of increments of the state variables of each iteration is calculated from the equation

$$[-\nabla_X J(X)] = \left[\frac{\partial \nabla_X J(X)}{\partial X}\right] \Delta X \tag{6.11}$$

where $\nabla_x J(X)$ is the gradient of objective function $J(X)$, the matrix in the right hand side is the matrix of second derivatives [the Jacobian matrix of the first derivatives of $J(X)$] and ΔX is the vector of state variable increments.

The systems of equations in the hierarchical model intended for the solution of this problem are based on the Lagrange function composed for this model

$$L = \sum_{I} F_I + \sigma^t (S_{bJ} + S_{bK}) \tag{6.12}$$

where S_{bJ} and S_{bK} are injections of the subsystems J and K obtained for the boundary node b. I is the index of the subsystems.

This function consists of the objective functions of the subsystems and the terms added in order to take into account the equality constraints of the boundary variables of the subsystems. The vector σ^t in (6.12) is the transposed vector of Lagrange multipliers, belonging to these constraints.

It is noteworthy that the added terms in (6.12) do not change the problem because at any solution these terms are equal to zero.

Therefore the problem of minimizing (6.12) consists of the calculation of the internal and boundary variables of the subsystems and Lagrange multipliers σ which provide a minimal numerical value to (6.12). In the iteration algorithm this problem is reduced to the calculation of the increments of these variables and multipliers at each iteration.

The function (6.12) can be presented as a sum of Lagrange functions of the subsystems. Each of these functions looks as shown below

$$L_I = F_I + \sigma^t S_{bI} \tag{6.13}$$

The FM algorithm for the solution of the SE problem based on the formulations above consists of the following steps.

1. Setting all state variables in the subsystems to the starting values.
2. Calculation of the state variables at the boundary nodes, corresponding to the values of the state variables in the subsystems of the previous step.
3. Calculation of the gradient of the Lagrange function (6.13) for each subsystem with respect to the internal state variables.
4. Calculation of the gradient of the Lagrange function (6.12) with respect to boundary state variables.
5. At this step the fulfillment of the conditions

$$\begin{aligned} \max \nabla x_i L_s \le \varepsilon \\ \max \nabla x_b L_s \le \varepsilon \end{aligned} \tag{6.14}$$

is checked to identify the finalization of the iteration process, where x_i is internal variables, x_b is boundary variables, L_s is the Lagrange function of the entire system and E is accuracy of calculations.

6. Formation of a linear system of equations for each subsystem representing it on the first iteration and subsequent determination of the FCs of the subsystem from this linear system. This system of equations is presented below

$$\begin{vmatrix} -\nabla_{X_i} L_I \\ -\nabla_{X_b} L_I \\ -\nabla_{\sigma} L_I \end{vmatrix} = \begin{vmatrix} w_{ii} & w_{ib} & w_{i\sigma} \\ w_{bi} & w_{bb} & w_{b\sigma} \\ w_{\sigma b} & w_{\sigma b} & w_{\sigma\sigma} \end{vmatrix} \begin{vmatrix} \Delta X_i \\ \Delta X_b \\ \Delta \sigma \end{vmatrix} \tag{6.15}$$

where the matrix on the right hand side is the Jacobian matrix of the first derivatives of the objective function. The FCs of the subsystems from (6.15) are determinated by successive elimination of increments of the internal variables in this system. The numerical values of all elements in the two lower vectors on the left hand side of (6.15) should be set equal to zero before elimination of the internal variables in (6.15). After this elimination it has the following form

$$\begin{vmatrix} \nabla'_{X_i} L_I \\ \nabla'_{X_b} L_I \\ \nabla'_{\sigma} L_I \end{vmatrix} = \begin{vmatrix} w'_{ii} & w'_{ib} & w'_{i\sigma} \\ 0 & w'_{bb} & w'_{b\sigma} \\ 0 & w'_{\sigma b} & w'_{\sigma\sigma} \end{vmatrix} \begin{vmatrix} \Delta X_i \\ \Delta X_b \\ \Delta\sigma \end{vmatrix} \tag{6.16}$$

in which w'_{ii} is an upper triangular matrix. It follows from (6.16) that the FCs of the subsystem are:

$$\begin{aligned} \left(\Delta \frac{\partial L_S}{\partial X_b}\right)_I &= w'_{bb}\Delta X_b + w'_{b\sigma}\Delta\sigma - \nabla'_{X_b} L_I \\ \left(\Delta \frac{\partial L_S}{\partial \sigma}\right)_I &= w'_{\sigma b}\Delta X_b + w'_{\sigma\sigma}\Delta\sigma - \nabla'_{\sigma} L_I \end{aligned} \tag{6.17}$$

7. Formation and solution of the SCE. The unknown variables in this system are the increments of boundary state variables $\Delta X_b$and the Lagrange multipliers $\Delta\sigma$. The basic equations for formation of the SCE in this algorithm can be obtained by differentiation of the Lagrange function (6.12) with respect to the boundary variables and the Lagrange multipliers σ. These basic equations look as shown below

$$\begin{aligned} \left(\Delta \frac{\partial L_S}{\partial X_b}\right)_J + \left(\Delta \frac{\partial L_S}{\partial X_b}\right)_K &= -\nabla_{X_b} L_S \\ \left(\Delta \frac{\partial L_S}{\partial \sigma}\right)_J + \left(\Delta \frac{\partial L_S}{\partial \sigma}\right)_K &= 0 \end{aligned} \tag{6.18}$$

Substitution of the derivatives in (6.18) with the FCs (6.17) gives the following SCE:

$$|A| \begin{vmatrix} \Delta X_b \\ \Delta\sigma \end{vmatrix} = \begin{vmatrix} -\nabla_{X_b} L_S \\ 0 \end{vmatrix} \tag{6.19}$$

where A is square matrix. The gradient on the right hand side of this equation consists of the first derivatives of L_S with respect to the boundary variables. It should be calculated from the following formula:

$$\nabla_{X_b} L_S = (\nabla_{X_b} L_S)_J + (\nabla_{X_b} L_S)_K \tag{6.20}$$

The solution of the SCE (6.19) completes this step of the algorithm.

8. Calculation of the increments of internal state variables Δx_i in the subsystems. The vectors ΔX_b and $\Delta\sigma$ determined in the previous step should be split into

sub-vectors corresponding to the boundary nodes of each subsystem. For each subsystem the sub-vectors of ΔX_b and $\Delta\sigma$, belonging to the boundary nodes should be substituted into the upper part of the transformed system (6.16).Successive back substitution into the upper part equations of (6.16) is used in order to obtain the values of the increments of all internal variables in the subsystems.

9. Calculation of the new values of the internal state variables in the subsystems in accordance with the general expression

$$X_i^{(k+1)} = X_i^{(k)} + \Delta X_i^{(k)} \tag{6.21}$$

At this step also new values of the Lagrange multipliers σ and the boundary state variables x_b should be calculated according to the following equations:

$$\begin{aligned} \sigma^{(k+1)} &= \sigma^{(k)} + \Delta\sigma^{(k)} \\ \theta^{(k+1)} &= \theta^{(k)} + \Delta\theta^{(k)} \\ U^{(k+1)} &= U^{(k)} + \Delta U^{(k)} \end{aligned} \tag{6.22}$$

10. Return to step 2.

In some cases a part of the nodes of the original system can be chosen as boundary nodes in the hierarchical model. This is possible if the measurements in the neighborhood of these nodes only include measurements of active and reactive power flows on transmission lines adjacent to these nodes.

In these cases the Lagrange function of the hierarchical model of the power system gets the following form

$$F_S = \sum_I F_I \tag{6.23}$$

In this situation the FM algorithm for the solution of the SE problem consists of the same steps as the algorithm presented above with the only difference that the equations (6.15) should be replaced as shown below. The system of equations representing each subsystem is:

$$\begin{vmatrix} \nabla_{X_i} F_I \\ \nabla_{X_b} F_I \end{vmatrix} = \begin{vmatrix} w_{ii} & w_{ib} \\ w_{bi} & w_{bb} \end{vmatrix} \begin{vmatrix} \Delta X_i \\ \Delta X_b \end{vmatrix} \tag{6.24}$$

This system should be formed in step 6 of the algorithm. The system of equations obtained after elimination of the internal variables from (6.24) is written below:

$$\begin{vmatrix} \nabla'_{X_i} F_I \\ \nabla'_{X_b} F_I \end{vmatrix} = \begin{vmatrix} w'_{ii} & w'_{ib} \\ 0 & w'_{bb} \end{vmatrix} \begin{vmatrix} \Delta X_i \\ \Delta X_b \end{vmatrix} \tag{6.25}$$

in which w'_{ii} is an upper triangular matrix. Hence, a FC follows for the subsystem:

$$\left(\Delta\frac{\partial F_S}{\partial X_b}\right) = w'_{bb}\Delta X_b - \nabla'_{X_b}F_I \tag{6.26}$$

In this case the basic equation for the formation of the SCE looks as follows:

$$\left(\Delta\frac{\partial F_S}{\partial X_b}\right)_J + \left(\Delta\frac{\partial F_S}{\partial X_b}\right)_K = -\nabla_{X_b}F_S \tag{6.27}$$

In this algorithm the Formation of the SCE is carried out in the same way as above. The resulting SCE is as follows:

$$B\Delta X_b = -\nabla_{X_b}F_S \tag{6.28}$$

where B is square matrix and the gradient on the right hand side should be calculated according to (6.20).

6.4.2 Solution Methods

Two methods are addressed for the solution of the coordination problem. In the first method a special module is created that constructs FCs after the procedure of SE of every subsystem. In the second method such a module is not needed.

Method 1:
The solution of the coordination problem with the first method can be divided into two stages. At first the voltage phases at all system nodes are brought to conformity with each other. Then the exact coordination of the solutions obtained for the subsystems is performed.

The coordination problem of the first method is solved in the following two steps:

1. The SE problem is solved in each subsystem. The results are the voltage phases and magnitudes at all nodes. The functional characteristics are constructed in each subsystem and then transferred to the upper level. A set of linear equations is solved on the upper level. The voltage phases and magnitudes at the boundary nodes are set equal to the values obtained by the solution of these equations. The values of the state vector components are transferred to the subsystems. The values of the voltage phases at the internal nodes of all subsystems are adjusted.
2. The voltage vectors at all system nodes are coordinated relative to each other. The solution of the coordination problem with this first method is obtained faster, provided that one of the following requirements is fulfilled:

   (a) a common slack node is used for all subsystems
   (b) PMU measurements are available in every subsystem

Method 2:
The second coordination method does not require a special module for the construction of the FCs, but at least one PMU is needed in each subsystem. If this requirement cannot be satisfied, the common slack node is fixed.

There are two main possibilities for the construction of the distributed hierarchical algorithm for SE on the basis of functional characteristics.

In the first possibility the boundary nodes are inserted in the system model in accordance with the FM principles in the form of a set of subsystems. These are inserted in addition to the nodes of the initial model. These boundary nodes are places on the branches connecting the subsystems adjacent to each other. In this case the first part of the branch is considered as belonging to one subsystem and the second part to the other.

In the second possibility the boundary nodes are nodes of the initial system model that correspond to the creation of a set of equations for the considered system as a whole.

Application of phasor measurements (PMU data) in combination with SCADA measurements improves the quality of the SE solution essentially and opens new opportunities for decomposition of the SE problem: (1) The placement of PMUs at the boundary nodes of the subsystems significantly simplifies the solution of the coordination problem and does not require iterative calculation of the subsystems; (2) representation of nodes with PMUs as slack nodes of the subsystems provides the possibility of coordinating the phase angles during the calculation of the subsystems; (3) measurements from PMUs placed near the boundary nodes enhances essentially the efficiency of the BDD in the boundary regions and increases the accuracy of the obtained estimates.

6.4.3 Illustrative Example

The hierarchical algorithm for the solution of the SE problem with the FM method is demonstrated on the network shown in Fig. 6.3. In this example the SE problem consists of the determination of the state variables, i.e. the phase angles θ at all buses of this network. All input data for the problem is presented below.

If the hierarchical algorithm is employed for SE this network is partitioned into two subsystems, I and J, as depicted in Fig. 6.3. The objective function for this network which should be minimized in the SE is written below

$$\begin{aligned} F_S = {} & [P_1 - 8.70\theta_1]^2 + [P_2 - 8.70(\theta_2 - \theta_1)]^2 + [P_3 - 8.70(\theta_3 - \theta_2)]^2 \\ & + [P_4 - 8.70(\theta_3 - \theta_4)]^2 + [P_5 - 8.70\theta_4]^2 \end{aligned} \tag{6.29}$$

where $P_1 - P_5$ are power flow measurements on the transmission lines.

The measurement data in this example are (in MW): $P_1 = -105$, $P_2 = 80$, $P_3 = 100$, $P_4 = 50$, $P_5 = 30$. The parameters of all 5 lines are: $x_1 = x_2 = x_3 = x_4 = x_5 = 20\ \Omega$, and the voltages are U = 100 kV.

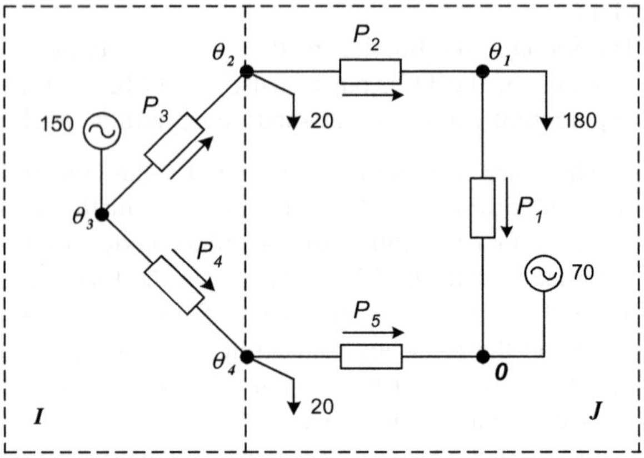

Fig. 6.3 Test system with 5 nodes

If this SE problem is solved by the basic algorithm of the Newton method, the estimated state variables can be calculated from equation (6.10) which takes the following form:

$$\begin{vmatrix} -3220.11 \\ -348.12 \\ 2610.9 \\ -348.12 \end{vmatrix} = \begin{vmatrix} 302.96 & -151.48 & 0 & 0 \\ -151.48 & 302.96 & -151.48 & 0 \\ 0 & -151.48 & 302.96 & -151.48 \\ 0 & 0 & -151.48 & 302.96 \end{vmatrix} \begin{vmatrix} \Delta\theta_1 \\ \Delta\theta_2 \\ \Delta\theta_3 \\ \Delta\theta_4 \end{vmatrix} \tag{6.30}$$

The system of equations for the subsystems I and J are written below according to (6.24):

$$\begin{vmatrix} -3220.11 \\ 0 \\ 0 \end{vmatrix} = \begin{vmatrix} 302,96 & -151.48 & 0 \\ -151.48 & 151.48 & 0 \\ 0 & 0 & 151.48 \end{vmatrix} \begin{vmatrix} \Delta\theta_1 \\ \Delta\theta_{b_1} \\ \Delta\theta_{b_2} \end{vmatrix} \tag{6.31}$$

$$\begin{vmatrix} 2610.9 \\ 0 \\ 0 \end{vmatrix} = \begin{vmatrix} 302.96 & -151.48 & -151.48 \\ -151.48 & 151.48 & 0 \\ -151.48 & 0 & 151.48 \end{vmatrix} \begin{vmatrix} \Delta\theta_2 \\ \Delta\theta_{b_1} \\ \Delta\theta_{b_2} \end{vmatrix} \tag{6.32}$$

After elimination of the internal variables the following systems of equations are obtained according to (6.25):

$$\begin{vmatrix} -3210.11 \\ -1609.93 \\ 0 \end{vmatrix} = \begin{vmatrix} 302.96 & -151.48 & 0 \\ 0 & 75.74 & 0 \\ 0 & 0 & 151.48 \end{vmatrix} \begin{vmatrix} \Delta\theta_1 \\ \Delta\theta_{b_1} \\ \Delta\theta_{b_2} \end{vmatrix} \tag{6.33}$$

$$\begin{vmatrix} 2610.9 \\ 1304.83 \\ 1304.83 \end{vmatrix} = \begin{vmatrix} 302.96 & -151.48 & -151.48 \\ 0 & 75.74 & -75.74 \\ 0 & -75.74 & 75.74 \end{vmatrix} \begin{vmatrix} \Delta\theta_2 \\ \Delta\theta_{b_1} \\ \Delta\theta_{b_2} \end{vmatrix} \tag{6.34}$$

The FCs of the subsystems in accordance with (6.26) are found from the lower parts of (6.33) and (6.34). These FCs are:

$$\begin{vmatrix} \frac{\partial F_I}{\partial \theta_{b_1}} \\ \frac{\partial F_I}{\partial \theta_{b_2}} \end{vmatrix} = \begin{vmatrix} 75.74 & 0 \\ 0 & 151.48 \end{vmatrix} \begin{vmatrix} \Delta\theta_{b_1} \\ \Delta\theta_{b_2} \end{vmatrix} + \begin{vmatrix} 1609.93 \\ 0 \end{vmatrix} \quad (6.35)$$

$$\begin{vmatrix} \frac{\partial F_J}{\partial \theta_{b_1}} \\ \frac{\partial F_J}{\partial \theta_{b_2}} \end{vmatrix} = \begin{vmatrix} 75.74 & -75.74 \\ -75.74 & 75.74 \end{vmatrix} \begin{vmatrix} \Delta\theta_{b_1} \\ \Delta\theta_{b_2} \end{vmatrix} + \begin{vmatrix} -1304.83 \\ -1304.83 \end{vmatrix} \quad (6.36)$$

The expressions from the right hand side of the FCs (6.35) and (6.36) are substituted into the general equation (6.27) in order to obtain the left hand side of the SCE (6.28).

Numerical values of the first derivatives of L_s with respect to the boundary variables are calculated from the input data of each of the two subsystems. It gives:

$$\frac{\partial F_I}{\partial \theta_{b_1}} = -1{,}392.48 \frac{\partial F_J}{\partial \theta_{b_1}} = 1{,}740.6 \quad (6.37)$$

$$\frac{\partial F_I}{\partial \theta_{b_2}} = -522.18 \frac{\partial F_J}{\partial \theta_{b_2}} = 870.3 \quad (6.38)$$

Then according to (6.20) we obtain:

$$\frac{\partial F_S}{\partial \theta_{b_1}} = 348.1 \frac{\partial F_S}{\partial \theta_{b_2}} = 348.1 \quad (6.39)$$

The SCE resulting from the operations above is:

$$\begin{vmatrix} -348.12 \\ -348.12 \end{vmatrix} = \begin{vmatrix} 151.48 & -75.74 \\ -75.74 & 227.22 \end{vmatrix} \begin{vmatrix} \Delta\theta_{b_1} \\ \Delta\theta_{b_2} \end{vmatrix} + \begin{vmatrix} 305.1 \\ -1304.83 \end{vmatrix} \quad (6.40)$$

The values of the boundary variables, found from these SCE are $\theta_{b1} = -2.6493$ and $\theta_{b2} = 3.329$. Substituting these values into the upper equation of (6.34) in subsystem J and solving this equation we get $\theta_2 = 8.975$. The same operation with the upper equation of (6.33) in subsystem I gives $\theta_1 = -11.952$.

The solution of the system (6.30) of the complete network results in exactly the same values of state variables as those obtained above by the hierarchical FM algorithm. The values of the power flows in the distributed SE are

- in subsystem J:

$$P_1 = 8.7 \times \theta_1 = 8.7 \times (-11.952) = -103.98$$
$$P_2 = 8.7 \times (\theta_{b1} - \theta_1) = 8.7 \times (-2.6493 + 11.952) = 80.93$$
$$P_5 = 8.7 \times \theta_{b2} = 8.7 \times 3.329 = 29.51$$

- and in subsystem I:

$$P_3 = 8.7 \times (\theta_2 - \theta_{b1}) = 8.7 \times (8.975 + 2.6493) = 101.31$$
$$P_4 = 8.7 \times (\theta_2 - \theta_{b2}) = 8.7 \times (8.975 - 3.329) = 49.1202$$

6.5 Outlook and Conclusions

The distributed data processing for the decomposition of the SE problem is an effective method of solving the SE, enhancing the quality of the results and the reliability of the computational procedure of EPS SE.

The algorithm has been developed to solve the SE problem for large interconnected power systems. It includes structural and functional decomposition of the problem and its solution by the TEs method (or by any other method). The structural decomposition is carried out by dividing the calculated scheme into subsystems, whose boundaries are nodes or tie lines. The functional decomposition is performed in accordance with the problems to be solved by the SE: bad data detection, SE on the basis of the quadratic and robustness criteria.

The two-stage algorithm has been developed to divide the calculated scheme into subsystems with installation of PMUs at the boundary nodes. At the first stage the scheme is divided into large subsystems with PMU installations at the boundary nodes. At the second stage of the decomposition each subsystem is divided into areas corresponding to the nodal voltage class, which considerably decreases the inhomogeneity of the calculated system.

The proposed two-stage algorithm for structural decomposition of the SE problem allows

- to process data for local subsystems of considerably lower dimension simultaneously
- to decrease the impact of inhomogeneity of the calculated system and measurements when calculating the subsystems of one voltage level. Essentially to simplify the solution of the coordination problem that does not require iterative calculations of the subsystems
- to reduce the time for the solution of the full scheme.

The functional decomposition of the SE problem allows

- to enhance the efficiency of bad data detection methods and correspondingly to increase the accuracy of the estimates
- to organize a flexible selection of the method to solve different SE problems for each subsystem
- to accelerate the measurement processing and thus to reduce the computation time of the SE for the whole system.

The application of synchronized measurements of highly accurate phasors received from PMUs allows for the solution of the following problems arising at the SE problem decomposition:

1. PMU installations at the boundary nodes of the subsystems simplify to the maximum extent the coordination problem solution and do not require iterative calculations of the subsystems.
2. Setting nodes with PMUs as slack nodes of the subsystems provides for a coordination of the voltage phases at the calculation of the subsystems.
3. Measurements from PMUs installed near the boundary nodes essentially enhance the efficiency of bad data detection in the boundary areas and the accuracy of the obtained estimates.

The results of experimental calculations have shown that the use of PMU data in combination with SCADA measurements considerably improves the result of the SE problem and enhances the efficiency of the algorithms.

The FM method is based on exact analytical transformations associated with two concepts which were introduced in this method: FC and SCE. When the FM algorithm is used to solve linear problems then it is possible to obtain the solution in a direct way (within a single iteration), otherwise if the hierarchical algorithm operates with nonlinear problems the same convergence properties are achieved as with basic (one level) algorithms from which they are derived. It is shown that these properties fully apply to a distributed SE algorithm. If PMU devices are installed in boundary nodes the dimension of the upper level problem with SCE can be decreased in distributed SE.

References

1. B. Ayuev, P. Erokhin, Y. Kulikov, in *IPS/UPS Wide Area Measuring System*, CIGRE, Session (27 Aug–1 Sep 2006)
2. A. Phadke, Synchronized phasor measurements. A historical overview, in *Proceedings of IEEE/PES Transmission and Distribution Conf*erence, vol. 1 (2002), pp. 476–479
3. Kazakov PN, Development of capabilities of phasor measurement devices and SMART-WAMS complexes, in *Proceedings of Modern Tendencies of Relay Protection System and Power System Automation Development* (Saint Petersburg, 2011)
4. A. Gamm, I. Kolosok, Test equations and their use for state estimation of electrical power system, in *Power and Electrical Engineering: Scientific Proceedings of Riga Technical University*, (RTU, Riga, 2002) pp. 99–105
5. O. Soukhanov, S. Shil, Application of functional modeling to the solution of electrical power systems optimization problems. Int. J. Electr. Power. Energy Syst. (2) (2000)
6. A. Abur, A. Gomez Exposito, *Power System State Estimation: Theory and Implementation* (Marcel Dekker, New York, 2004)
7. A. Monticelly, Electric power system state estimation. Proc. IEEE **88**(2), 262–282 (2000)
8. F.C. Schweppe, E. Handschin, Static State Estimation in Power Systems. IEEE Proc. **62**, 972–982 (1974)

9. K.A. Clements, O.J. Denison, R.J. Ringle, A multy-area approach to state estimation in power system networks, in *Proceedings of IEEE Power Engineering Society Meeting*, Paper C72 (San Francisco, CA, 1972), pp. 465-473
10. Y. Wallach, E. Handschin, An efficient parallel processing method for power system state estimation. IEEE Trans. Power App. Syst. **100**(1), 4402–4406 (1981)
11. S. Iwamoto, M. Kusano, V.H. Quantana, Hierarchical state estimation using a fast rectangular-coordinate method. IEEE Trans. Power Syst. **4**(3), 870–879 (1989)
12. A. El-KleibA, J. Nieplocha, H. Singh, D.J. Maratukulam, A decomposed state estimation technique suitable for parallel processor implementation. IEEE Trans. Power Syst. **7**(3), 1088–1097 (1992)
13. A.Z. Gamm, Y.A. Grishin, Distributed information processing in automated power system control systems, in *Proceedings of Vth International Workshop "Distributed Information Processing"* (Novosibirsk, Russia, 1995) pp. 243–247 (in Russian)
14. D.M. Falcao, F.F. Wu, L. Murphy, Parallel and distributed state estimation. IEEE Trans. Power Syst. **10**(2), 724–730 (1995)
15. R. Ebrahimian, R. Baldick, State estimation distributed processing. IEEE Trans. Power Apparatus Syst. **15**(4), 1240–1246 (2000)
16. I. Kolosok, E. Korkina, A. Paltsev, Decomposition of power system state estimation problem with the use of PMU data for large dimension schemes, in *Proceedings of the International Workshop "Liberalization and Modernization of Power Systems: Coordinated Monitoring and Control towards Smart Grids"* (SEI, Irkutsk, 2009) pp. 28–35
17. I. Kolosok, A. Paltsev, Application of the multi-agent approach to decomposition of power system SE problem, in *Collected papers of the III International Theoretical and Practical Conference "Power System: Management, Competition, Education"* vol. 1, (UGTU-UPI, Ekaterinburg, 2008) pp. 354–359 (in Russian)
18. I.N. Kolosok, E.S. Korkina, A.S. Paltsev, PMU placement for decomposition of power system state estimation problem, in *Proceedings Of International Conference "ELEKTROENERGETIKA-2010"*, (Slovakia, High Tatras, 2010) pp. 66–72
19. P.J. Huber, *Robustness in Statistics* (Wiley, New York, 1981)
20. P.J. Rousseeuw, A.M. Leroy, *Robust Regression and Outlier Detection* (Willey, New York, 1987)
21. L. Mili, V. Phaniraj, P.J. Rousseeuw, Least median of squares estimation in power systems. IEEE Trans. Power Syst. **6**(2), 511–523 (1991)
22. N.A. Manov, Y.Y. Chukreev, M.I. Uspensky et al., *New Information Technologies in Problems of On-line Control of Power Systems* (UroRAN, Ekaterinburg, 2002), p. 205. (in Russian)
23. D. Kroshko, D. Novitsky, O. Sukhanov, *Hierarchical Algorithms for Solving Power System State Estimation Problems* (Elektronnoye modelirovanie, Kiev, 2007). (in Russian)
24. A. Gomez-Esposito, de la A.J. Villa, C. Gomez-Quiles, P. Rousseaux, T. Van Cutsem, A taxonomy of multi-area state estimation methods. Electr. Power Syst. Res. **81**, 1060–1069 (2011). www.elsevier.com/locate/epsr
25. V. Venikov, O. Sukhanov, *Cybernetic Models of Electric Power Systems* (Energoizdat, Moscow, 1982), p. 328. (in Russian)
26. O. Sukhanov, Y. Sharov, *Hierarchical Models in the Analysis and Control of Power System Operation Conditions* (Izdatelsky dom MEI, Moscow, 2007), p. 312. (in Russian)

Chapter 7
Dynamic State Estimation

Anna Glazunova

7.1 Introduction

The procedure of state estimation is required to obtain electric power system (EPS) state variables on the basis of measurement data. It is differentiated into static and dynamic state estimation (SE). Static state estimation, which is discussed in Chap. 6, uses only a single snapshot. By replacing the static SE algorithms by a dynamic algorithm, the quality of the results can be enhanced and the solution of the on-line SE problem can be accelerated. Dynamic state estimation, which is the topic of the present chapter, makes it possible to:

- Consider interrelations among time-dependent state parameters.
- Apply the values of unmeasured state vector components that were calculated at the previous time instant (in the previous cycle).
- Increase the robustness of the algorithm against failures and data losses.
- Forecast of the state vector $x = (\delta, U)$.

Dynamic state estimation is performed on the basis of Kalman filters. The lead time can vary from several seconds (an interval between the acquisition of measurement data (snapshots) at the time instants k and k + 1) to several minutes. Snapshots provide information on measured values of the EPS state parameters at a certain time instant. This information is provided to the dispatching center by the supervisory control and data acquisition (SCADA) system and by Phasor Measurement Units (PMUs).

The state vector $x = (\delta, U)$ is used to calculate all state parameters. Forecasting of all state vector components at one $(k + 1)$ or several $(k + m)$ time instants ahead makes it possible to calculate all the values of all state variables of these snapshots.

A. Glazunova (✉)
Energy Systems Institute, Irkutsk, Russia
e-mail: glazunova@isem.sei.irk.ru

U. Häger et al. (eds.), *Monitoring, Control and Protection of Interconnected Power Systems*, Power Systems, DOI: 10.1007/978-3-642-53848-3_7,

7.2 Background Information

The procedure of state estimation is required to obtain the EPS state variables on the basis of measurement data [1, 2]. The measurements employed in EPS state estimations are the following measurements obtained from SCADA system and PMUs: Magnitudes of nodal voltages U_i; Active and reactive nodal generation $P_{g(i)}$, $Q_{g(i)}$; Active and reactive nodal loads $P_{l(i)}$, $Q_{l(i)}$; Power flows through transformers and lines P_{ij}, Q_{ij}; Voltage phases angles δ_i at the nodes of the system in which PMUs are placed; Magnitudes of currents I_{ij} through the lines connected to the PMU nodes; The angles φ_{ij} between current and voltage on these lines.

The vector of measurements is composed as follows: $\bar{y} = \left(P_i, Q_i, P_{ij}, Q_{ij}, U_i, \delta_i, I_{ij}, \varphi_{ij}\right)$, with $P_i = P_{l(i)} + P_{g(i)}$ and $Q_i = Q_{l(i)} + Q_{g(i)}$. It is recommended for the dynamic SE algorithm that all measurements are divided into two groups. The first group consists of the measurements that represent the state vector components $\bar{y}^I = (\delta_i, U_i)$, and the other group includes all the remaining measurements $\bar{y}^{II} = \left(P_{g_i}, Q_{g_i}, P_{l_i}, Q_{l_i}, P_{ij}, Q_{ij}, I_{ij}, \varphi_{ij}\right)$.

7.3 Modeling

This section explains the selection of the dynamic model. For short-term forecasting of state parameters simple dynamic models can be used. Furthermore, the measurement model is shown. In addition, this section considers the Kalman filter construction and presents the objective function of the dynamic state estimation.

7.3.1 Dynamic Model

The dynamic model has a strong impact on the efficiency of dynamic state estimation. Too complex models do not always improve quality and reliability of SE results. For short-term forecasting of state parameters we use simple dynamic models. Random fluctuations of state parameters are represented as a stationary Gaussian process in this model and all changes in state parameters are considered as random fluctuations and are taken into account in the vector $\xi_{F_k} \in N(0, \sigma^2_{F_k})$ as follows:

$$x_{k+1} = F_k x_k + \xi_{F_k} \tag{7.1}$$

$$M(\xi_{F_k}) = 0 \tag{7.2}$$

$$cov(\xi_{F_k}) = M(\xi_{F_k}, \xi_{F_k}) = \sigma^2_{F_k} \tag{7.3}$$

$\sigma^2_{F_k}$ is the noise variance of the dynamic model and F_k is a transition matrix between two EPS states.

To determine the noise variance of the dynamic model a sample is made up of s rows. Each row consists of $2n - 2$ elements, where n is the number of nodes in the system. For the elements of this sample we use the difference between measurements and their estimates Δ_k. The model of noise variance is determined by the following equation:

$$cov(\xi_{F_k}) = \sigma^2_{F_k} = cov(\Delta_{kj}, \Delta^T_{ki}) \tag{7.4}$$

In this equation $\sigma^2_{F_k} \neq 0$ for $i = j$ and $\sigma^2_{F_k} = 0$ for $i \neq j$. Furthermore, $\Delta_{kj} = \bar{x}_j - \hat{x}_j$ is the difference between the j-th measurement of state vector component and its estimate for the k-th snapshot.

The EPS can be represented by a Markov processes, i.e. the estimates of state parameters at a time instant $k + 1$ are determined by the estimates of state parameters at a time instant k with additional noise. Noise at a time instant k contains statistic information. The value of the noise variance is determined by recurrence relations.

7.3.2 Model of Measurement Data

Measurements coming from the SCADA system and from PMUs represent a sum of a true value and normal noise:

$$\bar{y} = y_{true} + \xi_y, \xi_y \in N\left(0, \sigma^2_y\right) \tag{7.5}$$

In this equation the measurement variance σ^2_y is determined by the metrological path. The difference between the two types of measurements lies in their accuracy.

7.3.3 Kalman Filter

A Kalman filter is used for dynamic SE. The Kalman filter is a recurrence method for solving the state estimation problem by the least squares method. Recurrence relations are constructed, in which the estimates of the k-th time instant are used to obtain the estimates at the time instant $k + 1$.

Based on a priori knowledge of the system behavior, the state space representation for a discrete time-variant system is described by the following equations [3]:

$$x_{k+1} = F_k x_k + \xi_{F(k)} \tag{7.6}$$

$$y_k = H_k x_k + \xi_{y(k)} \tag{7.7}$$

Equation (7.6) presents the dynamic model of the process $x(t)$. Equation (7.7) provides the dependence of the measurements $\bar{y}$ on the state vector x $(\bar{y}(x))$. After transformation of these equations we can obtain a covariance matrix of the forecast errors:

$$M_{k+1} = F_k P_k F_k^T + W_{F(k)} \tag{7.8}$$

$W_{F(k)}$ consist of $\sigma^2_{F(k)}$. The non-diagonal elements of this matrix are zero. P_k– the covariance matrix of estimation errors of state vector components. The transition matrix F is assumed to be a unit matrix because the considered time frames are small enough and the system changes are very slowly.

In dynamic state estimation the covariance matrix of noise from the dynamic model and of the state vector estimates are calculated with the recurrence relations. The elements of the covariance matrix of noise from the dynamic model are determined by the following equations:

$$\mu_{(k)j} = (1-\alpha)\mu_{(k-1)j} + \alpha \Delta_{(k)j} \tag{7.9}$$

$$\sigma^2_{F(k)} = (1-\alpha)\sigma^2_{F(k)} + \alpha\left(\Delta_{(k)j} - \mu_{(k)j}\right)^2 \tag{7.10}$$

In these equations $0 < \alpha < 1$ is the memory coefficient and μ_j are the mathematical expectations. The covariance matrix of the state vector component estimates is calculated as follows:

$$P_k = M_{k-1} - K_k H_k M_{k-1} \tag{7.11}$$

$H_k = \frac{\partial y}{\partial x}$ is the matrix of derivatives and $K_k = M_{k-1} H_k \left(H_k^T M_{k-1} H_k + R_k\right)^{-1}$ is the filtering coefficient.

7.3.4 Objective Function of Dynamic State Estimation

The objective function of the dynamic SE is composed as follows:

$$J(x) = (\bar{y} - y(x))^T R_y^{-1} (\bar{y} - y(x)) + (\bar{x} - x)^T * M^{-1} * (\bar{x} - x) \tag{7.12}$$

$\bar{x}$ are the measurements or pseudo measurements of state vector components, R_y is the measurement error covariance matrix whose diagonal elements are equal to the measurement variances $r_{ii} = \sigma_i^2$.

It is very important that all state vector components are used in the objective function. They are considered as measurements if they are measured and as pseudo measurements if they are not measured. For the weighting coefficients in the second term of the objective function the inverse matrix of the covariance matrix M^{-1} is used, which makes it possible to take into account the accuracy of the estimates obtained in the previous time instants (P) as well as the model noise (W).

Due to the nonlinear dependence $y(x)$, the problem is solved iteratively by use of the Newton method. The initial condition at each iteration is result of the EPS state estimation (values of state vector components) and the covariance matrix characterizing the estimation errors. The algorithm processes new measurement values by taking into account the measurement errors, and specifies the initial conditions. The adjusted initial conditions are the output data of the filter.

For a good performance of the dynamic SE tuning of the Kalman filter is necessary. The filter tuning implies the stabilization of the diagonal elements of the covariance matrix W_k of the dynamic model.

7.4 Applications of Dynamic State Estimation

This section introduces two problems of energy power systems that can be solved by use of dynamic state estimation. The main algorithm of dynamic state estimation is described. Formulas for the calculation of the estimate quality and the forecast accuracy are presented.

7.4.1 Filtering of Random Measurement Errors

In case of highly accurate voltage phase and magnitude measurements there is no need to estimate these components. In this case the gain matrix of the SE has the dimension $(2n - 2 - N_{acr}) \times (2n - 2 - N_{acr})$, where n is the number of nodes in the system and N_{acr} is the number of accurate state vector component.

Algorithm of dynamic SE

The main algorithm of dynamic state estimation is introduced in this section. The first few snapshots are used in order to tune and adjust the Kalman filter. Figure 7.1 shows the flowchart of the Kalman filter tuning. The Kalman filter is tuned each time a change of the topology happens. At the first snapshot the static state estimation is performed. Then initial approximations of the covariance matrix W are set. At the subsequent k_{tune} snapshots the dynamic SE is performed in which the covariance matrices W, P are calculated by Eqs. (7.4) and (7.15) respectively. A detailed description of all steps of the dynamic state estimation is provided below.

1. Read out of snapshot data at the time instant $k = 1$. The static state estimation is performed.
2. The measurements of these state vector components are used as initial approximations. If no measurements are available, then the state estimation results of the previous snapshot ($\hat{x}_{k-1}$) are used.

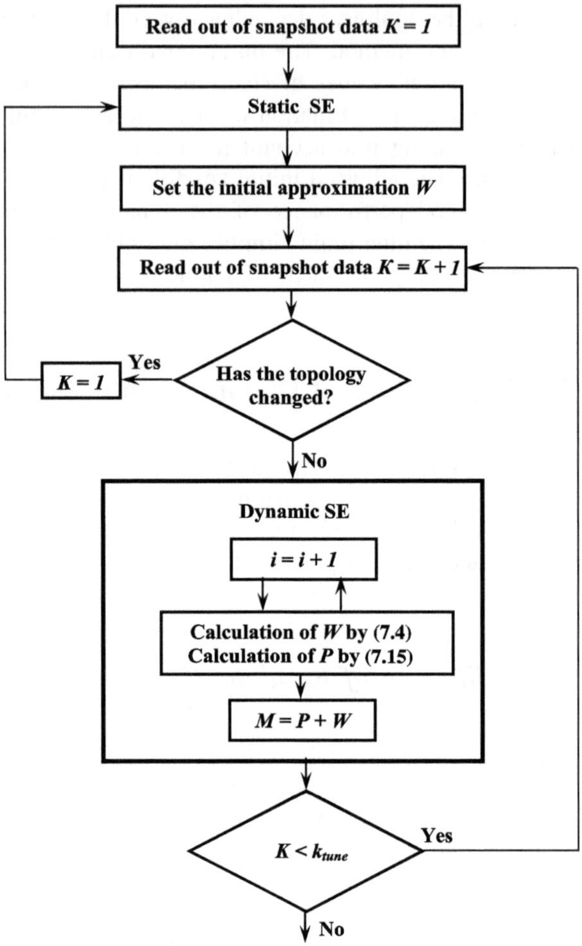

Fig. 7.1 Flow chart of the filter tuning. k_{tune} is the number of snapshots that are used to tune the Kalman filter

3. Calculation of the matrix of derivatives:

$$H_k^i = \frac{\partial y}{\partial x} \tag{7.13}$$

$$y_k^i = y(x_k^i) \tag{7.14}$$

In these equations i is the iteration counter.

4. Calculation of the estimation error covariance matrix of the state vector components by use of Eq. (7.11) or by use of the following equation:

$$P_k = \left(H_k^{\mathrm{T}}(R)^{-1}H_k\right)^{-1} \tag{7.15}$$

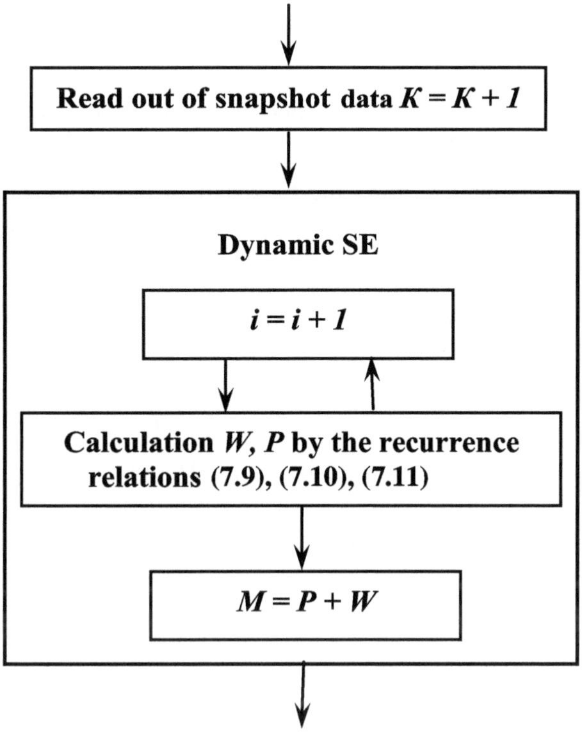

Fig. 7.2 Flow chart of the filter adjustment

The snapshots equation (7.15) (compare Fig. 7.1) is used to generate k_{tune}, while in other cases Eq. (7.11) (compare Fig. 7.2) is used.

5. Calculation of a correction value for state vector components:

$$\Delta x_k^i = P_k * M_{k-1}^{-1} * (\bar{x}_k - x_k) + P_k H_k^T * R_{y,k}^{-1} * \left(\bar{y} - y(x_k^i)\right) \tag{7.16}$$

6. Calculation of the state vector component values

$$x_k^{i+1} = x_k^i + \Delta x_k^i \tag{7.17}$$

7. If $\Delta x_k^i < \xi$, go to Step 8. Otherwise go to Step 3.
8. Calculation of the steady state conditions.
9. Calculation of the covariance matrix W_k of the noise from the dynamic model by use of Eqs. (7.9) and (7.10). If a change of the topology happens we use Eq. (7.4).
10. Calculation of the forecast error covariance matrix

$$M_k = P_k + W_k \tag{7.18}$$

11. Read out of the snapshot data at the time instant k. The input conditions $\hat{x}_k, M_k$ are set.
12. In case of detected topology changes set $k = 1$ and go to Step 1.
13. Go to Step 2.

Figure 7.2 shows the flowchart of the Kalman filter adjustment. This is performed by calculating the covariance matrices W and P with recurrence relations (7.9), (7.10) and (7.11). Adjustment of the Kalman filter is performed after each cycle of the dynamic SE.

The quality of the estimates is determined with the following value:

$$\varphi_k = \sum_{i=1}^{m} \frac{(\bar{y}_i - y_i(\hat{x}))^2}{\sigma_i^2} \tag{7.19}$$

m is the number of measurements. The smaller $_{\varphi k}$ is, the better is the filtering of the random errors.

7.4.2 Forecasting

The state vector is used to calculate all state variables. Forecasting of all state vector components for some time ahead makes it possible to forecast the values of all state variables.

Forecasting of state vector components is performed as dynamic SE if there are no new measurements available. The measurement variance in Eq. (7.12) is then assumed to be equal to infinity.

The values of the forecasts of all state vector components are calculated by the following equation:

$$\tilde{x}_{k+1} = \bar{x}_k + P_k * M_k^{-1} * \Delta\tilde{x}_k \tag{7.20}$$

The forecast error in the first forecasting cycle is determined by the following equation:

$$\Delta\tilde{x}_k = \hat{x}_k - \bar{x}_k \tag{7.21}$$

$\hat{x}_k$ are the estimates of the state vector components of the previous snapshot and $\bar{x}_k$ are the measurements of the previous snapshot. In the subsequent snapshots the forecast error is calculated by the following equation:

$$\Delta\tilde{x}_k = \hat{x}_k - \tilde{x}_k \tag{7.22}$$

$\tilde{x}_k$ is the forecast of the state vector components of the previous snapshot.

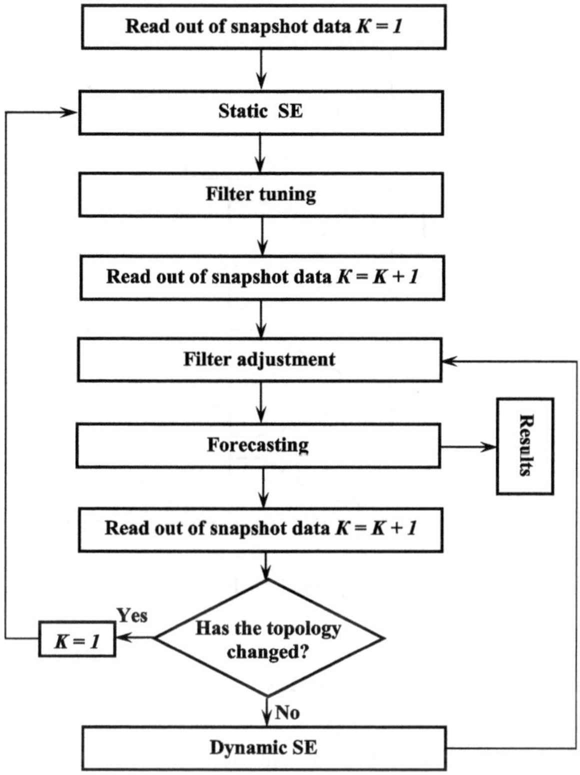

Fig. 7.3 Flowchart of state vector forecasting

Figure 7.3 shows a flowchart of the forecasting process of state vector components. At the first snapshot the static state estimation is performed. At the subsequent m snapshots the filter is tuned to execute the forecast. Furthermore, the state vector components of the next snapshot are forecasted. If needed, all parameters of the forthcoming state are calculated, and the obtained information is transferred to the respective programs. With arrival of the next snapshot the procedure of dynamic state estimation is started. Based on the current state estimation the Kalman filter is adjusted.

The forecast accuracy is determined according to the following equation:

$$\varphi_x = \frac{1}{k}\sum_{1}^{k}\left(|\tilde{\mathrm{x}}| - |\hat{\mathrm{x}}|\right) \tag{7.23}$$

φ_x is the average error for forecasting the voltage magnitude and phase angle with regard to estimates of the total number of snapshots.

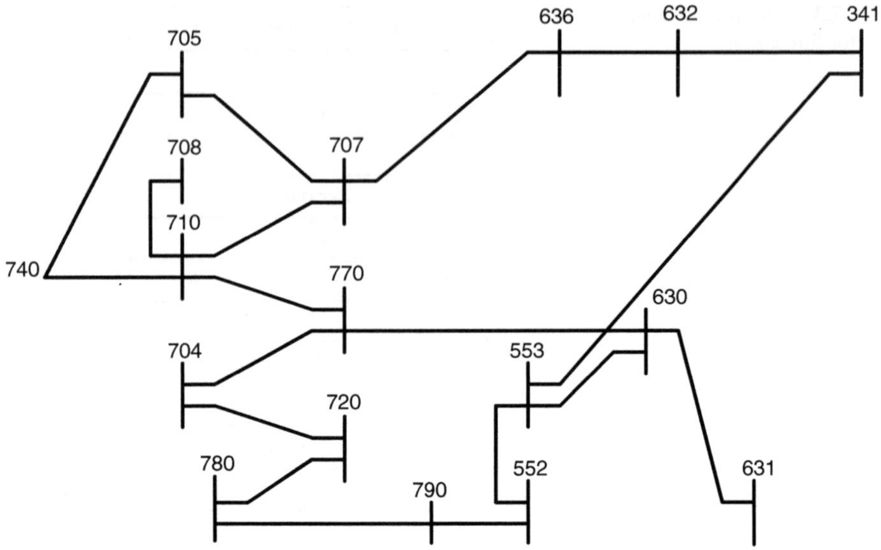

Fig. 7.4 The calculation model of the interconnected electric power system

7.5 Numerical Results

This section demonstrates the ability of dynamic state estimation to filter random measurement errors and to perform a short-term forecast of voltage magnitude and phase angle in a particular system. For these purposes a database of measurement snapshots is created. Numerical results are obtained for each node of the system. These results are shown for several nodes.

7.5.1 Test System

The technique for state estimation and for forecasting of state vector components is validated with the scheme shown in Fig. 7.4, which presents part of the actual network. This system consists of 17 nodes and 21 lines with several closed loops. There are 46 measurements in this system.

7.5.2 Database Creation

In order to create a test case, a virtual database of measurement snapshots is generated. The database is created by time-domain simulations on the basis of the active and reactive load curves at node 341. The curves, which are illustrated in

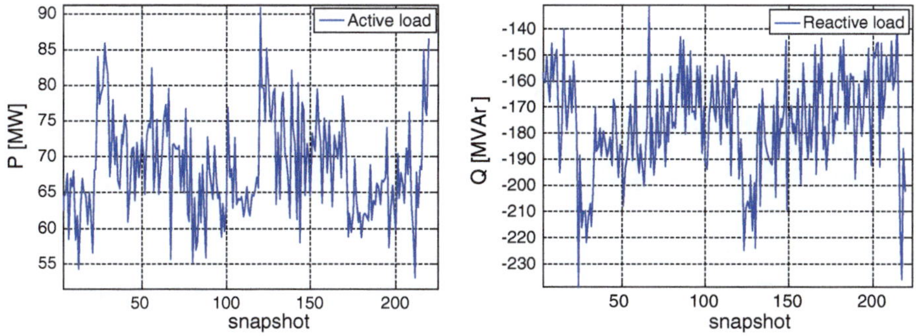

Fig. 7.5 Load measurements at node 341

Fig. 7.5, have two maximums (and minimums) (snapshots 20–30 and snapshots 120–130). The frequency of repetition is 96 snapshots. Every database consists of 46 measurement curves.

Each snapshot is created in the following way: The steady state situation is calculated at a certain load value. The error x_{run} is generated by use of a random number and then superimposed on the steady state $y_{i(ss)}$. The measurement values are calculated according to the following expression:

$$y_{i(sim)} = y_{i(ss)} + x_{run}\sqrt{\sigma^2_{i(y)}} \tag{7.24}$$

$\sigma^2_{i(y)}$ is the measurement variance and $x_{run} \rightarrow N(0, 1)$. The Calculated measurement values are called snapshots. Two virtual databases are created. The measurements in these two databases differ by the value of the variance of the simulated measurement errors. Measurements of the second group (y^{II}, compare Sect. 7.3) are measured with the following precision: $\bar{P} = P_{true} \pm 15$ MW, $\bar{Q} = Q_{true} \pm 45$ MVAr.

In database 1 the components of the state vector x are measured with high accuracy $\bar{U} = U_{true} \pm 0.75$ kV $\bar{\delta} = \delta_{true} \pm 0.017°$. In database 2 the components of the state vector x are measured with low accuracy. $\bar{U} = U_{true} \pm 5$ kV $\bar{\delta} = \delta_{true} \pm 0.085°$.

Figures 7.6 and 7.7 show the voltage magnitude and the phase angles obtained at different values of the measurement error variance.

7.5.3 Kalman Filter Tuning

Static SE is used to receive the first snapshot, which is applied as initial approximation for the dynamic SE. At first several snapshots are used to perform the filter tuning. Statistic information is accumulated in these snapshots for the calculation of the covariance matrix W_F. Figure 7.8 presents the stabilization process of the diagonal elements of the covariance matrix.

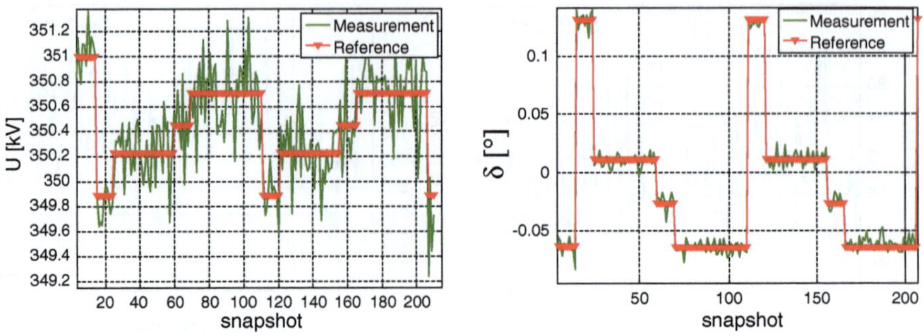

Fig. 7.6 Voltage magnitude and phase angle at node 341 with accuracy according to data base 1

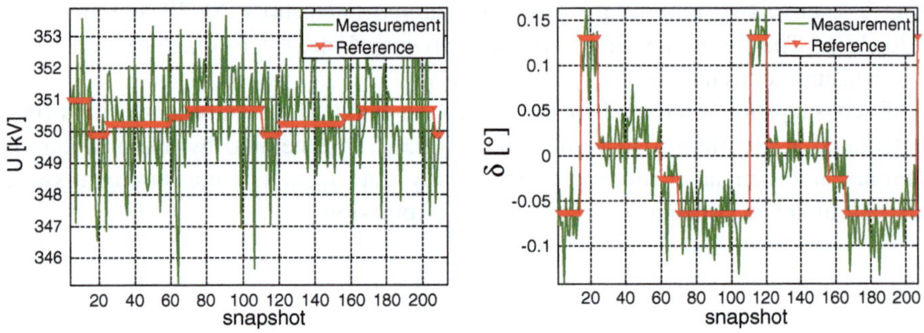

Fig. 7.7 Voltage magnitude and phase angle at node 341 with accuracy according to data base 2

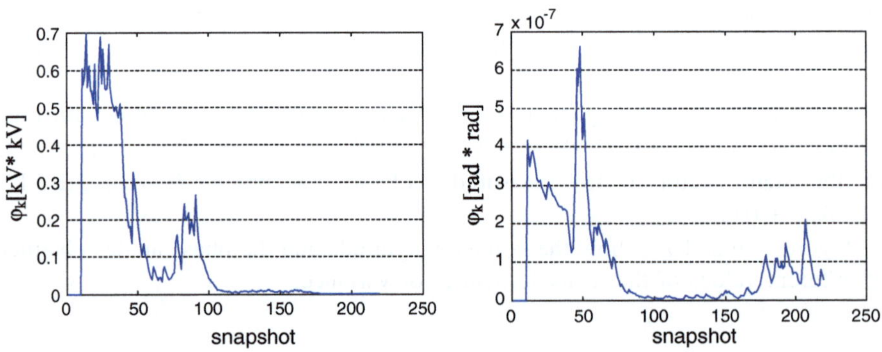

Fig. 7.8 Stabilization process of the diagonal elements of the covariance matrix

7.5.4 Filtering of Random Errors

The dynamic state estimation (filtering of random errors in the measurements) is carried out for both accuracy datasets. Figure 7.9 presents the reference values, the measurements and the estimates of voltage magnitude and phase angle with data

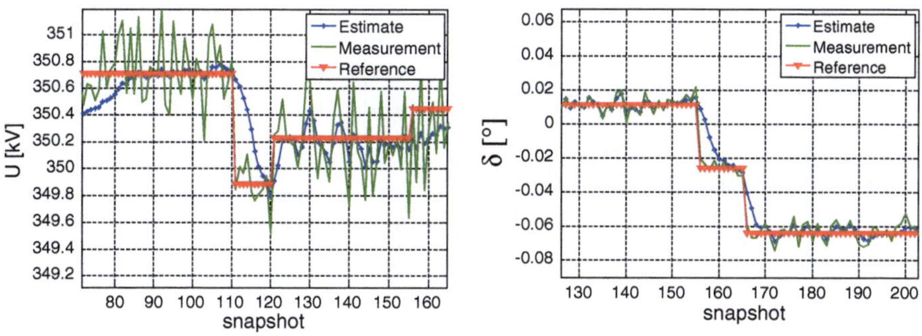

Fig. 7.9 References, measurements and estimates of the voltage magnitude and phase at node 341 with data base 1

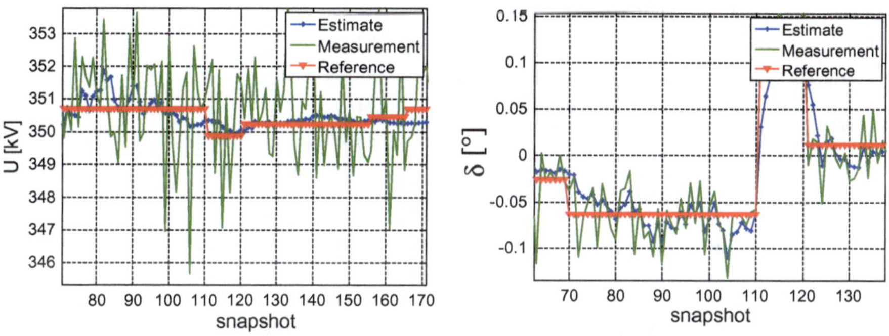

Fig. 7.10 References, measurements and estimates of the voltage magnitude and phase at node 341 with data base 2

base 1 at 200 snapshots at node 341. From Fig. 7.9 it can be seen that the estimates of the phase angle are close to the measurements. This shows that there is no need to estimate these components.

Figure 7.10 presents the reference values, the measurements and the estimates of voltage magnitude and phase angle with data base 2 at 200 snapshots.

From Figs. 7.9 and 7.10 it can be derived that the estimates are close to the reference values. This means that the filtering of random errors in measurement improves the quality of these measurements.

7.5.5 Forecasting the State Vector Components

For the forecast analysis all state vector components are divided into two groups. Measured components belong to the first ground and unmeasured components belong to the second group.

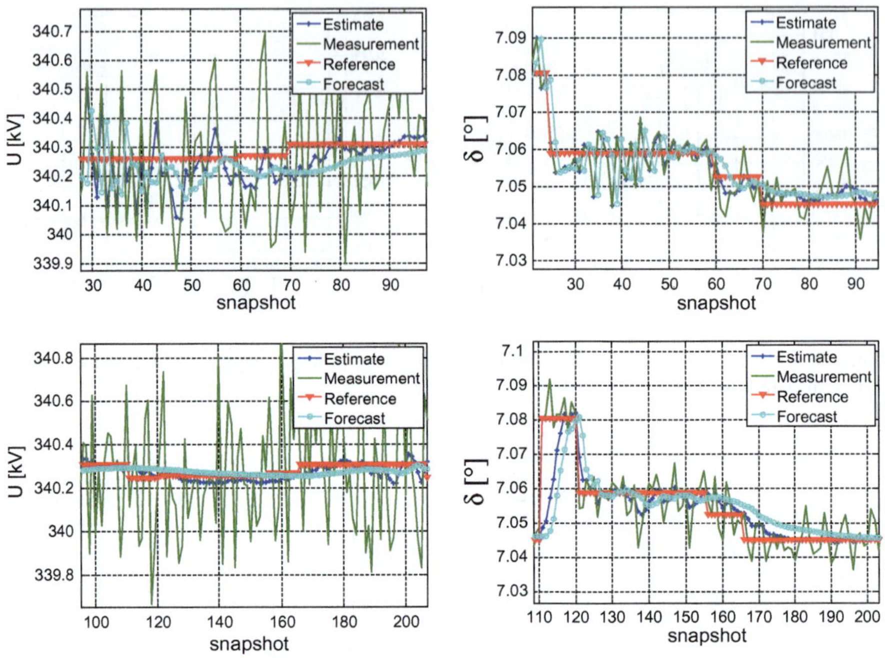

Fig. 7.11 Forecast of the accurate state vector component

At first the forecast of measured state vector components is presented for both data bases. Measurements are applied as initial approximations of the state vector components. Figure 7.11 presents the forecast values of voltage magnitude and phase angle for several snapshots with data base 1.

Figure 7.12 presents the average forecast error that was calculated at 200 snapshots according to Eq. (7.23). Table 7.1 presents the average forecast error for the voltage magnitude and phase angle respectively of 200 snapshots according to Eq. (7.23). Figure 7.12 and Table 7.1 show that the forecast is made with high accuracy, because the average forecast error does not exceed the accuracy of the measurements. The voltage magnitude is measured with an accuracy of ±0.75 kV. The voltage phase angle is measured with an accuracy of ±0.017°.

Figure 7.13 presents the forecast values of voltage magnitude and phase angle for several snapshots with data base 2. Table 7.2 presents the average forecast error of 200 snapshots. Figure 7.13 and Table 7.2 show that under the considered conditions the forecast is made with high accuracy, because the average forecast error does not exceed the accuracy of the measurements. The voltage magnitude is measured with an accuracy of ±5 kV. The voltage phase angle is measured with an accuracy of ±0.085°.

In the following the forecast of unmeasured state vector components is presented. In this case the estimates of the state vector components from the previous cycle are applied as initial approximations of the new state vector components.

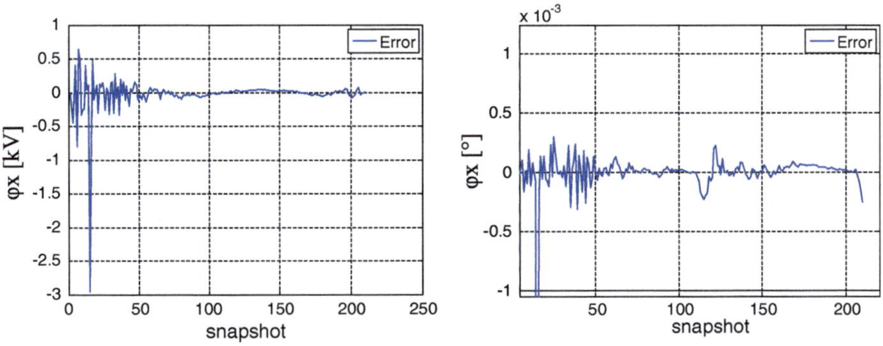

Fig. 7.12 Average forecast error at 200 snapshots

Table 7.1 Average forecast error of accurate state vector components

Component	Average forecast error (%)	
U	0.0856 kV	0.02
δ	0.0064°	0.085

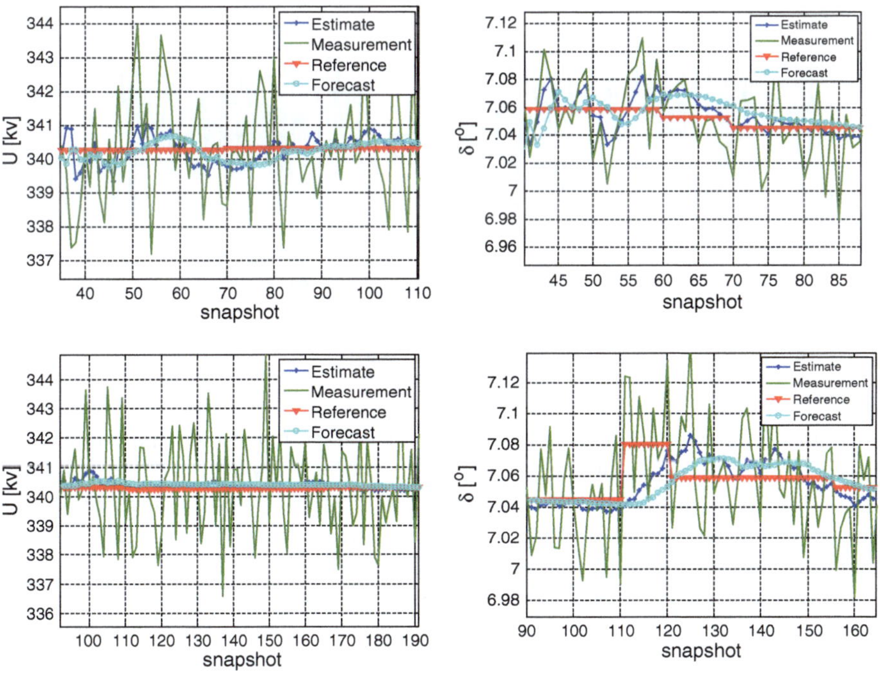

Fig. 7.13 Forecast of the inaccurate components

Table 7.2 Average forecast error of inaccurate state vector components

Component	Average forecast error (%)	
U	0.29 kV	0.08
δ	0.012°	0.17

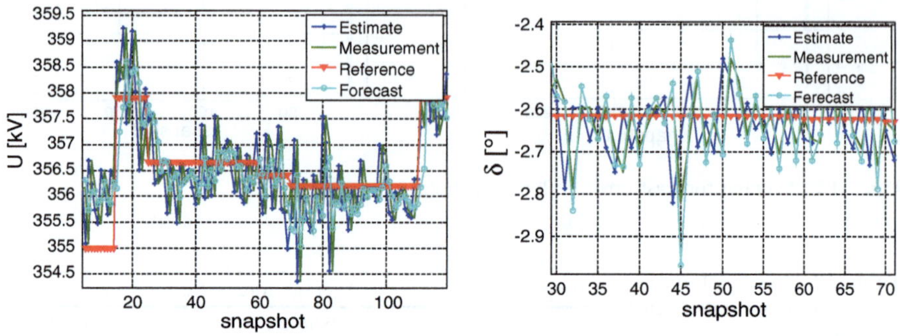

Fig. 7.14 Forecast of the state vector components

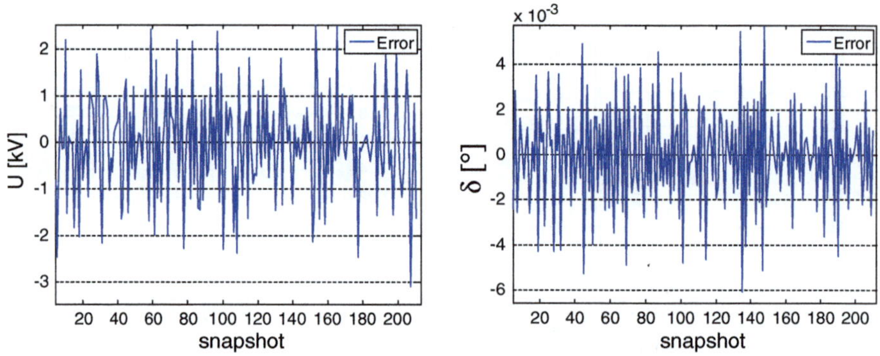

Fig. 7.15 Forecast error of the state vector components

Figure 7.14 presents the forecast values of voltage magnitude and phase angle, based on the following variances: $\sigma_P^2 = 25\ \text{MW}^2$, $\sigma_{Pij}^2 = 25\ \text{MW}^2$, $\sigma_Q^2 = 100\ \text{MVAr}^2$, $\sigma_{Qij}^2 = 100\ \text{MVAr}^2$. Figure 7.15 indicates the average forecast error of 200 snapshots. Table 7.3 presents the average forecast error of the unmeasured state vector components at 200 snapshots.

Analysis of Fig. 7.15 and Table 7.3 shows that the forecast is accurate enough, because the average forecast error of the unmeasured state vector components does not exceed the measurement accuracy of the measured state vector components.

Table 7.3 Average forecast error of the unmeasured state vector components

Component	Average forecast error (%)	
U	1.3 kV	0.37
δ	0.085°	3.66

7.6 Conclusions

This chapter suggests approaches for short-term forecasting of the parameters of expected operating conditions with use of Kalman filter based algorithms of dynamical state estimation. The lead time can vary from several seconds to several minutes. The state vector $x = (\delta, U)$ is used to calculate all state parameters. Forecasting of all state vector components at one $(k + 1)$ or several $(k + m)$ time instants ahead makes it possible to calculate all the values of all state variables of these snapshots. In addition, dynamic state estimation is used for filtering random errors in measurements. This chapter demonstrates the ability of the dynamic state estimation algorithm to filter random measurement errors and to perform short-term forecast of voltage magnitude and phase angle in a test system.

The filtering of random measurement errors of the state vector components is presented on the basis of measurements of the second group (compare Sect. 7.2) $\bar{y}^{II} = (P_{g_i}, Q_{g_i}, P_{l_i}, Q_{l_i}, P_{ij}, Q_{ij})$.

If the measurements have high accuracy, then there is no need to estimate these components. The analysis of all the considered situations illustrates that the suggested forecasting technique yields acceptable accuracy:

- For measured and unmeasured state vector components;
- With different random errors in the measurements.

Measurements from PMUs that are also applied as state vector components and precise measurements improve the forecast results. The forecast error of the voltage magnitude in comparison with the estimate does not exceed 1.3 kV. The forecast error of the voltage phase angle in comparison with the estimate does not exceed 0.09°.

The disadvantage of this tool is the need to update existing information when the network configuration is changed. The dynamic state estimation algorithm starts from the beginning.

References

1. A.Z. Gamm, Statistical methods for state estimation of electric power systems. M.: Nauka, 220 pp. (1976, in Russian)
2. A. Gamm, Y. Grishin, A. Glazunova, I. Kolosok, E. Korkina, New EPS state estimation algorithms based on the technique of test equations and PMU measurements, in *Proceedings of the International Conference PowerTech'2007*, Lausanne, 2007
3. M.B. Do CouttoFilho, J.C.S. de Souza, Forecasting-Aided state estimation–part I: Panorama. IEEE Trans. Power Syst. **24**(4) (2009)

Chapter 8
Inter-TSO Solutions for Monitoring and State Estimation

Anna Mutule, Karlis Brinkis, Oleg Kochukov and Kay Görner

8.1 Introduction

The main goal of this section is to test and validate the state estimation algorithms which are described in Chaps. 6 and 7. For this purpose a special test case, the Baltic Electrical Ring was developed. The test case represents a real network with more than 100 nodes. Furthermore, there are created several scenarios of the network operation, which are tested with the state estimation algorithms. In order to verify the accuracy, the results are compared with SCADA data.

In order to complete state estimation, it is also necessary to create an information technology complex. A prototype of such complex is presented in Sect. 8.3. In each sub-system standard software of static State Estimation can be applied. In the course of the research we applied software OTSENKA and MUSTANG. All other algorithms except the static state estimation algorithm are implemented in MATLAB.

The objectives of the state estimation prototype also include:

- Development of methods and algorithms for solving the state estimation (SE) problem on the basis of the Test Equations method and new metering devices (PMU, smart power meters etc.);

A. Mutule (✉) · K. Brinkis · O. Kochukov
Institute of Physical Energetics, Riga, Latvia
e-mail: amutule@edi.lv

K. Brinkis
e-mail: info@bleea.lv

O. Kochukov
e-mail: olegsk@edi.lv

K. Görner
TU Dortmund University, Dortmund, Germany
e-mail: kay.goerner@tu-dortmund.de

U. Häger et al. (eds.), *Monitoring, Control and Protection of Interconnected Power Systems*, Power Systems, DOI: 10.1007/978-3-642-53848-3_8,

- Development of methods for dynamic state estimation to forecast and monitor operating conditions on the basis of new systems and technologies (PMU, artificial intelligence (AI) methods, etc.);
- Enhancing the efficiency of methods for bad data detection in measurements on the basis of AI methods, dynamic approaches and non-quadratic criteria;
- Development of distributed algorithms for solving the SE problem on the basis of decomposition methods for SE problem and PMU measurements.

Future objectives of the state estimation prototype could include automatic and fast data input from PMU and operational state forecast.

8.2 Test Case for Inter-TSO Network Monitoring and State Estimation

To test operational state calculation methods and algorithms for application of emergency automation, the Baltic Electrical Ring model was chosen. This model is interesting for many reasons that are described in more detail in Sects. 8.2.1 and 8.2.2.

8.2.1 Selection Reason of the 330 and 750 kV Electrical Ring Network

The electrical ring is sufficiently long (around 2,500 km) with a large number of power plants—more than 10, which have capacities exceeding 900 MW and which are connected to the 330 and 750 kV network.

Breaking the electrical ring, especially by disconnection of the branch between Moscow and St. Petersburg, as well as between Belarus and Smolensk, leads to maximal active power flow at the remaining part of the electrical ring transmission network that in several cases can exceed maximal permissible values, causing blackouts [1].

The electrical ring is equipped with emergency control schemes, which operate with emergency power reserves (start-up of hydro generators, switch-off of electrical energy consumers or hydro generators, which operate in pump-mode, etc.), prevent asynchronous states of the system and in this way prevent blackouts. These actions refer to a power system or power system union with active generation power deficit.

It is clear that if the electrical ring is opened between Moscow and St. Petersburg and in pre-emergency state considerable active power flow is turned to Moscow, emergency control schemes of Leningrad power system have to limit generation power, for example by switching-off a 500 MW thermo generator of Leningrad NPP etc. This causes a decreasing transit power flow through the remaining branch of the electrical ring.

If in accident situations selective break automatics is used, which defines the angle between the two voltage levels of the model and, when this angle exceeds the adjusted value, the automatics initiate a break of the electrical ring at the optimal point. Usually selective break automatics triggers, if the angle δ between the equivalent electromotive force (EMF) in the electrical ring reaches 120–140°. It is worth mentioning that for all power system distance protections automatic blocking at oscillations is used, therefore triggering is not possible at asynchronous operation state.

8.2.2 Modeling Scheme of the Baltic Electrical Ring

The modeling scheme of 330 ÷ 750 kV electrical network for the electrical ring contains Estonian, Latvian and Lithuanian power systems with three 330 kV electrical links to Kaliningrad power system, united power system of Belarus and Smolensk power system, as well as the equivalent of the united power system of Russian central territory, which is connected to Kalininskaya Nuclear Power Plant (NPP) and Smolenskaya NPP. The equivalent of North-West united power system is connected to Kirishi power plant. Electrical Ring also includes parallel 330 kV electrical lines Chudovo–Novgorod–Pskov power plant connecting Velikoreck and Polock (Fig. 8.1).

At the 330 kV bus bars of Kirishi power plant (Russian North-West united power system) the equivalent electrical link is created with the power system of Kareliya and Kola, but at the 330 kV bus burs of Vostochnaya the equivalent electrical link is created through Viborg substation AC–DC–AC converter (±1,200 MW) with Finland.

The Estonian 330 kV network includes an equivalent electrical link through Harku substation AC–DC–AC converter (±1,200 MW) with Finland. The remaining 220 kV network Pjussi–Rakvere–Kizja is not modeled, because it is going to be liquidated in the nearest future.

The combined cycle gas and steam turbine power plant (450 MW) of Kaliningrad power system is connected to the 110 kV network next to the 330 kV substation Centralnaya, so the power plant is modeled at the 330 kV bus bars of this substation.

The model considers the part of Belarus 220 kV power network, which is connected to Berjeza generation units. Berjeza power plants, which are merged into one equivalent, are connected with equivalent 220 kV electrical links to Rosja, Baranovichi and Slutsk 330 kV substations.

From the bus bars of Gomel and Mozirija substations the model considers 330 kV electrical links from the 330 kV bus burs of Chernobyl NPP to the equivalent of the united power system of Ukrainian central territory. It should be taken into consideration, that electrical links to Ukrainian united power system are used only in accident situations, for example, in 2008 during the period, when 2,400 MW of power at Lukomla power plant was lost due to an accident. In normal operating state they are not connected to Gomel and Mozirija substations. In all

Fig. 8.1 The Baltic electrical ring

330 kV nodes it is possible to model static load curves, which are defined two times per year within the unitary program (maximal winter and minimal summer consumption) and one time per year in Russia, Ukraine, Belarus and the Baltic countries.

The nodes of the model are sorted in folders and subfolders that allow convenient operation with output data and results, forming and filtering them by power system, as well as by voltage class indication. Generation nodes in each Baltic power system model are separated from the others. In each generation node it is possible to use a number of operating generation unit and an amount of generated power that corresponds to its actual values defined by the operating state of the system, for example, in maximal or minimum consumption hours.

For modeling power plant generators and its capacity for steady run calculations, the reactance Xd is used, while for dynamic operating state the generator reactance X'd is used.

In the power system of Latvia the model considers only a small part of the 110 kV network, including the part of the 110 kV network that shunts the 330 kV network, as well as the 110 kV electrical links from Kegums HPP (240 MW) to the 330 kV substation, because the power plant is located close to the 330 kV substation.

High capacity power plants, which are connected to the nodes of the electrical ring, are: Smolenskaya NPP (3,000 MW), Kalininskaya NPP (3,000 MW),

Leningraskaya NPP (4,000 MW), Kirishi TPP (1,800 MW), Baltiyskaya TPP (1,000 MW), Estonian TPP (1,400 MW), Plavinu HPP (<900 MW), Lithuanian Elektrenu TPP (1,800 MW), Kruonio HPP (±900 MW), as well as Lukomla TPP (2,400 MW). It is important that the last block of Ignalina NPP (1,350 MW) was stopped on December 31, 2009, but the actual lifetime without any investments could be at least 3 ÷ 5 years longer. All power plants which are out of the Baltic region were also modeled for steady run and dynamic operating state calculations analogically to the Baltic power plant operational states.

The great advantage of this mathematical model is the possibility of calculating power flows in inter-system electrical connection points even if measurement information at some points is lost. For the Baltic Ring model inter-system the electrical interconnection points are:

- Russian North-West united power system—Russian Center united power system
- Estonian power system—Russian North-West united power system
- Estonian (Pskov) power system—Latvian power system
- Latvian power system—Lithuanian power system
- Lithuanian power system—Kaliningrad power system
- Lithuanian power system—Belarus united power system
- Belarus united power system—Smolensk power system.

For the maintenance of reactive power balance it is necessary to use generator power in summer, because the static reactive power compensation facilities of the 330 kV network are not able to ensure maximal permissible level of the 330 kV network.

Totally, the model has 140 nodes, 180 links, involving power systems of six countries.

8.2.3 Analysis of Interconnection Points in Electrical Ring

Each sub-system of the Electrical Ring has electrical connections with other sub-systems, with definite maximum power flow, voltage class, number and length of lines, and repair schemes.

From the stability point of view, the main feature of a closed electrical ring is that stability loss is possible only at two variable electrical interconnection points (ECPs); in our case it is the combination Leningrad–Estonia and Smolensk Belarus, or the combination Estonia–Latvia and Smolensk–Belarus.

Combinations of this kind depend on the outage place where the power flow is limited by the condition that the summary flow of two corresponding ECPs has reached its maximum according to the static stability criterion. The advantage of an electrical ring is that the static and dynamic stability criteria for two ECPs are always higher than those for any single ECP.

When analyzing possible emergency situation, it should be taken into consideration that the mentioned Electrical Ring is non-uniform and involves repair schemes, especially in summer months.

Such an analysis should be done taking into account all possible operational states of power plants connected to the ring, as well as the maximum electricity consumption—for example, when one of the units of a nuclear power plant (Leningrad, Tver, or Smolensk NPP) is stopped due to a failure entailing a decrease of generation by 1,000–1,300 MW.

Owing to the non-uniformity of the ring in definite seasons of the year there could be different combinations of weakened ECPs, which requires a new approach for creation of adaptive, selective and reliable emergency automatics.

It should be mentioned that the ring in some ECPs is limited by the maximum power flow of 1,500–2,000 MW, and during maintenance schemes even by 600 MW. Therefore, in the case of maintenance schemes and short-circuit at one of the remaining transmission lines of a weakened ECP even possible breakage of the Electrical Ring was considered—for example, between Smolensk and Belarus, between Estonia and Leningrad, etc.

8.3 Prototype of State Estimation

This subsection describes the approach for creation of distributed and dynamic state estimation prototypes. Section 8.3.1 explains the main features of distributed state estimation and Sect. 8.3.2 discusses the main features of dynamic state estimation.

8.3.1 Prototype Description

In the course of the research we examined the collaborative operation of an interconnected large scale network. Several system operators need to have reliable information about the overall system state for secure operation of the interconnection and they need to have detailed information about their own sub-system.

A framework has been developed which enables testing of state estimation algorithm. Figure 8.2 displays its structure.

The scenarios of the 'Baltic Ring' model are based on the power systems of the Baltic countries and their neighbors, Russia and Belarus, as described in Chap. 2. Scenarios can be obtained by real SCADA measurement or simulated by software. Scenarios are calculated by MUSTANG software, which has been developed at the Institute of Physical Energetics (IPE). The calculations are regarded as real because reference scenarios are compared with real SCADA data (see Sect. 8.4). Besides the scenarios of the Baltic Ring, other scenarios can also be calculated, as for example the full scheme of the interconnected network of ENTSO-E Continental Europe and IPS/UPS.

Simulation results or, in case of real scenarios, real SCADA measurements provide initial data for state vector and topology information. The preparation of

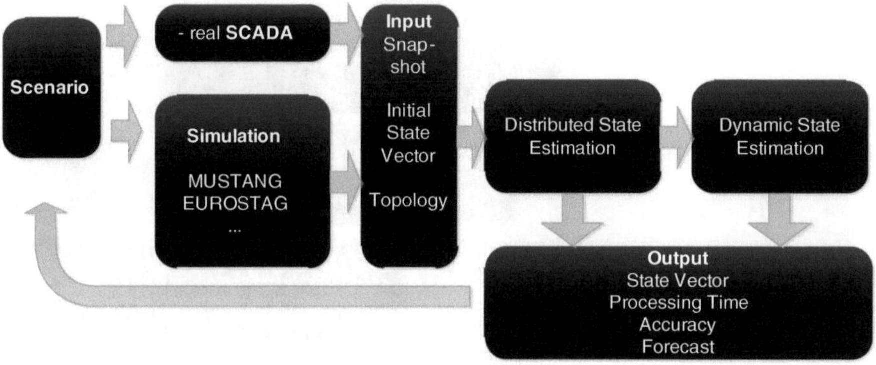

Fig. 8.2 Prototype framework for state estimation algorithms

data requires a common data protocol containing a consistent data set and a snapshot of measurement or simulation results respectively.

The distributed State Estimation provides a first decomposition of the full network. Here we receive the results on how the system is decomposed into sub-systems and static state estimation. Dynamic state estimation is applied in the selected sub-systems. Several results of static state estimation are necessary for dynamic state estimation in order to tune internal data processing algorithms such as Kalman filter. Both, Distributed and Dynamic State Estimation are described in detail in Sects. 8.2.1 and 8.2.2 respectively. A short summarization is given in the following Sects. 8.2.2 and 8.2.3.

The procedure of evaluation is as follows: At first scenarios are developed and simulations are performed. Then, this data is prepared for the execution of the State Estimation algorithms. The results such as state vector, forecast in case of dynamic state estimator and accuracy and processing time are finally evaluated.

8.3.2 Distributed State Estimation

Figure 8.3 displays the structure of the prototype. Beginning from topology information, the full network is divided into sub-systems. Boundary nodes and timelines are selected. For each sub-system a slack node must be selected. Time synchronized measurements are beneficial because of their accuracy and the measurement of the angle. If a sub-system does not contain any time synchronized measurements, then the angle must be determined by additional calculations taking into account additional variables such as power flows. Very often a system is already divided into sub-systems because of political borders or technical structure such as division in different voltage levels.

State Estimation is executed on the basis of quadratic and robust criteria for each sub-system. A priori bad data detection is performed in order to improve the quality

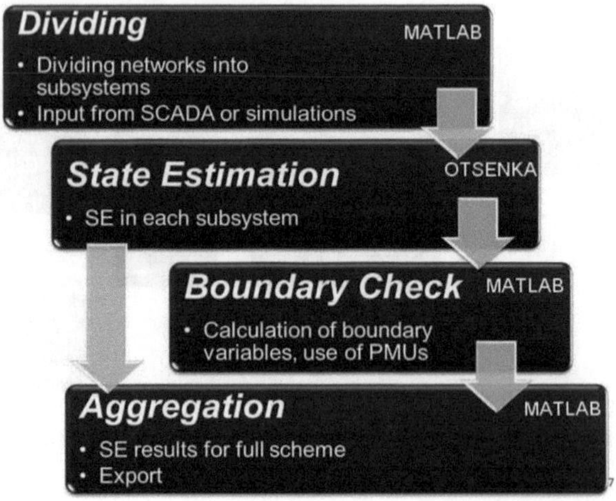

Fig. 8.3 Prototype of distributed state estimation

of the initial state vector. In each sub-system standard software of static State Estimation can be applied. In the course of the research we applied OTSENKA and MUSTANG software which has been developed in ESI, Irkutsk and IPE, Riga. MUSTANG is used for model data simulation and OTSENKA for state estimation [3, 4]. All other algorithms except the static state estimation algorithm are implemented in MATLAB.

The State Estimation results contain boundary values. These values must be compared with the boundary values of neighboring sub-systems. If the values don't match then state estimation of the affected sub-systems must be performed again taking into account additional values or update of measurement accuracy for the covariance matrix.

8.3.3 Dynamic State Estimation

First, the term 'Dynamic State Estimation' is defined as follows:

> Dynamic State Estimation makes it possible to take into account the interrelation among state variables changing in time.

Besides the measurement model, Dynamic State Estimation considers a signal model as well. The applied dynamic model determines largely the efficiency of dynamic state estimation. Models containing too many variables, i.e. nodes do not always improve the quality and reliability of the SE results. For short-term forecasting of state parameters we use small scale dynamic models. In the small scale dynamic models random fluctuations of state parameters are represented as a stationary Gaussian process (white noise) and all changes in state parameters are considered as random fluctuations and are taken into account.

Furthermore a Kalman Filter is applied, which is a recursive method of solving the state estimation problem by the least squares method. The recursive relations are

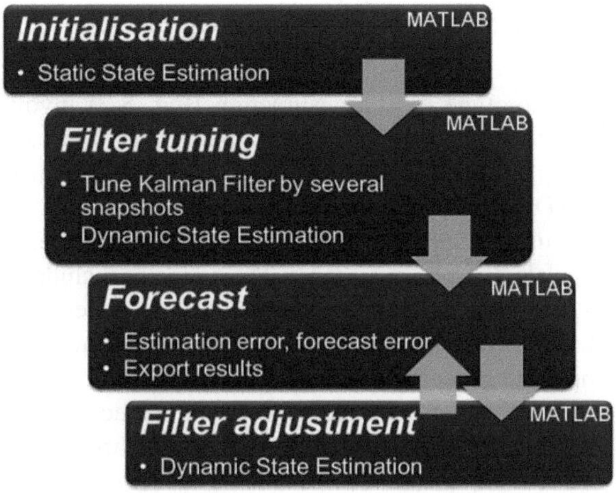

Fig. 8.4 Prototype of dynamic state estimation

constructed, in which the estimations of the k-th time instant are used to obtain the estimates at the time instant k + 1.

Based on the description of the Kalman filter, the forecast error covariance matrix for the time instant k + 1 is obtained from past snapshots. Additional consideration is given to forecast error results in a new objective function of Dynamic State Estimation.

Due to the nonlinear dependence, the objective function is solved iteratively by Newton's method. The initial conditions for each cycle are the state estimation of the system (values of state vector components) and the covariance matrix which characterizes the estimation error. The algorithm processes newly coming measurement vectors by taking into account the measurement errors, and specifies the initial conditions. The adjusted initial conditions are the filter output data.

Figure 8.4 displays the structure of Dynamic State Estimation. The program is written in the MATLAB environment. Dynamic State Estimation is divided into four parts.

For initialization, static state estimation is performed. Several snapshots from static state estimation are furthermore applied for tuning the Kalman Filter. The objective function is then solved. After that a forecast can be provided based on a trend calculation. Such a forecast is beneficial in case of slow response of static state estimation induced by bad or missing data.

Dynamic State Estimation and the correction of forecast and estimation error are performed for each time step.

8.4 Testing the Efficiency of the Suggested Algorithm for Distributed State Estimation

This subsection focuses on data simulation issues and comparison of calculated data and SCADA measurements.

8.4.1 Testing of Simulator Efficiency

The prototype of state estimation (Fig. 8.2) involves Mustang software in Data Simulation module. In order to show the efficiency of Mustang as simulator, multiple operational states were tested—calculated values were compared with SCADA data provided by Latvian TSO.

Mustang software is able to perform static and dynamic process modeling in large electric power systems. The software includes possibility to model emergency automation, which is necessary in large generation unit disconnection modeling. Static operational state calculation is made using Newton–Rawson method with improved borderline state convergence in accordance with Matveev method. Linear algebraic equation system is solved in accordance with Gauss method, which includes optimization of prior unknown variables excluding sequence [2].

For comparison with calculated data from Mustang software, the Latvian TSO prepared SCADA measurements. Several characteristic operational states were calculated and an example was chosen with largest difference between calculated values and SCADA measurements.

Figure 8.5 shows the calculated values of power flows between the systems (green color) for April 26, 2010 and SCADA measurement (red color).

From the calculation results it can be concluded that the power flow difference does not exceed 5 % and is mostly 2 ÷ 3 %, therefore Mustang simulations have high accuracy and the simulated data could be used for State Estimation.

8.4.2 Comparison of Calculated Power Flow and Voltage Values in OTSENKA Software with SCADA Measurements in Electrical Ring

This testing stage includes the calculation of four operational states and comparison of different calculated values with SCADA and MUSTANG-simulated parameter values.

The operational state descriptions are given in Table 8.1.

Due to an accident in the Baltic Ring all turbo generators (8 × 300 MW) in Lukomla power plant are disconnected (operational state 1), as well as all outgoing 330 kV electrical lines (7 in total) are disconnected after 1–1.5 min (operational state 2). After triggering of emergency automation in Belarusskaya substation, hydro power plants at the river Daugava were launched with total capacity of 750 MW (in fact, HES generation increased by 600 MW) (operational state 3).

In the morning of 25 June, 2008 during the peak hours one 300 MW block was disconnected after a short-circuit at the Belarus Lukomla power plant, and eight minutes later—the remaining seven 300 MW blocks with the total power of 2,100 MW were tripped as well. The loss of 2,400 MW generation and

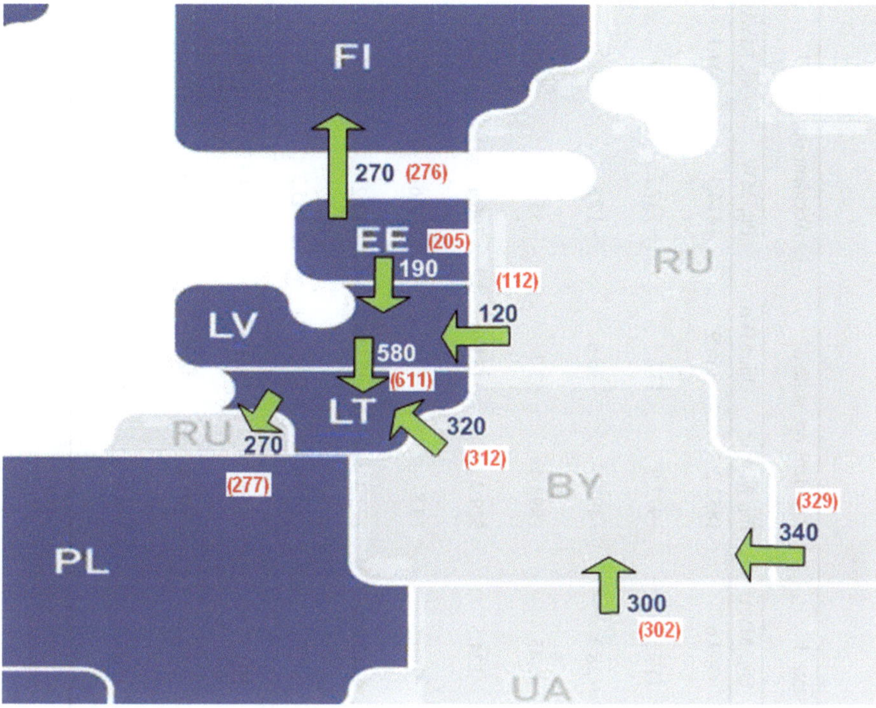

Fig. 8.5 Calculated values of power flows (MW) between the systems (*green color*) for April 26, 2010 and SCADA measurement (*red color*)

Table 8.1 Descriptions of operational states

Op. state 0.	Normal operational state
Op. state 1.	Disconnection of power plant Lukomla (2,100 MW) from the power system of Belarus, lines between power plant and grid remain in operation
Op. state 2.	Disconnection of power plant Lukomla (2,100 MW) from the power system of Belarus, lines between power plant and grid are disconnected
Op. state 3.	Disconnection of power plant Lukomla (2,100 MW) from the power system of Belarus, lines from power plant are disconnected, power plant Plav. HES in LAT increased generation by 600 MW

disconnection of TLs in Belarus caused a considerable weakening of the electric ring Moscow–St. Petersburg–Estonia–Latvia–Lithuania–Belarus–Smolensk–Moscow in the Belarus territory. At the remaining 330 and 750 kV transit TLs the power flow rose from the maximum of 800–2,050 and 1,400 MW, respectively, causing considerable overload (1.4 of the permissible) of the 750/330 kV 1,000 MVA transformer at the Belarus substation. At the 750/330 kV transformer the overload current emergency automatics was triggered (the setting was 15 % over Pnom) with a 5 min. delay, which impacted a cascade of the hydro power plants at Daugava

Table 8.2 Power flows in initial and accident operational states

System 1 System 2	Initial op. state		Accident op. state 1		Accident op. state 2		Accident op. state 3	
	Pij, MW	Qij, MVAr	Pij, MW	Qij, MVAr	Pij, MW	Qij, MVAr	Pij, MW	Qij, MVAr
Centr Sev-Zap	772.2	−481.7	1651	−369.6	1853.2	−346.9	1414.3	−404.1
Estonia Sev-Zap	−460	−59.6	−866.5	117.5	−972.0	177.5	−737.2	54.9
Estonia Pskovskaya	−59	−32.9	−39.2	−38.3	−49.6	−36.6	−17.6	−39.7
Latvia Estonia (Through Pskov)	−158	−116.2	−734	17.45	−893.4	62.7	−559.8	−36.4
Latvia Lithuania	−196	−45.8	374.7	−124.5	529.7	−133.6	796.5	−144.0
Lithuania Belarus	−701.5	−89.6	−128.5	−146.5	19.8	−10.6	282.6	−136.7
Lithuania Kaliningradskaya	273.3	−106.9	272.57	−112.8	272.3	−115.9	272.2	−118.0
Belarus Pskovskaya	57.7	−59.1	−180.4	−0.72	−193.8	3.9	−145.3	−8.7
Belarus Smolenskaya	−835	−365	−2126.7	104.17	−1908.7	−141.1	−1663.5	−163.6
Belarus Ukraine	−297	72.8	−297.9	−37.6	−297.9	−54.5	−297.9	−32.1
Estonia Finland	270	150	270	150	270.0	150.1	270.0	150.1

Table 8.3 Comparison of real and calculated active power flows in different operational states (part)

Node 1 Node 2	Initial op. state		1 op. state		2 op. state		3 op. state	
	Active flow (MW)		Active flow (MW)		Active flow (MW)		Active flow (MW)	
	Calc.	Real	Calc.	Real	Calc.	Real	Calc.	Real
CENTR KalinNPP	−2,204	−2,205	−995	−996	−877	−877	−1,367	−1,367
SmolNPP Roslavlj	411	412	715	716	591	592	564	564
Vostochnaya Juzhnaya	−476	−477	−455	−456	−499	−499	−462	−463
LAES-33 Gatchina	577	577	690	690	716	716	659	658
Kirishi Chudovo	−398	−398	−378	−378	−376	−377	−380	−381

(Plavinu, Keguma and Rigas) with a total power of 750 MW (actually, within this time up to 1,000 MW total could be reached), which decreased the overload of ECPs in the Electrical Ring. Due to the advantages of the Electrical Ring, dynamic stability break did not happen.

Power flows between the systems in the described operational states are presented in Table 8.2.

The changes in power flow at the interconnection between Belarus and Ukraine are insignificant. Comparing all scenarios, however some connections have considerable changes in power flows, for example at the interconnections Belarus–Smolenskaya and Centr–Sev-Zap.

The parameters of scenario 3 show that the operational state is stable, despite 2,400 MW generation capacity and seven 330 kV transmission lines in Belarus power system are out of operation.

However, such overloadings of the electrical network in the northern and southern part are not permissible, because they can cause damage to some elements that can lead to a decrease in the capacity at the connection point causing a dynamic stability break at multiple connection points.

Table 8.3 provides a comparison of calculated and real active power flows in the modeled network for operational states that are given in Table 8.1. Though, it is just a summarization of the obtained results. The full table includes also reactive power flows and 140 nodes in total. Initial data for the computer model included injections in nodes, voltages and voltage angles.

The results in Table 8.3 show that the calculated values are very close to real SCADA values in all operational states. Based on these results it can be concluded that the suggested algorithms for distributed state estimation have a high accuracy.

8.5 Conclusions

The use of Distributed State Estimation enables the following advances:

- Minimization of information transferred among control centers of subsystems operating in parallel and the coordination center;
- Decrease in SE execution time for the full scheme;
- Simplification to the maximum extent of the problem of solution coordination for individual subsystems, in some cases—elimination of iteration calculations for subsystems;
- Decrease in inhomogeneity of the calculated schemes of subsystems, providing a guaranteed convergence of the computational process;
- Enhance of efficiency of bad data detection algorithms and accuracy of SE results.

The main advance of Dynamic State Estimator can be summarized as follows:

- Forecast and estimation of date during dynamic changes
- Loss of data is compensated by the use of pseudo measurement
- The accuracy is increased
- The calculation time depends on the scale of the (sub-) system.

Power system research with pre-estimation of accident situations requires well-adjusted and appropriate mathematical models, which ensure state calculations of static and dynamic operation. Electrical ring 330 ÷ 750 kV Estonia–Latvia–Lithuania (Kaliningrad)–Belarus–Smolensk–Moscow–St. Petersburg–Estonia was chosen as a test case for testing the efficiency of state estimation algorithm. The network is appropriate by its size and the number of countries involved. This network involves five countries and their TSOs with huge experience in operative and technological management.

The event that was chosen for SE efficiency testing is Lukomla power plant disconnection, including three further operational states. It is real event that meets the requirements of test case event, including having SCADA data provided by Latvian TSO as input data, having major part of the network involved in the event, having a large number of objects involved in the events, namely power plant, electric lines and transformer substations.

Mustang software was chosen to be simulation tool in the SE prototype and calculations have proved high accuracy of simulated results. SCADA and simulated voltages, angles and power flows difference is less than 3 % in all cases.

State estimation performed with Otsenka software proves high accuracy of the offered state estimation algorithms. The calculated and real value difference is relatively small in initial operational state, as well as in accident operation state scenarios, which were based on real SCADA data.

The dynamic state estimation prototype has a high efficiency. However the model used was smaller as compared to the model for Distributed State Estimation testing. Also, small angles in the results of calculations indicate an operational state with lower power flow values, which means that the accuracy is not at maximum, though it is sufficient.

The new monitoring and state estimation technologies provide opportunities for TSOs, such as new communication channel, Dynamic State Estimation, accident investigation and prevention of blackouts.

References

1. K. Brinkis, V. Kreslinsh, A. Mutule, I. Oleinikova, Z. Krishans, O. Kochukov. Fulfilment of criteria of electricity supply reliability in the Baltic region. Latvian J. Phys. Tech. Sci. N 6 (2011)
2. V. Ivanovs, V. Rimarevs, K. Briņķis, J. Gurevičs, Patent Nr 12960. A method for control of static and dynamic stability processes in a large electric system, a software for its realization and application for adaptive regulation of the system. Latvian Republic Patent Department Publication (2003), p. 24
3. A.Z. Gamm, I.N. Kolosok, Test equations and their use for state estimation of electrical power system/power and electrical engineering. in *Scientific Proceedings of Riga Technical University* (RTU, Riga, 2002), pp. 99–105
4. Y.A. Grishin, I.N. Kolosok, E.S. Korkina, L.V. Em, State estimation of electric power system for new technological system. in *Proceedings of the 1999 International Conference PowerTech99*, Budapesht

Part IV
Control of Interconnected Power Systems

Chapter 9
Technologies for the Control of Interconnected Power Systems

Angelo L'Abbate and Ulf Häger

9.1 Introduction

Since the early stages of development of interconnected electric power systems throughout the world, the issue of how to best perform network control functions has been faced by the respective system operators. The key parameters that need to be continuously handled and controlled by operators in order to keep their HVAC (High Voltage Alternating Current) system security at both steady-state and dynamic level include: line current magnitude, nodal voltage amplitude, nodal voltage angular difference, active power flow, reactive power flow, system frequency.

In the past, prior to the electricity industry liberalization, vertically integrated utilities, jointly operating generation and transmission systems, had at their disposal several means of controlling the different parameters in their respective system: in addition to switching devices (lines, reactors, capacitors), changing transformer taps and operating topology modifications, one of the main control options included the possibility to directly regulate generating units outputs towards a power redispatch. Nowadays, in a liberalized power sector, the unbundling between transmission and generation functions has certainly restricted the system observability and control possibilities in the hands of TSOs (Transmission System Operators). Moreover, the fast growing penetration of variable RES-E (Renewable Energy Sources of Electricity)—especially wind and solar—generation and the increasing deployment of downstream DG (Distributed Generation) stress even more the need for an advanced, faster and multi-variable power system control features.

Currently, the technology progress makes it possible for the TSOs to have at their disposal power electronics-based system controllers that may play a further

A. L'Abbate (✉)
RSE S.p.A., Milan, Italy
e-mail: angelo.labbate@rse-web.it

U. Häger
TU Dortmund University, Dortmund, Germany
e-mail: ulf.haeger@tu-dortmund.de

U. Häger et al. (eds.), *Monitoring, Control and Protection of Interconnected Power Systems*, Power Systems, DOI: 10.1007/978-3-642-53848-3_9,

crucial role for the future development and operation of the electricity grids. Among these advanced devices, one can distinguish the categories of FACTS (Flexible Alternating Current Transmission System) and HVDC (High Voltage Direct Current) technologies with their respective differences. So far, at least in the European system, mechanically-regulated devices like the PSTs (Phase Shifting Transformers) have been generally implemented by TSOs for static power flow control, especially at cross-border level, while HVDC links have been used to couple asynchronous systems and/or for long submarine interconnections. In other systems outside Europe, the situation is however different for several reasons, and this may also impact on future evolution of the pan-European grid.

The technologies of PST, FACTS and HVDC families can be grouped and categorized as PFCs (Power Flow Controllers), for having the feature of power flow control capability, with their differences. The active redirection of power flows offered by PFCs also allows exploiting lines closer to their thermal limits towards a more optimized utilization of the existing grid.

Considering the vast technical literature existing on the PFC technologies under investigation, the following sections briefly focus on the main technical features of the most mature and promising PFC technologies.

9.2 Control Technologies: PSTs

A PST is a transformer with complex turn ratio, able to control the active power flow by handling the voltage phase angle between two nodes in the system. The phase angle difference between the PST terminal voltages can be obtained by connecting a transformer in series with the transmission line (boosting transformer). The active and reactive power, injected into the transmission line by this boosting transformer, must be absorbed from the network by a shunt (excitation) transformer. Then, neglecting losses, the PST neither produces nor absorbs active and/or reactive power; it acts as a "power transfer accelerator" [1]. Figure 9.1 shows a basic scheme of a PST.

PST is a simple, robust and reliable technology. The theory behind power flow control, the operational principles of PST and the different topologies and categories of PST are detailed in the literature (see for example [1–5]).

The PSTs, as controllers of power flows, are used for different scopes, such as: parallel lines load sharing; total transmission capacity increase; system reliability improvement by mitigation of post-contingency overloads and of undesired power flows; RES-E integration facilitation; grid congestion removal; transmission losses reduction [6]. However, it has to be noted that these devices do not eliminate losses: the shifting of power to other lines can cause losses in other parts of the grid. Depending on their constructive characteristics PSTs may have a very high total efficiency (up to 99.8 % ca. at full load and maximum phase angle) [4].

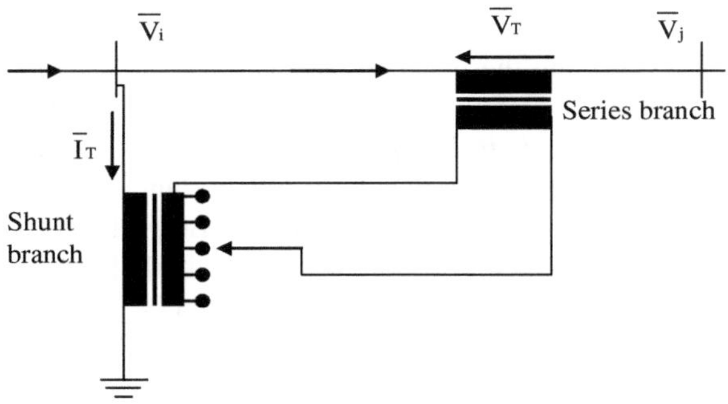

Fig. 9.1 Basic scheme of PST

From the electrical point of view two types of PST can be distinguished [1, 2].

The first one is a PST with equal input and output voltage magnitude but with a phase shift between these voltages: this type of PST is known as Phase Angle Regulator (PAR). For the PAR the controllable parameter is the voltage shift angle.

The second type of PST is known as Quadrature Boosting Transformer (QBT). The phasor of the injected voltage of the QBT series branch is shifted by a constant angle (equal to 90° in most cases) with respect to the input voltage vector. The controllable parameter of the QBT is the magnitude of the injected voltage: in this case, in addition to the phase angle shift, there is a change in the output voltage magnitude, depending on the complex turn ratio.

PSTs are well proven and reliable devices: they are extensively utilized and they have been implemented by TSOs worldwide, especially in the last twenty years, to solve local load flow problems in HVAC networks. Also, given socio-environmental constraints often hindering the structural grid development, the use of PSTs can be a convenient option as these devices can be commissioned in a relatively short time with respect to a HVAC line expansion [3].

In general, preventive and curative control strategies are implemented for power flow controllability. In the preventive mode, the permanent phase shift allows redistributing the power flows between several HVAC lines and relieves network stresses in case of line outage. In the curative mode, the phase shift is small (sometimes down to zero) in normal operation but is automatically controlled to reduce the power flow on the overloaded lines and to avoid a tripping out [3, 6].

Today, some hundreds PSTs are estimated to be globally installed [6]. In Europe, there exist several PST applications for static power flow control, especially at cross-border network ties. For example, at the Netherlands-Germany grid border, there are five PSTs installed, which are operated by three different TSOs. With natural flow the interconnectors at the Dutch-German border are utilized unevenly, leading to reduced cross-border capacities compared to the sum of capacities of each single line. By the construction of PSTs the redistribution of the natural power flow on the interconnectors has been possible, leading then to equalize the loading of the transmission lines. Furthermore, at the Netherlands-

Belgium grid border, three PSTs have been installed in order to reduce the loop flows through the networks of the two countries [5, 7, 8].

In addition to the PSTs implemented at the Netherlands-Belgium and Netherlands-Germany grid borders (see also Chap. 10), there are different PSTs currently installed in the networks of e.g. UK, Ireland, France, Austria, Italy, as well as at the France–Spain, France–Italy, France–Belgium, Slovenia–Italy, Switzerland–Italy, Germany–France grid borders, for example [3, 6, 9].

In the European system, further PSTs are under construction or planned to be installed at the Italy–Austria, Czech Republic–Germany, Germany–Poland, Germany–Denmark grid borders, for instance [10, 11].

Outside Europe, the growth of the PST market is expected to be largely driven by developing economies, such as China, India, Brazil and Saudi Arabia, with upcoming long-distance transmission network projects. On the other hand, the PST market in certain developed countries, such as the USA, Canada, Japan and Australia, is expected to decline in the mid-long term future, also due to the concurrence provided by emerging, more advanced FACTS and HVDC technologies (see Sects. 9.3, 9.4).

Further applications of PSTs may consider the installation of PSTs in series with reactive elements for specific targets like: substation uprating, substation reserve sharing, network decoupling, HVAC transmission line de-icing. All these applications rely on conventional PSTs and reactive elements interconnected in order to meet unusual objectives for PSTs [12]. Behind these applications there has also been the definition of the concept of a technology, the Interphase Power Controller (IPC), which is a series-connected device consisting of two parallel branches, each with an impedance in series with a PST element. Developing IPC technology has aimed at creating a family of new power flow controllers that would overcome the limitations on system operation caused by high short-circuit levels. Different types and topologies of IPCs are described in literature (see [13] for example). Some implementations are reported in the USA and in Canada [12, 13].

9.3 Control Technologies: FACTS

The definition and the categorization of FACTS concept date back to the late 1980s when the Electric Power Research Institute (EPRI) in the USA started to investigate and develop these advanced controllers. FACTS technology consists in a collection of devices that individually or in coordination with others offer the possibility to control one or more of the interdependent parameters that influence the operation of HVAC transmission networks. These parameters include the line series impedance, the shunt impedance, the line current, the nodal voltage amplitude, the nodal voltage angular difference, for instance [1, 14, 15].

The design of the different schemes and configurations of FACTS devices is based on the combination of traditional power system components (such as transformers, reactors, switches and capacitors) with power electronics elements (such as various types of transistors and thyristors). In recent years, the nominal current rating of thyristors has increased, revolutionizing the power electronics industry with high-power applications of up to thousands of MW. The development of semiconductor technology has made it possible to manufacture new types of thyristors, such as Gate Turn-Off (GTO) thyristors, Integrated Gate Commutated Thyristors (IGCT) and Insulated Gate Bipolar Transistors (IGBT). Promising thyristors are those depending on Metal Oxide Semiconductor (MOS) integrated circuits, such as the MOS Controlled Thyristor (MCT), the MOS Turn-Off Thyristor (MTO) or the Emitter Turn-Off Thyristor (ETO). The increasing progress in thyristor-based technology has resulted firstly in the development of HVDC transmission systems (see Sect. 9.4) as an alternative to long-distance HVAC transmission. Subsequently, this proven HVDC technology served as the basis for the implementation and utilization of FACTS controllers [16].

In addition to the rapidly increasing development of power electronics technology, several other driving factors related to electricity market liberalization may contribute to make FACTS utilization necessary and profitable for power systems purposes. Also, power flow patterns, often dictated by market decisions but also more frequently influenced by variable RES-E generation, are generally more unpredictable: therefore, the uncertainties in generation and network planning and operation require modern transmission systems to be as flexible as possible. In this view, FACTS can provide useful support to grid operators in system control. Besides, open access to the transmission grid is leading to generally higher utilization of transmission systems, which may result in more frequent network congestion. For example, in the European power system, this occurs in particular on cross-border and inter-zonal interconnections.

The traditional solution to address network congestion consists in increasing transmission capacity by building new overhead HVAC lines. However, implementing this solution nowadays is becoming more and more restrained by socio-environmental, economic, and political obstacles. Therefore, an effective way to cope with this situation is to use existing transmission structures more efficiently, by freeing paths that are 'occupied' by undesired power flows, i.e. loop flows, thereby effectively utilizing these HVAC lines and preventing possible system congestion thanks to FACTS technologies.

Moreover, due to their fast response features, FACTS devices can offer several advantages for controlling variable RES-E generation units, such as wind power plants, facilitating their integration into the system.

FACTS devices, able to address all these needs, can then make utility networks more reliable, more controllable and more efficient. More specifically, the utilization of FACTS devices can allow [1, 14, 15]:

- controlling active and reactive power flows smoothly and rapidly up to a certain level;
- reducing undesired active and reactive power flows in the system and thereby network losses;

- increasing the loading capability of the transmission lines to levels closer to their thermal limits without violating security constraints;
- improving steady-state and transient stability;
- reducing series voltage drops on the lines;
- enhancing system damping in the presence of oscillations;
- shifting the power flow from congested transmission lines to free parallel paths in a rapid and precise way;
- controlling voltage and improving power quality.

Furthermore, FACTS controllers, differently from mechanical devices, such as transformer tap changers or shunt capacitor switches that have enabled the HVAC power system to be controlled so far, are not subject to mechanical wear: this is an important advantage of FACTS devices in addition to high flexibility and speed.

The main drawback of FACTS technology so far has been related to its costs, which are higher than those ones of the corresponding mechanical devices: this factor, in addition to higher FACTS complexity, has slowed down the integration of FACTS devices and prevented their more widespread employment in transmission systems.

However, since all FACTS controllers are built using similar technologies, their deployment can benefit from economies of scale linked to volume production: the cost of these devices is decreasing with the development of high-power electronics and increasing economies of scale [16].

A detailed description of the different types of FACTS can be found in numerous specific literature sources (like [1, 14, 15]). The scope of this section will be limited to the most promising and broadly used types of FACTS.

In general, FACTS devices can be classified, according to their connection, as [1, 17]:

- **Shunt controllers**. Among shunt controllers the main devices are the Static VAR Compensator (SVC) and the Static Synchronous Compensator (STATCOM) (also known as Static Condenser, STATCON). Other devices of this type are BESS (Battery Energy Storage System) and SMES (Superconducting Magnetic Energy Storage).
- **Series controllers**. The series controllers category includes devices such as the Thyristor Controlled Series Capacitor (TCSC) and the Static Synchronous Series Compensator (SSSC).
- **Combined controllers**. Devices such as the Thyristor Controlled Phase Shifting Transformer (TCPST), the Interline Power Flow Controller (IPFC), the Dynamic Flow Controller (DFC) and the Unified Power Flow Controller (UPFC) belong to this third category of FACTS.

Another possible classification of FACTS is based on the power electronics technology used for the converters, that is thyristor-based or voltage source-based. Figure 9.2 displays the two types of FACTS classification with the main devices typologies [6, 15, 16].

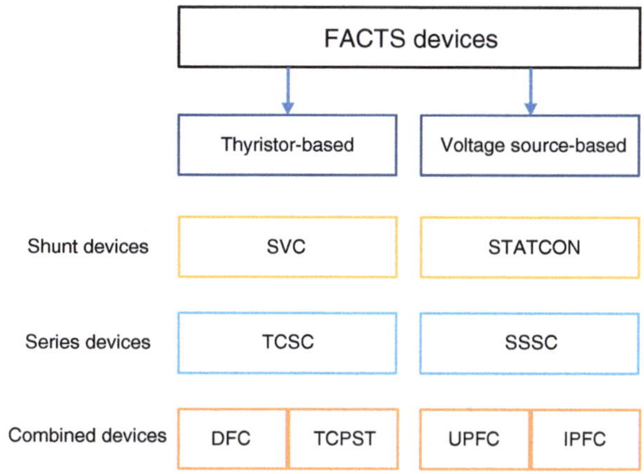

Fig. 9.2 Two types of FACTS classification (adapted from [15])

Unlike thyristor-based devices, voltage source-based controllers are able to impose shunt or series voltage at the point where power is injected into the system in order to achieve defined control targets.

9.3.1 Shunt Controllers

The SVC (Static VAR Compensator) is basically composed of a combination of thyristor controlled reactors (TCRs), thyristor switched capacitors (TSCs) and fixed capacitors or reactors; additional filters are used to cut harmonics. Figure 9.3a shows a SVC scheme with TSCs and a TCR [1, 14].

The primary role of SVCs is to regulate the amount of reactive power compensation to the actual system needs and then to control voltage at the HV (high voltage) bus both continuously and rapidly. SVCs can also be used for stability improvement, as they can have a very positive impact on power oscillations damping, i.e. for steady-state stability [18, 19]. Flexible and continuous reactive power compensation is made feasible by using thyristor-switched shunt elements operating in both the capacitive and inductive regions. These features can be very important also for controlling wind farms' reactive output [15, 16].

SVCs were first applied in the USA in the 1970s, long before the concept of FACTS was formulated. The first application was the EPRI-Minnesota Power & Light and Westinghouse project commissioned in 1978, in which SVCs enabled a 25 % power increase along the line where they were installed. Worldwide, there is a steady increase in the number of installations. Most recently, SVCs have been ordered or installed in Chile, Canada, USA, Mexico, South Africa, Peru, France, UK, Norway and Finland, among others [16]. In the case of Europe, the largest number of SVCs can be found in the UK. The number of SVCs installed worldwide (at industrial and utility level) is currently estimated at more than 800 devices

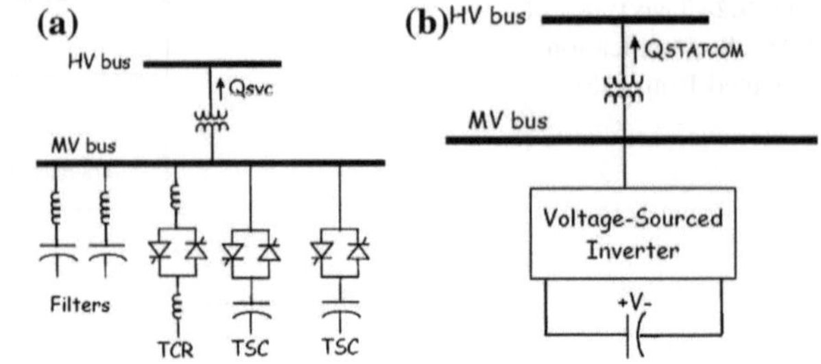

Fig. 9.3 Examples of shunt FACTS [**a** SVC, **b** STATCOM]

for a total installed power of over 90 GVA [16]. Installations of SVCs in Europe are expected to further increase, especially due to the growing RES penetration. For example, new SVCs are planned or under design in Germany and France, whereas projects for SVCs are under study in Poland and in Italy [10, 11].

Further developments of SVC equipment have resulted in installations of relocatable SVCs (RSVCs) in some substations in South Africa and in the UK. In this way it is possible to fully exploit the potential of these devices to adapt to changed needs in the power system. To this end SVC installations need to be compact in order to make relocation possible within 3–6 months. It is evaluated that relocation might occur up to five times in the 40-year operating life of an RSVC [1, 16].

For further details about different aspects and properties of SVCs the reader is referred to [1, 14–16, 20] and to the references therein, among others.

The Static Synchronous Compensator (STATCOM or SSC) represents a further development of the SVC. In the literature this device is also called a Static Condenser (STATCON), Static VAR Generator (SVG), GTO-SVC or Advanced SVC (ASVC) [1, 14].

After two experimental installations of converter-based VAR compensators during the 1980s, GTOs with greatly increased ratings have become available, and a ±80 MVAR installation, using 4,500 V, 3,000 A GTOs, has been set up in Japan. In the USA, in 1995, a STATCOM rated for ±100 MVAR was commissioned at the Sullivan substation of the Tennessee Valley Authority (TVA) power system. In this case the GTOs are rated for 4,500 V and 4,000 A to control a 161 kV bus voltage. The number of STATCOMs installed worldwide is currently estimated at about 20 devices deployed or under construction in different countries, such as the USA, Japan, China, Chile and the UK (the only European country having STATCOM applications so far) for a total installed power of over 1,500 MVA [16].

In contrast to the SVC, a STATCOM does not use capacitor or reactor banks to produce reactive power. The basic electronic building block for a STATCOM is a voltage-sourced converter that inverts the DC voltage at its input terminals into a three-phase set of AC output voltages: the device is then able to generate a (nearly sinusoidal) synchronous voltage. This capability can be further enhanced using an external energy storage device. Figure 9.3b shows a scheme of a STATCOM [14].

The reactive and active power, generated or absorbed by the STATCOM, can be controlled independently of each other. The real power that the synchronous voltage source exchanges at its AC terminals with the AC system must be supplied to or absorbed from its DC terminals by the energy storage device. On the other hand, the reactive power generation or absorption is developed internally in the STATCOM, and the DC energy storage device plays no role in it.

Reference signals for reactive and active power injections respectively define the amplitude and the phase angle of the generated output voltage [16].

In terms of operation, it is worth noting that the STATCOM is faster than the SVC; STATCOM has an increased transient rating compared to the SVC in both the inductive and capacitive operating regions. The inherently available transient rating of the STATCOM is dependent on the characteristics of the power semiconductors used and the junction temperature at which the devices are operated.

The capability to exchange energy between the AC and DC systems may be used to improve system efficiency and prevent power outages. Also, in combination with fast reactive power control, dynamic real power exchange provides an extremely effective tool for transient and dynamic stability improvement, not only for voltage control and reactive compensation. Due to all these features devices like STATCOMs can then be very suitable for the control of wind farms.

STATCOM can be designed also to act as an active filter to absorb system harmonics. One of the advantages of using STATCOM, in comparison with SVC, is the space saving obtained by utilization of electronic converters instead of physical devices like reactances or capacities; this advantage is very important in order to simplify location changes. Cost of STATCOM can be higher due to the utilization of more sophisticated GTO thyristors. An additional cost is due to utilization of PWM (Pulse Width Modulation) control system to reduce harmonics [14–16, 20].

Installations of STATCOMs in Europe and worldwide are expected to further increase, especially in presence of growing RES penetration [15, 16].

The BESS (Battery Energy Storage System) is a chemical-based energy storage system that can be connected to a STATCOM; it can rapidly exchange active energy absorbed/supplied by the AC system. For applications on transmission systems like transient stability improvement, BESS have generally small sizes (a few tens of MWh), but it could deliver MWs with high MW/MWh ratio. The associated STATCOM can simultaneously absorb or deliver reactive power within the converter's MVA capability.

The SMES (Superconducting Magnetic Energy Storage) is a storage system including a superconducting coil and electronic converters that rapidly injects and/or absorbs real and/or reactive power or dynamically controls power flow in an AC system. The time delay during charge and discharge is quite short and power is available almost instantaneously; very high power output can be provided for a brief period of time. Although the DC current in the magnet does not change rapidly, an appropriate electronics interface for connection to a STATCOM can provide a rapid control of voltage across the magnet [1, 15].

9.3.2 Series Controllers

In terms of controlling active power flows, series elements have a much greater impact than shunt elements. Series elements are also very effective for damping power oscillations and for obtaining an improvement in transient stability, which can be achieved through modulation of active power, while they have very little influence on voltage control [1, 15, 18].

The Thyristor Controlled Series Capacitor (TCSC) can vary the series impedance continuously to levels below and above the line's natural impedance. This is a powerful way of increasing and controlling power transfer. TCSCs can respond rapidly to control signals to increase or decrease the capacitance or inductance, thereby damping those dominant oscillation frequencies that would otherwise create instabilities or unacceptable dynamic conditions during and after a disturbance [21]. This second-generation FACTS device, which is based on conventional thyristors, can greatly improve power flow control and also dampen subsynchronous oscillations [1, 14, 15, 22]. A basic scheme of a TCSC is shown in Fig. 9.4. The variation of the capacitance can be obtained by varying the TCR reactance connected parallel to the capacitance of the TCSC. A complete TCSC module consists of a series capacitor with a parallel path including a thyristor switch with surge inductor; it also includes a Metal-Oxide Varistor (MOV) for overvoltage protection and a by-pass breaker, typical of series capacitors. A complete TCSC system may include several such modules in series and be a part of an overall project to improve power system performance together with a conventional series capacitor bank [14, 20].

Almost all TCSC installations worldwide are located outside Europe, with the notable exception of a single European project (in Sweden). In 1991, in the USA, a multi-segment, mechanically-switched series compensation system was installed by American Electric Power (AEP) at its Kanawha River substation, with one phase of one segment augmented with a thyristor switch. In 1992, a single-module TCSC was built for the Western Area Power Administration (WAPA) and put into operation at the Kayenta substation, in the USA as well. This system is located at the mid-point of a 200-mile 230 kV line and increases power transfer on the line by 100 MW. A complete multi-module TCSC was installed at the US Slatt substation operated by the Bonneville Power Administration (BPA) in 1993.

A further project involving TCSCs has been carried out in Brazil, where a TCSC is used in combination with five conventional series capacitors on a 1,017 km-long 500 kV transmission system. In this application (operating since 1999) the TCSC system is utilized for damping and transient stability enhancements.

The latest TCSC projects have been carried out in India and China. The number of TCSCs installed worldwide is currently estimated at about 10–12 devices for a total installed power of 2,000 MVA ca. [16].

While in a controlled series compensation device the reactive power is produced or consumed by energy storage elements (capacitors and reactors), another

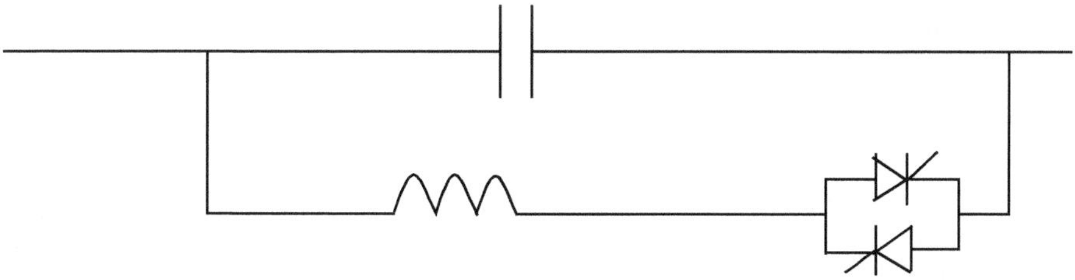

Fig. 9.4 Basic scheme of a TCSC

Fig. 9.5 Basic scheme of a SSSC

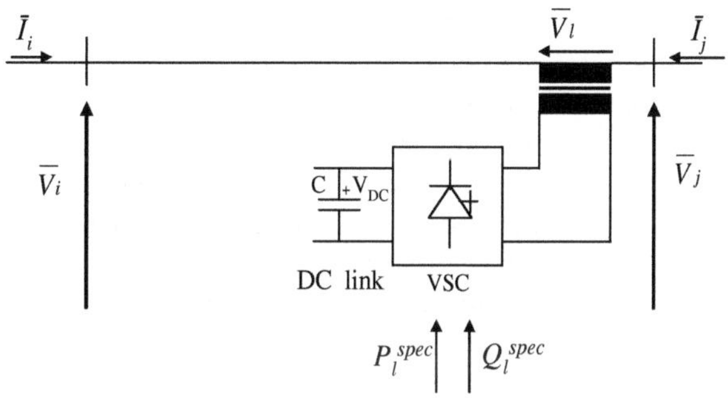

approach is possible using GTOs in a device which in the literature is generally called a Static Synchronous Series Compensator (SSSC), an Advanced Series Compensator (ASC) or a GTO-CSC. As a STATCOM is an improved SVC, an SSSC is similarly a development of a controlled series compensation device. SSSCs have so far been deployed mostly in North-American implementations. Recent developments related to SSSCs concern the testing and installation of a device in the Spanish 220 kV grid as part of the Spanish TSO plans for accommodating wind and other RES penetration [11]. The general structure of an SSSC is presented in Fig. 9.5 [16].

In contrast to the series capacitor, an SSSC can exchange active power with the AC system, by controlling the angular position of the injected voltage with respect to the line current. This capability of the SSSC allows simultaneous compensation of both the reactive and resistive components of the series line impedance. Also unlike the TCSC, the SSSC has immunity to resonance: in fact, the voltage drop across the relatively small inductive output impedance of the SSSC, provided by the leakage inductance of the series insertion transformer, is automatically balanced at the fundamental frequency when the SSSC provides capacitive line compensation. Thus, the effective output impedance versus frequency characteristic of the SSSC remains that of a small inductor at all frequencies except for its fundamental operating frequency. On the other hand, the SSSC has a very fast (almost instantaneous) response and thus it can be very effective in damping subsynchronous oscillations (which may be present due to existing series capacitors) [18]. Other details about SSSCs are given in [1, 14–16, 20, 23] and the references therein, among others.

9.3.3 Combined Controllers

These controllers combine several devices, some of which have been described in the previous Sects. 9.2, 9.3.1 and 9.3.2. Usually, they have both shunt and series capabilities, with the exception of the IPFC, which is a combination of several series controllers.

Combined devices have the big advantage of being able to simultaneously use the features of both types of device they are made of.

The TCPST (Thyristor Controlled Phase Shifting Transformer) is a device based on thyristor valves and PST technologies. Power electronics has revolutionized the use of PSTs in power system control, because the replacement of the traditional mechanical tap changers by thyristor valves (to obtain the TCPST) has increased the response speed of phase shifters and made them effective for the enhancement of both small-disturbance and transient stability [14].

Like for a PST, the phase angle difference between the TCPST terminal voltages can be obtained by a boosting transformer in series with the transmission line. The active and reactive power, taken to the transmission line by this boosting transformer (by injected voltage), must be absorbed from the network by a shunt (excitation) transformer. Figure 9.6 shows a basic scheme of a TCPST [16].

As for the PST, by neglecting losses, the TCPST neither produces nor absorbs active or reactive power.

From the electrical point of view two types of TCPST can be distinguished: TCPAR (Thyristor Controlled Phase Angle Regulator) and TCQBT (Thyristor Controlled Quadrature Boosting Transformer). These two types have the corresponding features of PAR and QBT, respectively (see also Sect. 9.2).

In fact, for the TCPAR, keeping equal magnitude for input and output voltages, the controllable parameter is the voltage shift angle.

For the TCQBT, in which there is a change in the output voltage magnitude, depending on the complex turn ratio, the controllable parameter is the magnitude of the injected voltage, whose phasor is shifted by a constant angle (equal to 90° in most cases) with respect to the input voltage vector [1].

Finally, concerning its impact on system problems, a TCPST is reasonably effective for load flow control and, like series FACTS devices, offers good performance for power oscillation damping and transient stability improvement (but in these applications it is less effective than an SSSC), while it has little influence on voltage control.

Further details about TCPSTs are given in [1, 14–16, 20] and in the references therein.

Like the TCPST, the Dynamic Flow Controller (DFC) is based on both thyristor controlled and conventional technologies. It combines a standard (mechanically switched) PST, a mechanically switched shunt capacitor (MSC), a multi-module, thyristor switched series capacitor (TSSC) and thyristor switched series reactors (TSSRs). It is in fact a hybrid device combining traditional PST and switched series compensation.

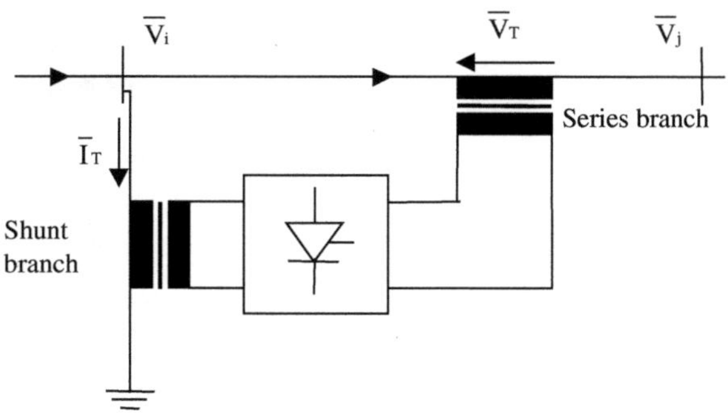

Fig. 9.6 Basic scheme of a TCPST

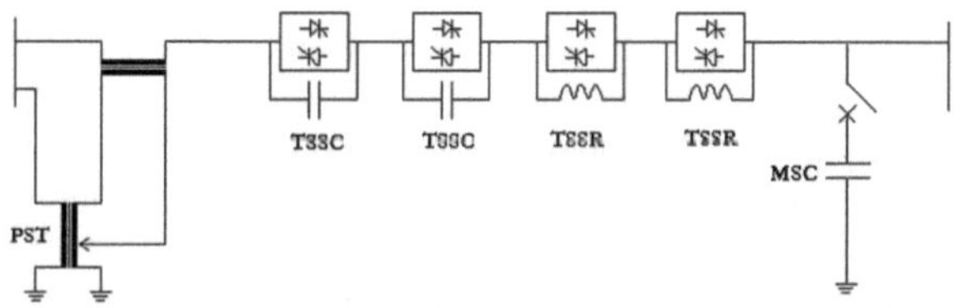

Fig. 9.7 Basic scheme of a DFC [15]

In terms of operation, the purpose of the MSC is to provide voltage support in the event of overload and other possible conditions, and the reactances of reactors and capacitors are selected on a binary basis to result in the desired stepped reactance variation [15]. Figure 9.7 shows a basic scheme of a DFC.

The switching of series reactors occurs, in principle, at zero-angle control to avoid producing any kind of harmonics. Nevertheless, the principle of phase-angle control used in the TCSC can be applied for continuous control as well.

This device is intended to be quite effective in terms of power flow control and reactive compensation, being seen as having some advantages in terms of cost-effectiveness, simplicity, maturity and robustness of the technologies of its sub-systems [15]; however, for the time being it is still a prototype device. Further details on the above mentioned points are given in [15] and in the references therein.

Developments in FACTS research have led to a new device: the Interline Power Flow Controller (IPFC) [14]. This arrangement consists of two (or more) series voltage source converter-based devices (SSSCs) [24] installed on two (or more) lines and connected at their DC terminals. Thus, in addition to serially compensating the reactive power, each SSSC can provide real power to the common DC link from its own line. The IPFC thus enables the problem of controlling different transmission lines at a given substation to be solved. In fact, the under-utilized lines make available surplus power which can be used by other lines for real power control. This capability makes it possible to equalize both real and reactive power flow between the lines, to transfer power demand from overloaded to under-loaded lines, to compensate against resistive line voltage drops and the corresponding reactive line power, and to increase the effectiveness of the compensating system

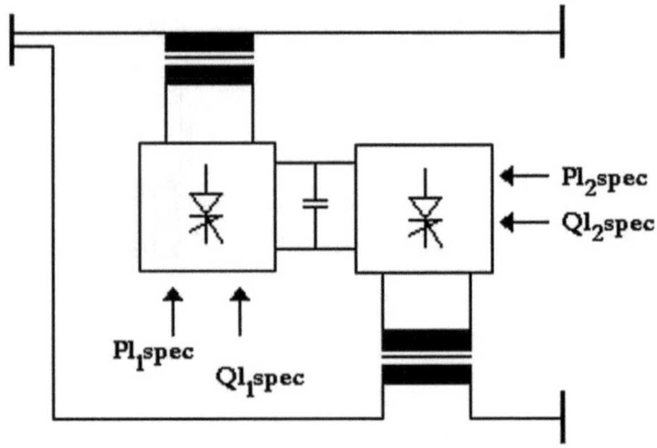

Fig. 9.8 Basic scheme of an IPFC

for dynamic disturbances (transient stability and power oscillation damping). Therefore, the IPFC provides a highly effective scheme for power transmission at a multi-line substation [16].

The IPFC is a multi-line FACTS device. In the case of two lines controlled by an IPFC the basic diagram is that one shown in Fig. 9.8.

The first phase of the convertible static compensator deployed, as a pilot project, at the Marcy substation operated by the New York Power Authority (NYPA), with the support of the EPRI, was completed in 2001 and had the IPFC among its possible multiple configurations [24]. The purpose was to establish further control concepts for all the voltage source converter-based FACTS devices and to provide benefits to the New York transmission system, allowing additional system flow for a variety of loading patterns and contingencies. The multi-line device can be used in eleven configurations: STATCOM 1, STATCOM 2, both STATCOMs, SSSC 1, SSSC 2, both SSSCs, STATCOM 1 + SSSC 2, STATCOM 2 + SSSC 1, IPFC, UPFC 1, UPFC 2. The analysis carried out ex-post showed an improvement in the power flow and, in particular, in the congested Utica-Albany segment, increased stability and reliability [16, 24].

The UPFC (Unified Power Flow Controller) is the most powerful and versatile (and costly) FACTS device, able to independently and autonomously control voltage amplitude, active and reactive power flow. This device (see Fig. 9.9) results from the combination of a STATCOM (converter 1 for the shunt part) and a SSSC (converter 2 for the series part), interlinked via a common DC capacitor [1, 14, 15]. Converter 1 is used primarily to provide the active power demand of converter 2 at the common DC link. Converter 2 itself generates the reactive power demand corresponding to series voltage injection and, therefore, the transmission system is not burdened by reactive power flow due to the operation of the UPFC. In fact, since converter 1 can also generate or absorb reactive power at its AC terminal, independently of the active power it transfers to (or from) the DC terminal, it follows that, with proper controls, it can also fulfill the function of an independent STATCOM. That is, it can provide reactive power compensation for

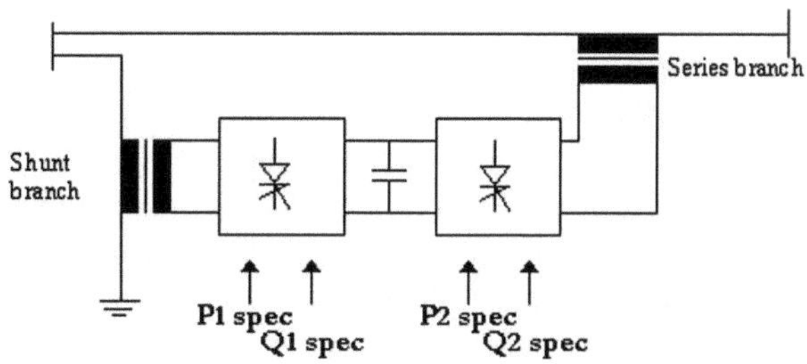

Fig. 9.9 Basic scheme of an UPFC

the transmission line and thus performs indirect voltage regulation at the input terminal of the UPFC. In addition, the UPFC can operate as a series impedance compensator when the shunt element is out of service and as a static VAR source when the series element (SSSC) is out of service [1, 14–16].

The main functions of the UPFC, which combines the features of a STATCOM, SSSC and TCPST, are [1, 14–16, 20]:

- Voltage regulation with continuously variable in-phase/anti-phase voltage injection. Functionally, this is similar to that obtainable with a transformer tap-changer having infinitely small steps.
- Series reactive compensation. This is similar to, but more general than, controlled series capacitive and inductive series compensation. This is because the UPFC injected series compensating voltage (as for the SSSC) can be kept constant, if desired, independent of line current variation, whereas the voltage across the series compensating (capacitive and inductive) impedance varies with the line current. This is a clear advantage of VSC-based devices compared to thyristor-based devices.
- Phase shifting. The UPFC can function as a perfect phase shifter. From the practical viewpoint, in contrast to conventional phase shifters, the AC system does not have to supply the reactive power that the phase shifting process demands, since it is actually generated by the UPFC converter.
- Multi-function power flow control, executed by simultaneous terminal voltage regulation, series capacitive line compensation and phase shifting.
- Enhancement of transmission capacity, transient stability, power oscillation damping, and voltage stability. For its speed and control characteristics, the UPFC is the most complete and powerful FACTS device in performing those steady-state and dynamic functions.

There are currently three UPFC solutions implemented worldwide: two are in the USA, one is in South Korea. The first installation of this device was carried out in 1998 at the Inez station by American Electric Power (AEP) in eastern Kentucky, in a joint effort with the EPRI and Westinghouse. In this application, the UPFC employs two GTO-based converters, each rated ±160 MVA, connected by a common DC link [14, 25]. The second implementation concerns the convertible static compensator project, previously mentioned, developed by the New York Power Authority (NYPA) and the EPRI at the Marcy substation: the UPFC results from two out of eleven possible configurations of this multi-line system [24]. The third UPFC application is in South Korea, where Korea Electric Power Corporation (KEPCO) has installed an 80 MVA UPFC at its 154 kV Kang-Jin substation. The device has been operational since October 2002 [26].

Recent developments in the FACTS field are those relating to SSSCs (see Sect. 9.3.2), aiming at less device complexity and reduced technology cost by deploying transformer less controllers with more advanced thyristors [16].

9.4 Control Technologies: HVDC

Worldwide, transmission and distribution of electricity mostly operate in AC. However, for some specific applications, DC is preferred over AC: this is the case of HVDC, which is a well-established technology, presenting no limitation on link length, differently from AC systems. The first HVDC installations date back to 1950s; nowadays, HVDC technologies are worldwide deployed counting on a long operational experience. In fact, this power electronics-based technology exhibits characteristics that over the years have made it widely attractive over HVAC transmission for several applications, such as: long-distance overhead transmission (several hundreds to few thousands kilometers); asynchronous systems interconnection via back-to-back (BTB) installations; medium/long-distance submarine/underground cable transmission; bulk power transport [16, 27].

Thanks to its response features, the HVDC technology can deliver several benefits to the transmission system, such as [27]:

- transfer capacity enhancement
- power flow control
- transient stability improvement
- power oscillation damping
- voltage stability and control
- rejection of cascading disturbances.

Classic HVDC technology, developed over the years, has been exploiting the significant progress of power electronics components evolving from mercury arc-based to thyristor-based converters. The basic principle of power systems coupling by an HVDC link is the conversion of AC of any frequency to DC and back to AC as illustrated in Fig. 9.10.

It is important to distinguish between two general groups of HVDC systems depending on the distance between the rectifier and the inverter. If both converters are installed in the same location, this device is an HVDC Back-to-Back (BTB) station. If the DC circuit bridges longer distances and the converters are installed in different locations this represents an HVDC long-distance transmission system. Usually, these links connect only two points, i.e. only two converters, one acting as inverter and the other one operating as rectifier, exist. Projects with HVDC long-

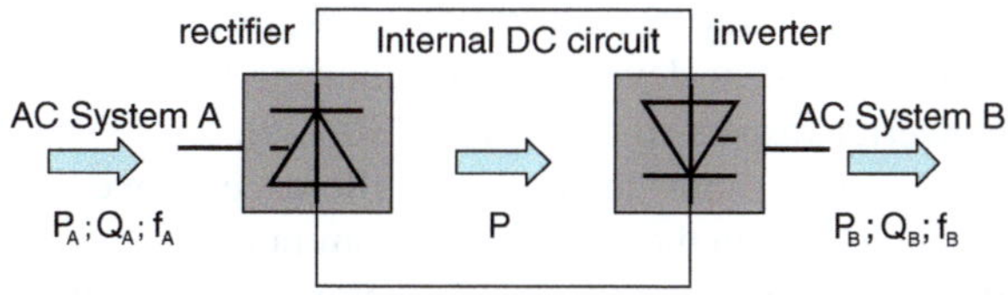

Fig. 9.10 Basic principles and scheme of a HVDC system

distance transmission systems connecting more than two points, so-called multi-terminal HVDC systems, have been implemented in the past in a very limited amount, being the Sardinia-Corsica-Italian Peninsula (SA.CO.I.) and the Quebec-New England links the only multi-terminal HVDC interconnections currently in operation around the world. Today, however, multi-terminal HVDC systems are again more and more investigated [6, 16].

The system characteristics of an HVDC system differ from the ones of AC transmission systems. The most important differences are:

- Asynchronous interconnection: Coupling of two independent AC systems with different nominal voltage and/or frequency and/or operational rules.
- Controllability of active and reactive power flow.

The most important difference between an AC and DC link is the possibility to control the active power transmitted via the tie. The controllability of the power flow could be used to improve the operating conditions of the coupled systems. Loop flows are avoided and would not additionally load the lines in the interface area. This allows to optimize the use of existing equipment and to maximize the transfer capability.

Several severe disturbances in the past have shown how minor failures affected a whole interconnected system. A DC link would act as a firewall, avoiding cascading effects and such spreading of disturbances from one system to another. This decoupling character of a DC link is its major advantage. In case of system disturbances, neither a dynamic interaction between the coupled systems nor the consideration of a specific margin which reduces the installed interface capacity has to be regarded compared to the synchronous coupling. Additionally, DC links allow the control of the power flow which discharges the grid under overload and emergency situations [28].

Adding up on HVDC traditional features, recent advances in power electronics, related to the availability of fully controllable solid-state components like GTO thyristors, IGCT thyristors and IGBT transistors (as seen also for FACTS), may lead to a further deployment of HVDC technology in the transmission grids. This is the case of the very promising self-commutated Voltage Source Converter (VSC)-based HVDC, which represents the state-of-the-art technology for connection of offshore wind farms and for multi-terminal applications. The key advantages of VSC-HVDC with respect to the line-commutated Current Source Converter (CSC)-based HVDC (the classic or conventional HVDC) are the possibility to feed reactive power into a network node and provide a smoother voltage support.

VSC technology is available for two-, three- or multi-level converters. Multi-level VSC is a recent development, requiring less filtering and having lower losses per converter respect to two-level VSC [29].

Going in more details, one difference between CSC-HVDC and VSC-HVDC technologies, in the case of line-commutated converter implementation, consists in that CSC is based on using a reactor as "smoothing" element for keeping the DC current constant, while VSC is based on using a capacitor as "smoothing" element for keeping the DC voltage constant. One of the main features of the CSC technology is that switching frequency is defined by the system frequency. In addition, CSC-HVDC also differs from VSC-HVDC since it is a robust technology with lower losses and higher rating.

Furthermore, CSC-HVDC and VSC-HVDC differ in their ability to provide reactive power support for the power grid. Neglecting losses, while the CSC rectifier receives active power from and the CSC inverter injects active power into the AC power grid, both rectifier and inverter consume reactive power from the connected AC network in order to perform a reliable commutation. In case of VSC-HVDC transmission, depending on the magnitude and the phase angle of the required voltage, the VSC converter can function in four different operation modes, i.e. the rectifier or the inverter operation mode, and therein, the power factor to be leading or lagging. Ideally, in a capability chart of active (P) and reactive (Q) power transmission, a VSC-HVDC link behavior can be represented by a four-quadrant operation characteristic. Hence, VSC-HVDC provides for reactive power exchange between DC and the AC systems and therefore contributes to voltage stability [27, 29].

There are however limits restricting the VSC-HVDC capability. Figure 9.11 provides an example of the operating range of a VSC-HVDC transmission system [30]. There are three main factors that limit the capability of the VSC-HVDC converters seen from a power system stability perspective. These are [29, 30]:

1. Maximum current. The maximum current that can flow through the valves, when multiplied by the AC voltage, yields the maximum MVA power circle in the power plane. If the AC voltage decreases, the MVA power capability reduces accordingly.
2. Maximum DC voltage. Reactive power is mainly dependent on the voltage difference between the AC voltage the VSC can generate and the AC network voltage. If the AC network voltage is high, the difference between the maximum DC voltage and the AC voltage is low. The reactive power capability is then moderate, but increases with decreasing AC voltage.
3. DC link ampacity. The maximum DC current that can flow through the link provides also a limitation.

Considering then the three limitations above reported, Fig. 9.12 provides an example of the resulting operating range of a VSC-HVDC transmission system in more realistic terms [30].

Fig. 9.11 Example of capability limits of VSC-HVDC transmission system [30]

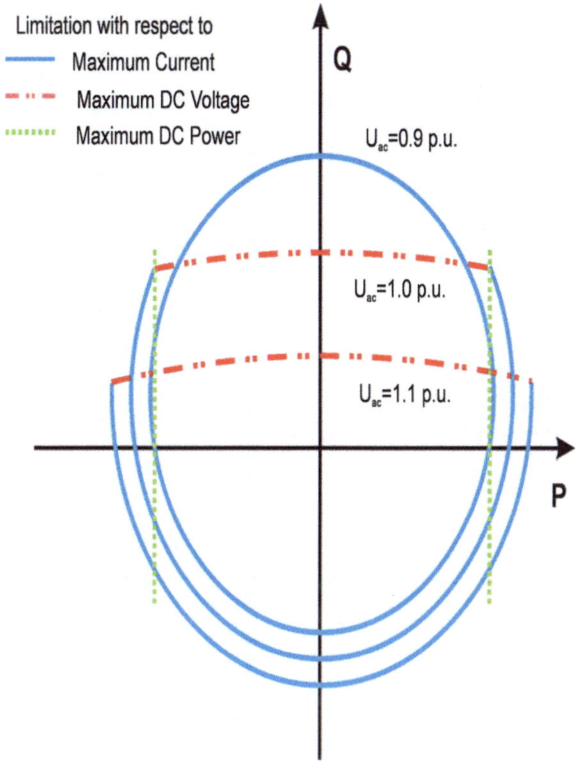

Several revolutionary changes have occurred lately in the HVDC power industry. For the line-commutated CSC-HVDC ongoing developments, the current world record example has concerned the installation of ±800 kV, 6,400 MW CSC-HVDC overhead lines (OHLs) in China [30, 31], and these ratings will be further overcome in a close future. In fact, large line-commutated CSC-HVDC OHLs projects in China (rated 7,200 MW at ±800 kV) and in India (multi-terminal link rated 8,000 MW at ±800 kV) have been already planned. Further developments foresee reaching voltage ratings of 1,000 kV for CSC-HVDC OHLs [30, 31].

On the other hand, self-commutated VSC-based HVDC technologies give nowadays the possibility to transfer up to 1,000 MW per transmission system if the operation voltage is ±320 kV [16]. New prototypes of VSC-HVDC devices at higher voltage levels (up to ±400–500 kV) for higher power transfer (up to 1,500–2,000 MW) are under preparation/testing. These circumstances allow recognizing several new structures of transmission systems. Sometimes such structures are named as hybrid technologies. One of the applications of the hybrid technologies, when AC and DC are used in couple, is segmentation of power grids in order to prevent cascading outages and soften the effects of disturbances [28, 32]. The needs of the hybrid AC-DC grid structure for future power systems

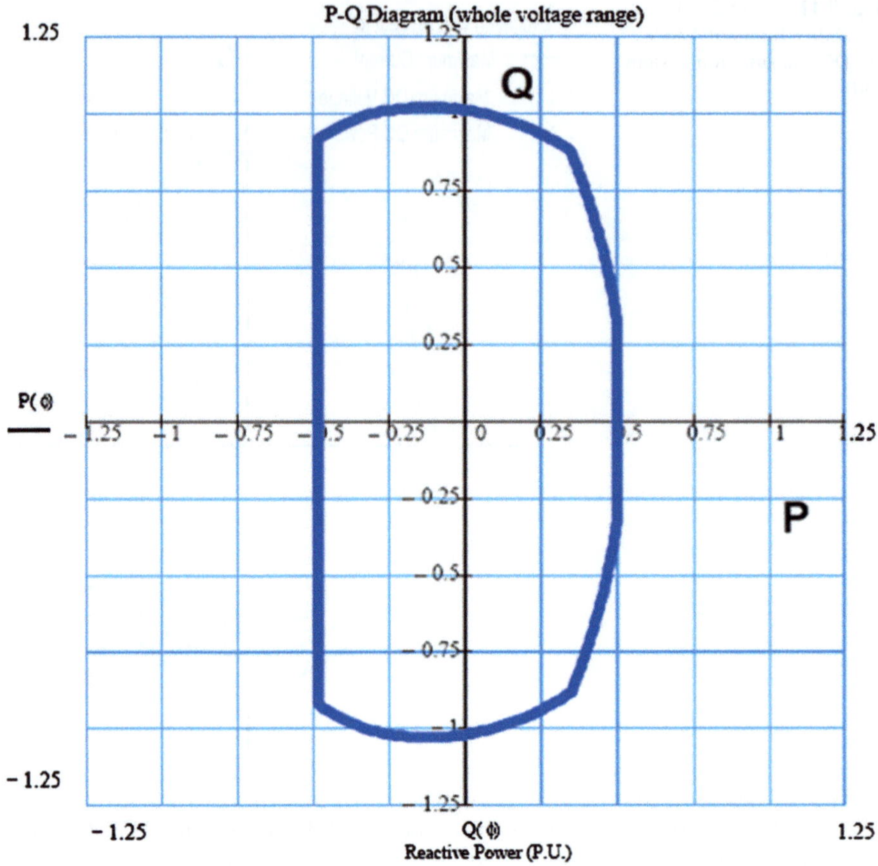

Fig. 9.12 Example of operating range of VSC-HVDC transmission system [30]

are discussed in [29, 32]. The embedding of VSC-HVDC transmission system in meshed AC grid is improving efficient power delivery. The hybrid AC/DC transmission system could effectively manage the overall power grid operation security and efficiency. Technology advantages of VSC-HVDC incorporated in meshed AC are:

- power flow control flexibility,
- fast response to disturbances,
- multi-terminal configurations.

Several examples of operation of HVDC links exist in Europe and around the world. Table 9.1 illustrates selected applications of HVDC in Europe (as of 2012).

In Europe there is at the moment only one HVDC BTB link in operation, located between Finland and Russia. A planned HVDC BTB link is the one between Lithuania and Poland (located in Alytus, Lithuania). Others could be

Table 9.1 Examples of HVDC systems in operation in Europe (as of 2012)

Name	Countries connected	Rating (MW)
Konti-Skan	Denmark–Sweden	300
Skagerrak 1, 2 + 3	Denmark–Norway	1,000
BalticCable	Germany–Sweden	600
Kontek	Denmark–Germany	600
GRITA	Greece–Italy	500
SwePol	Sweden–Poland	600
Fenno-Skan 1	Sweden–Finland	500
Fenno-Skan 2	Sweden–Finland	800
Vyborg BTB	Finland–Russia	1,420
Estlink 1	Estonia–Finland	350
CrossChannel	France–UK	2,000
SA.CO.I.	Sardinia–Corsica–Italy	300
SA.PE.I.	Sardinia–Italy	1,000
NorNed 1	Netherlands–Norway	700
Moyle interconnector	Scotland–Northern Ireland	450
BritNed	UK–Netherlands	1,000
Romulo	Balearic–Spain	400
Volgograd-Donbass	Russia–Ukraine	720

potentially installed in the future at the ties between the Russian system (IPS/UPS) and the continental European system (ENTSO-E CE). Earlier studies have identified possible locations for HVDC BTB systems in the interface zone between ENTSO-E CE and IPS/UPS in Rzeszów (Poland), Zakhidnoukrainska (Ukraine) and Isaccea (Romania) [28].

Table 9.2 summarizes the main features, benefits and the drawbacks for CSC-HVDC and VSC-HVDC technologies (as of 2012) [16, 27, 29].

Looking at the features of HVDC technologies, for the transmission of bulk power over long distances, currently CSC-HVDC is the best suited HVDC technology. On the other hand, VSC-HVDC is useful for the interconnection of remote offshore wind farms to the main power grid since VSC-HVDC is not dependent on a specified Effective Short-Circuit Ratio (ESCR) or on reactive power support at the connection points in order to perform a reliable commutation process. Furthermore, its independency of external commutation enables VSC-HVDC to perform a black start which helps TSOs to overcome the problem of how to start-up connected offshore wind farms and to re-energize network sections that suffered from a system blackout. VSC-HVDC can provide additional capacity for the point-to-point power transmission over short-to medium-long distances within a power grid while also providing fast reactive power flow and voltage control at its terminals and thereby contributing to power system and voltage stability. For all these features VSC-HVDC is the most promising technology for multi-terminal DC applications and also for forming the backbone of potential offshore grids implementation. HVDC provides a transfer corridor of fixed power rating at all times during network operation which is especially beneficial during network

Table 9.2 Comparison of CSC-HVDC and VSC-HVDC

	CSC-HVDC	VSC-HVDC
System description		
System ratings in operation	±500 kV, 1,000 MW (cable)	±200 kV, 400 MW (cable)
	±800 kV, 6,400 MW (OHL)	350 kV, 300 MW (OHL)
System ratings planned/available	±600 kV, 2,200 MW (cable)	±500 kV, 1,400 MW (cable)
	±800 kV, 8,000 MW (OHL)	±350 kV, 700 MW (OHL)
Future trend of system ratings	Towards higher ratings	Towards higher ratings
Operational experience	>50 years	>10 years
Lifetime	30–40 year	30–40 years[a]
Converter losses (at full load, per converter)	0.5–0.8 %	0.9–1.3 %
Availability (per system)	>98 %	>98 %
System capabilities		
Reactive power injection possibility	No	Yes
Easy meshing	No	Yes
Limitation in cable line length	No	No
Black start capability	No	Yes

[a] estimated values, not enough experience yet

disturbances. Furthermore, while VSC-HVDC is able to directly inject reactive power into a network node, this reactive power is no longer transported to the considered network node through neighboring AC transmission lines. This frees transmission capacity in the vicinity of the considered network node which was formerly occupied by reactive power and can now be used for active power transmission. Therefore, the positive effect of a VSC-HVDC transmission line on neighboring network paths has to be taken into account by power flow analyses in order to perform a complete economic assessment [16].

In conclusion, considering the fast developments of VSC-HVDC technology and its applications, especially in Europe but not only, the following prospects on VSC-HVDC can be taken [16, 28]:

- VSC-HVDC technology is now emerging as a robust and economical alternative for future transmission grid expansion. In particular, embedded VSC-HVDC applications in meshed AC grids could significantly improve overall system performance, enabling smart operation of transmission grids with improved security and efficiency.
- VSC-HVDC transmission also offers a superior solution for many challenging technical issues associated with integration of large-scale renewable energy sources such as offshore wind power. The technology is under continuous development rapidly into higher voltage, higher power and more flexibility.

9.5 Reliability and Availability of FACTS and HVDC

The reliability of a transmission system describes the degree of performance of its elements and can be addressed by the aspects of security and adequacy of the analyzed system [33]. It is expressed in percent and takes overall account of the probability that each transmission component adequately performs under the required operating conditions for the considered period of time.

In the reliability analysis some indices are essential and they are recalled in the following [34]:

Mean Time To Failure (MTTF). It is the average time from the instance a component/device/system enters a working state until a component/device/system enters a failed state. This may also be defined as the component/device/system uptime.

Mean Time To Repair (MTTR). It is the average time it takes to restore a component/device/system to a specified working condition from the instance the component/device/system failed. This may also be defined at the component/device/system downtime.

Availability. It is the probability of finding the component/device/system in the operating state at some time into the future. The availability of a component with two states can be calculated by

$$A = \frac{Uptime}{Uptime + Downtime} = \frac{MTTF}{MTTF + MTTR} \tag{9.1}$$

Failure rate. It is the number of times a component/device/system is expected to fail per unit of time or the number of times a component/device/system is expected to fail per unit of time the component/device/system is in a working condition. It is represented by λ and it can be calculated by two methods. In one case the number of failures for a sample of components for a specified period of time is recorded without suspending time for a component upon failure, whereas another approach considers suspending time when a component enters a failure state.

Reliability analyses normally specify the failure rate and MTTR. Therefore, if the reliability survey has calculated the failure rate for a component without suspending time for failed components, the failure rate can be related to MTTR and MTTF by the expression

$$MTTF = \frac{1}{\lambda} - MTTR \tag{9.2}$$

The availability of a transmission system is the counterpart to reliability (also given in percent): it takes into account planned (e.g. regular maintenance) and unplanned (e.g. valve failure) outages. These statistical values are subject to a number of non-deterministic influences and cannot be mathematically derived from the station design or other design parameters but rather be observed from operational experience. However, the overall reliability or availability of a

Table 9.3 Reliability indices values for selected HVAC assets [34–36]

Component	MTTF (years)	MTTR (hours)	Availability
AC Transformer	7.73	230	0.99662
Overhead AC line (single)	2.04	11	0.999385
Overhead AC line (double)	0.66	24	0.995885
Capacitor	1.42	100	0.992036
SSSC	1.41	200	0.984043
UPFC	2.05	200	0.988986

Table 9.4 Estimated reliability indices values for selected HVDC assets [38, 39]

Component	MTTF (years)	MTTR (hours)	Availability
Onshore VSC converter transformer (100–300 kV)	19.88	1,008	0.99425
Onshore CSC converter transformer (500 kV)	4.54	116	0.99710
Offshore converter transformer (100–300 kV)	19.83	1,512	0.99137
Onshore converter reactor	7	24	0.99961
Offshore converter reactor	7	192	0.99688
Onshore VSC converter (100–300 kV)	1.9	12	0.99928
Onshore CSC converter (500 kV)	6.98	136	0.99778
Offshore VSC converter (100–300 kV)	1.9	60	0.99641
Undersea DC cable (100 km)	14.12	1,440	0.98849
Overhead DC line	4.98	200	0.995443
Onshore DC Switchyard	4.02	26	0.99926
Offshore DC Switchyard	4.02	98	0.99722

complete transmission system can be mathematically derived from the single reliabilities or availabilities of all the equipment involved.

In Table 9.3 some typical failure and repair times as well as average availability values for different AC components are provided.

It has to be however said that, as the increasing deployment of FACTS is rather recent, more experiences are generally needed to derive for the different FACTS a comprehensive picture of their availability behavior. A notable exception is the SVC, as it was the first device to be deployed on a commercial scale. Availability values found in the literature for SVC devices usually lie between 0.98 and 0.99 [16]. Other values provided in literature are the availability for TCPAR (0.986667) and the MTTR for STATCOM (60 h) [16, 35, 36].

Concerning the reliability figures of HVDC, regular technical reports (see e.g. [37]) provide the availability records for selected line-commutated CSC-HVDC transmission systems collected over the last twenty years. It has also to be stated that there is only a limited experience so far in the operation of self-commutated VSC-HVDC transmission systems (especially those ones in a maritime environment) yet, so that reported availability figures have to be handled with care. Further observational long-term studies in the field of VSC-HVDC are necessary to confirm the presently assumed values.

Table 9.4 gives account of some elements on the reliability indices values for different HVDC system components [38, 39].

Table 9.5 Main technical–economic features of selected FACTS and HVDC [16, 28]

Device	Transmission capacity increase	Power flow control	Transient stability improvement	Voltage stability improvement	Power oscillation damping	Investment costs
SVC	▲▲	▲	▲	▲▲▲	▲▲	▲
STATCOM	▲	▲	▲▲	▲▲▲	▲▲	▲▲
TCSC	▲▲▲	▲▲	▲▲▲	▲	▲▲	▲
SSSC	▲▲▲	▲▲▲	▲▲▲	▲	▲▲	▲▲
UPFC	▲▲▲	▲▲▲	▲▲▲	▲▲▲	▲▲▲	▲▲▲
CSC-HVDC	▲▲▲	▲▲▲	▲▲	▲	▲▲	▲▲▲
VSC-HVDC	▲▲	▲▲▲	▲▲▲	▲▲	▲▲▲	▲▲

9.6 Conclusions

In addition to the application of PSTs, widespread deployed, the role of advanced transmission technologies, like FACTS and HVDC, for improving power system control and performances is expected to largely increase: these devices will extensively impact on future transmission grids, in Europe and elsewhere. This trend is also more and more driven by the steadily growing penetration of RES and related consequences. On the other hand, considering that the deployment of FACTS and HVDC technologies requires a significant financial effort, striking a balance between gains and drawbacks will be of crucial importance.

Table 9.5 [16, 28] summarizes main characteristics of different FACTS and HVDC technologies.

References

1. CIGRE Task Force 38.01.06, *Load flow control in High Voltage Power Systems using FACTS controllers* (1996)
2. J. Verboomen, D. Van Hertem, P.H. Schavemaker, W.L. Kling, R. Belmans, Phase Shifting Transformers: principles and Applications, in *Proceedings of International Conference on Future Power Systems*, Amsterdam (Netherlands), 16–18 Nov 2005
3. P. Hurlet, J.-C. Riboud, J. Margoloff, A. Tanguy, French experience in phase-shifting transformers, in *Proceedings of 2006 CIGRE Session*, Paris (France), 27 Aug–1 Sep 2006
4. D. Van Hertem, J. Verboomen, S. Cole, W. Kling, R. Belmans, Influence of phase shifting transformers and HVDC on power system losses, in *Proceedings of 2007 IEEE PES General Meeting*, Tampa (USA), 24–28 Jun 2007
5. J. Verboomen, D. Van Hertem, P.H. Schavemaker, W.L. Kling, R. Belmans, Border-Flow Control by means of Phase Shifting Transformers, in *Proceedings of 2007 IEEE PowerTech Conference*, Lausanne (Switzerland), 1–5 Jul 2007
6. A. Vaféas, S. Galant, T. Pagano, Final WP1 report on cost/benefit analysis of innovative technologies and grid technologies roadmap report validated by the external partners, REALISEGRID Project Deliverable D1.4.2, May 2011 [Online]. Available: http://realisegrid.rse-web.it/content/files/File/Publications%20and%20results/Deliverable_REALISEGRID_1.4.2.pdf

7. U. Häger, J. Schwippe, K. Görner, Improving network controllability by coordinated control of HVDC and FACTS devices, REALISEGRID Project Deliverable D1.2.2, Apr 2010 [Online]. Available: http://realisegrid.rse-web.it/content/files/File/Publications%20and%20results/Deliverable_REALISEGRID_1.2.2.pdf
8. J. Warichet, J.-L. Leonard, J. Rimez, O. Bronckart, J. Van Hecke, Grid implementation and operational use of large phase shifting transformers in the Belgian HV grid to cope with international network challenges, in *Proceedings of 2010 CIGRE Session*, Paris (France), 22–27 Aug 2010
9. E.M. Carlini, G. Manduzio, D. Bonmann, Power Flow Control on the Italian network by means of phase-shifting transformers, in *Proceedings of 2006 CIGRE Session*, Paris (France), 27 Aug–1 Sep 2006
10. ENTSO-E, Ten-Year Network Development Plan 2010, Jun 2010 [Online]. Available: https://www.entsoe.eu/major-projects/ten-year-network-development-plan/tyndp-2010
11. ENTSO-E, Ten-Year Network Development Plan 2012, Jul 2012 [Online]. Available: https://www.entsoe.eu/major-projects/ten-year-network-development-plan/tyndp-2012
12. J. Brochu, F. Beauregard, R. Cloutier, A. Bergeron, L. Garant, F. Sirois, M.I. Henderson, Innovative Applications of Phase-Shifting Transformers supplemented with Series Reactive Elements, in *Proceedings of 2006 CIGRE Session*, Paris (France), 27 Aug–1 Sep 2006
13. J. Brochu, *Interphase Power Controllers*, (Polytechnic International Press, Montreal, 1999)
14. N.G. Hingorani, L. Gyugyi, *Understanding FACTS: Concepts and Technology of Flexible AC Transmission Systems*, (IEEE Press Inc., New York, 2000)
15. X.P. Zhang, C. Rehtanz, B. Pal, *Flexible AC Transmission Systems: Modeling and Control*, 2nd edn, (Springer, Berlin, 2011)
16. S. Rüberg, H. Ferreira, A. L'Abbate, U. Häger, G. Fulli, Y. Li, J. Schwippe, Improving network controllability by Flexible Alternating Current Transmission System (FACTS) and by High Voltage Direct Current (HVDC) transmission systems, REALISEGRID Project Deliverable D1.2.1, Mar 2010 [Online]. Available: http://realisegrid.rse-web.it/content/files/File/Publications%20and%20results/Deliverable_REALISEGRID_1.2.1.pdf
17. FACTS Terms and Definitions Task Force—FACTS Working Group—DC and FACTS Subcommittee, Proposed Terms and Definitions for Flexible AC Transmission System (FACTS), IEEE Trans. Power Delivery, **12**(4), 1848–1853 (1997)
18. CIGRE Task Force 38.01.07, *Analysis and control of power system oscillations* (1996)
19. N. Mithulananthan, C.A. Cañizares, J. Reeve, G.J. Rogers, *Comparison of PSS, SVC and STATCOM controllers for damping power system oscillations*, IEEE Trans. Power Syst. **18**(2), 786–792 (2003)
20. A. L'Abbate, Modeling of FACTS Devices and Applications in Liberalized Power Systems, Doctoral degree dissertation, Politecnico di Bari (Italy)/Universität Dortmund (Germany) (2004)
21. J.J. Paserba, N.W. Miller, E.V. Larsen, R.J. Piwko, *A thyristor controlled series compensation model for power system stability analysis*, IEEE Trans. Power Delivery **10**(3), 1471–1478 (1995)
22. CIGRE Working Group 14.18, *Thyristor Controlled Series Compensation* (1997)
23. CIGRE Working Group B4.40, *Static Synchronous Series Compensator* (SSSC) (2009)
24. B. Fardanesh, M. Henderson, B. Shperling, S. Zelingher, L. Gyugyi, C. Schauder, B. Lam, J. Mountford, R. Adapa, A. Edris, Convertible Static Compensator Application to the New York Transmission System, in *Proceedings of 1998 CIGRE Session*, Paris (France), 31 Aug–5 Sep 1998
25. B.A. Renz, A.J.F. Keri, A.S. Mehraban, J.P. Kessinger, C.D. Schauder, L. Gyugyi, L.J. Kovalsky, A.A. Edris, World's First Unified Power Flow Controller on the AEP System, in *Proceedings of 1998 CIGRE Session*, Paris (France), 31 Aug–5 Sep 1998
26. B.H. Chang, S.Y. Kim, J.S. Yoon, S.P. Moon, D.H. Baek, B.M. Kwak, J.B. Choo, Control Strategies Study for KEPCO UPFC Operation Automation in Korean Sub-Transmission System, in *Proceedings of 2006 CIGRE Session*, Paris (France), 27 Aug–1 Sep 2006

27. M.P. Bahrman, B.K. Johnson, *The ABCs of HVDC transmission technologies*, IEEE Power Energy Mag. **5**(2), 32–44 (2007)
28. D. Efimov, S. Palamarchuk, M. Vassilyev, A. Osak, I. Naumkin, S. Smolovik, E. Bompard, L. Krumm, A. Grobovoy, K. Görner, U. Häger, M. Osthues, A. Iaria, E. Gaglioti, A. L'Abbate, O. Soukhanov, R. Cherkaoui, M. Stubbe, X.-P. Zhang, T. Lomane, Requirements and technologies for the development, extension and interconnection of large power systems, ICOEUR Project Deliverable D1.4, Sep 2010 [Online]. Available: http://www.icoeur.eu/index.php/publications/26-deliverables.html
29. J. Arrillaga, Y.H. Liu, N.R. Watson, *Flexible Power Transmission—The HVDC Options*, (Wiley, New Jersey, 2007)
30. ABB, HVDC Reference Projects, 2012 [Online]. Available: http://www.abb.com/hvdc
31. Siemens, HVDC Reference Projects, 2012 [Online]. Available: http://www.energy.siemens.com/us/en/power-transmission/hvdc/
32. J. Pan, R. Nuqui, K. Srivastava, T. Johnson, P. Holmberg, Y.-J. Hafner, *AC Grid with Embedded VSC-HVDC for Secure and Efficient Power Delivery*, IEEE Energy 2030 Conference, Atlanta (USA), 17–18 Nov 2008
33. ENTSO-E, *Glossary of Terms*, 2010 [Online]. Available: https://www.entsoe.eu/fileadmin/user_upload/_library/resources/statistics/100903_Statistical_Glossary.pdf
34. R. Billinton, R.N. Allan, *Reliability Evaluation of Electric Power System*, 2nd edn. (Plenum Press, New York, 1996)
35. R. Billinton, M. Fotuhi-Firuzabad, S.O. Faried, S. Aboreshaid, Composite System Reliability Evaluation incorporating an HVDC Link and a Static Synchronous Series Compensator, in *Proceedings of 2012 IEEE Canadian Conference on Electrical and Computer Engineering*, Winnipeg (Canada), 12–15 May 2002
36. A. Rajabi-Ghahnavieh, M. Fotuhi-Firuzabad, R. Feuillet, Evaluation of UPFC Impacts on Power System Reliability, in *Proceedings of 2008 IEEE PES Transmission and Distribution Conference and Exposition*, Chicago (USA), 21–24 Apr 2008
37. M.G. Bennett, N.S. Dhaliwal, A. Leirbukt, A Survey on the Reliability of HVDC Systems Throughout the World during 2009–2010, in *Proceedings of 2012 CIGRE Session*, Paris (France), 26–31 Aug 2012
38. S. Zadkhast, M. Fotuhi-Firuzabad, F. Aminifar, R. Billinton, S.O. Faried, A.-A. Edris, *Reliability evaluation of an HVDC transmission system tapped by a VSC station*, IEEE Trans. Power Delivery **25**(3), 1962–1970 (2010)
39. A. Beddard, M. Barnes, *VSC-HVDC Availability Analysis*, (2011)

Chapter 10
Coordinated Power Flow Control

Ulf Häger

10.1 Introduction

Constructing new transmission lines is difficult due to environmental specifications and missing public acceptance. A temporary alternative is the integration of Power Flow Controllers (PFC) in order to dynamically redirect power flows. Chapter 9 has already introduced these devices as conventional Phase Shifting Transformers (PST) or as series controlled Flexible Alternating Current Transmission Systems (FACTS). In general, the installation of multiple PFCs with mutual influence to the transmission grid requires sophisticated coordination mechanisms in order to increase the overall transmission capacity and to avoid conflicts that may lead to unexpected behavior [1]. Several coordination methods based on optimal power flow (OPF) are known from the literature [2–5]. These methods can be used for coordinating PFCs in the security centers of the involved TSOs while the power system is in a normal operational state. Such methods take account of market constraints or $(n - 1)$ security constraints. The computation time of these methods to calculate a set of optimal PFC settings is very large, but can be reduced into the range of below 15 min under special circumstances [6].

This chapter is based on a novel approach for an autonomous multi-agent based real-time coordination system for PFCs which was developed and firstly introduced in [7, 8]. The main area of application of this approach is the coordination of PFCs during unforeseen events with dynamically changing power flows and fault/tripping scenarios. An analysis of the applicability of this method was carried out in [9] and robustness issues have been discussed in [10].

Section 10.2 provides an introduction of the autonomous multi-agent based real-time coordination system. Then, in Sect. 10.3 a test case is presented,

U. Häger (✉)
TU Dortmund University, Dortmund, Germany
e-mail: ulf.haeger@tu-dortmund.de

U. Häger et al. (eds.), *Monitoring, Control and Protection of Interconnected Power Systems*, Power Systems, DOI: 10.1007/978-3-642-53848-3_10,

comparing the agent coordination with an OPF method. In Sect. 10.4 the multi-agent coordination is further verified by analyzing the behavior on a large scale interconnection. A summary and conclusions are provided in Sect. 10.5.

10.2 Multi-agent System Structure for Distributed Coordination of PFC

In this approach all serial devices of the power system (transmission lines, transformers and PFCs) are equipped with agents. There are two kinds of agents, active agents and passive agents. Each PFC is equipped with an active agent. Each non-controllable power system device is equipped with a passive agent. An overview of the coordination system is given in Fig. 10.1.

The basis of this coordination system is that the active agents receive state messages containing recent information from passive agents about the current grid topology and about the needs for control actions.

Each active agent evaluates this information by carrying out a sensitivity analysis (including a distributed topology analysis), in order to estimate the influence of a control action of the corresponding PFC on power system devices. Based on the topology information each active agent sorts all power system devices into two zones: The loading of devices in the *ControlUpZone* is reduced by controlling the PFC upwards and the loading of devices in the *ControlDownZone* is reduced by the opposite control action.

To carry out the distributed coordination a weighting function is calculated for each power system device in the region of influence of the PFC in order to sort the control needs according to their significance, which depends on the loading of the device and the sensitivity of PFC control actions on this device. Based on the values of the weighting function, each active agent decides about the control direction of the next control action. If there are overloaded devices in the *ControlUpZone* and *ControlDownZone* then all control needs could only be compensated through simultaneous, yet contrary control actions. The overloading of the devices in the zone with the lowest weighting function value must be reduced by a second PFC that might be available.

Finally, the changes in PFC set-points are reflected by changed loadings (and states) of the power system devices, closing the control loop by passing new measurements from passive agents to active agents through the communication with state messages. This control loop also comprises a verification of the control actions in order to assure robustness of the control system.

The following sections provide details about the communication model (10.2.1), the principle communication among Agents (10.2.2), the communication rules (10.2.3), the assignment of the sensitivities (10.2.4) and the direction of impact (10.2.5), followed by a description of the distributed coordination (10.2.6) and the control of PFCs (10.2.7).

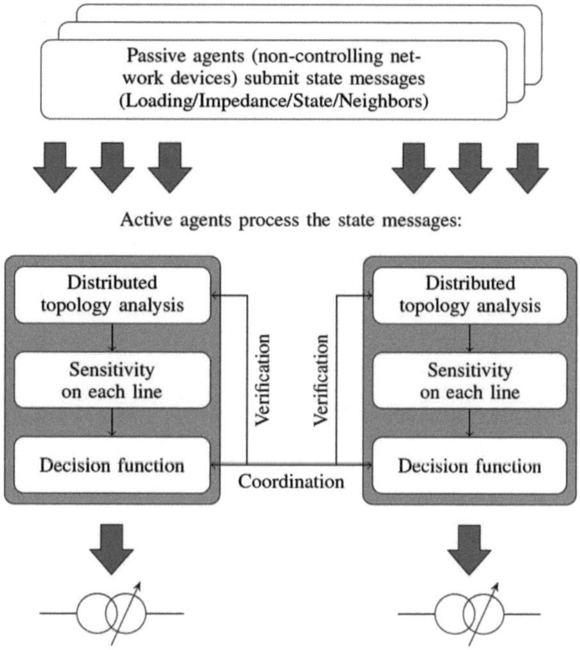

Fig. 10.1 General structure of the real-time coordination method of PFC devices based on a multi-agent system

10.2.1 Communication Model [8]

Communication rules are set up for all agents to enable an appropriate information exchange. State messages are submitted from non-controllable power system devices and propagated along the network topology containing information about loading, (cumulated) impedance and state of the power system devices. Controlling agents of PFCs evaluate the state messages to calculate the local (relevant) network topology, the sensitivities (estimated impact) of control actions on non-controllable power system devices and the demand for control actions (the criticality). Each non-controlling agent has the following knowledge about its element:

- Direction of the power flow through the device
- Loading of the device
- Impedance of the device
- State of the device (active or inactive)
- Neighboring devices physically connected to the connection buses of the own device.

10.2.2 Principle Communication Among Agents [8, 9]

The evaluation of the system state is done on the basis of *StateInformMessages* submitted by non-controlling agents along the power system topology. These

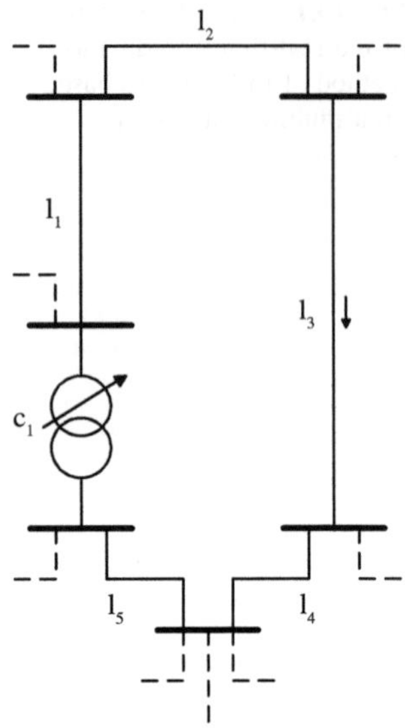

Fig. 10.2 Exemplary network situation for assigning the network topology

messages are submitted from one agent to the agents of the neighboring devices along the grid topology. Each device and each message has a unique identifier. The history of sending and forwarding agents is stored in each message. Each forwarding agent appends the following information to the message:

- Time stamp
- Identifier of the agent
- Loading and impedance of the corresponding power system device
- Direction of the power flow
- Identifier of the sending connection bus from which the message will be forwarded.

In the following a basic network situation, presented in Fig. 10.2, is used to illustrate this communication. It is explained how the controlling agent of PFC c_1 receives state information submitted by the non-controlling agent of transmission line l_3.

The agent of transmission line l_3 submits one message to every agent of its neighboring devices. These messages contain the above mentioned information about transmission line l_3. All agents of the devices physically connected to the sending connection bus receive this message and add the corresponding information of their own transmission line. The accumulated impedance of one message expresses the transmission path impedance. Subsequently, the messages are updated and forwarded further along the grid topology. Finally, the controlling agent of PFC c_1 receives one message from line l_3 at each connection bus of the PFC. The first message was submitted along the transmission lines l_3, l_2 and l_1, while the second message was transmitted along the transmission lines l_3, l_4 and l_5. By analyzing these two messages the controlling agent calculates that the transmission line l_3 is located on a transmission path connecting the two connection buses of the PFC.

10.2.3 Communication Rules [8, 9]

To process an incoming *StateInformMessage* a non-controlling agent will use the following rules:

- If the message has not traversed the current agent and thus the message is unknown, the agent will add the identifier and the impedance of its corresponding device. Subsequently, the message is forwarded to all neighboring devices connected to the opposite connection bus of the device (omitting the devices whose identifiers are listed in the message and have already been traversed).
- If the unique identifier of the current device is already listed in the identifier list of the message the *StateInformMessage* has been sent in a loop. The message is discarded.
- If the transmission path impedance of the incoming message is larger than an allowed maximum threshold for the transmission path impedance, then the message will be discarded. This ensures that messages only traverse agents within the vicinity of a PFC (in terms of impedance and impact of control actions) and reduces the total number of messages.

A controlling agent uses the same rules like a non-controlling agent to process and propagate incoming *StateInformMessages*. In contrast to a non-controlling agent, a controlling agent periodically evaluates the received messages. In this way the information about the area of influence of a PFC is permanently updated. The area of influence is defined by all devices on which the sensitivity for control actions of the PFC is above a certain boundary. In case of changes to the power system topology the information about control paths and parallel paths needs to be updated. Hence, *StateInformMessages* have to be submitted and evaluated periodically.

10.2.4 Assigning the Sensitivity

A controlling agent is required to have knowledge about the impact of a control action of its corresponding PFC on a certain power system device to be controlled. Two kinds of information are needed. On the one hand, the agent needs to know the sensitivity and, on the other hand, the direction of the impact on a device to be con-trolled (whether a directed control action of a PFC will increase or decrease the power flow through the device).

The sensitivity is the quotient between the power flow change through the affected device and the power flow change through the control path of the PFC. If the PFC c changes the power flow on the control path by $\Delta P(c)$ and this leads to a change of power flow $\Delta P(d)$ through the power system device d, the sensitivity between c and d is calculated according to [8, 9]:

$$sens(c,d) = \frac{\Delta P(d)}{\Delta P_{PFC}(c)} \tag{10.1}$$

In the distributed agent system it is impossible to assess the precise sensitivity by use of local information only. However, the sensitivity can be estimated by using the relevant topology information obtained through the communication of the agents (in the vicinity of PFCs until *StateInformMessages* are discarded due to the impedance threshold). Each controlling agent extracts continuously the topology information of all incoming *StateInformMessages*. A nodal admittance matrix is build, containing the topology data within the area of influence of a PFC. This process is carried out frequently with a time step T_{step} (in our experiments a typical interval for T_{step} is between 20 and 100 ms) by clearing the previous matrix and rebuilding it from scratch. The impedance x_{ij} of each power system device is appended to the rows and columns of the nodal admittance matrix corresponding to the connection buses i and j of the power system device (10.2)–(10.4). Shunt elements are neglected.

$$\underline{B} = \begin{bmatrix} b_{11} & \cdots & b_{1n} \\ \vdots & \ddots & \vdots \\ b_{n1} & \cdots & b_{nn} \end{bmatrix} \tag{10.2}$$

$$b_{ii} = \sum_{k=1}^{n_n} -\frac{1}{X_{L,ik}} \tag{10.3}$$

$$b_{ik} = \frac{1}{X_{L,ik}};\ i \neq k \tag{10.4}$$

The first two rows and columns of $\underline{B}$ correspond to the connection buses of the PFC c, associated with the controlling agent. Starting from this matrix of the dimension 2×2, one additional row and column is inserted for each bus (of power system devices discovered through the *StateInformMessages*) which does not exist in the nodal admittance matrix, yet. Since *StateInformMessages* are propagated along the power system topology, the same power system device can be recorded in several messages. The time stamp of each entry is used to filter the most recent information about a power system device which will be used for building the nodal admittance matrix.

After the creation of the nodal admittance matrix, DC power flow computation methods can be applied for the calculation of the sensitivities. Since $\underline{B}$ is singular it cannot be inverted. The pseudoinverse $\underline{B}^{\dagger}$ is used for the calculation of the sensitivity between c and d according to Eqs. (10.5)–(10.7), in which the power system device d is located between the connection nodes i and j:

$$\underline{P} = \begin{bmatrix} 1 \\ -1 \\ 0 \\ \vdots \\ 0 \end{bmatrix} \tag{10.5}$$

$$\underline{\delta'} = \underline{B}^{\dagger} \cdot \underline{P} \tag{10.6}$$

$$sens(c,d) = \frac{1}{X_{L,ik}} \cdot \delta'_i - \frac{1}{X_{L,ik}} \cdot \delta'_k \tag{10.7}$$

By injecting a positive loop flow between the first connection bus [first element of $\underline{P}$ in Eq. (10.5)) of the PFC and the second connection bus (second element of $\underline{P}$ in Eq. (10.5)], the resulting loading of the grid (DC power flow) is utilized to estimate sensitivities of control actions of PFC c on the different paths. If a path consists of several parallel power system devices, then the particular sensitivities on each power system device have to be calculated according to the proportion of the impedances of the parallel devices.

10.2.5 Assigning the Direction of Impact [8]

Each PFC can change the power flow on its control path in two directions by either controlling it upwards or downwards. To determine how a control action influences the power flow on a parallel path, the controlling agent of the PFC needs information about the direction of impact on a device to be controlled. That is to say which control action will relieve or stress the considered power system device. To implement this knowledge, each power system device within the area of influence of the PFC is grouped into a *ControlUpZone* or a *ControlDownZone*. To decrease the loading of all power system devices of the *ControlUpZone,* the PFC has to be controlled upwards and to decrease the loading of the power system devices of the *ControlDownZone,* the PFC has to be controlled downwards.

To support this kind of grouping, the non-controlling agents insert information about the power flow direction into the *StateInformMessages*. For each power system device it is defined that a positive power flow is directed from connection bus 1 to connection bus 2. To assign a power system device to one of the two zones, the controlling agent compares the direction of power flow through a power system device obtained by the *StateInformMessages* with the direction of power flow obtained by the sensitivity analysis (sign of the sensitivity).

- If both signs are equal, then the corresponding power system device is a serial element and will be included into the *ControlUpZone*.
- If both signs are not equal, then the corresponding power system device is a parallel element and will be included into the *ControlDownZone*.

10.2.6 Distributed Coordination [8]

The coordinated control system has to react when control actions of PFCs are required. For this purpose, a couple of parameters are defined for the distributed coordination.

The critical loading of a device is usually defined as 100 % and specified by the parameter *crit*. The parameter *high* is defined as a boundary for high loading of a device in order to initiate control actions before the maximum loading of a power system device is reached. If the loading of a device is higher than this value, then first control actions will be carried out. *Slope* is another parameter which is needed for the valuation of the loading of a power system device. Typical values for these parameters are given in Table 10.1.

Based on these boundaries different control actions are required according to the following ranging:

- **The loading of a power system device is below the boundary value *high***: There is no need for control actions. Possible deviations from the neutral position of the PFCs can be reversed.
- **The loading of a power system device is between the boundary *high* and 100 %**: The power system device is about to be overloaded. First precautious control actions should be taken.
- **The loading of a power system device is above 100 %**: The power system device is overloaded. Control actions have to be taken to reduce the loading within a reasonable amount of time.

The parameter *minSens* defines the lowest sensitivity of a power system device to be assigned to the area of influence of the corresponding PFC. Every controlling agent observes all power system devices within the area of influence of its corresponding PFC. The controlling agent has knowledge about the loading of each observed power system device and knows the sensitivity and direction of impact between its corresponding PFC and the observed power system devices. The fundamental decision criteria for taking control actions by the PFC *c* on the loading of a power system device *d* are the loading *load(d)* and the sensitivity *sensitivity (d, c)*. A weighting function uses this fundamental input data to determine one value $f_{weight}(d,c)$ for each power system device *d* within the area of influence of a PFC *c* expressing the urgency of a possible control action. For a detailed description of the weighting function please refer to [8]. In the following we use the values according to Table 10.1 for the parameters of the distributed coordination.

10.2.7 Control of PFCs [8]

All devices in the *ControlUpZone* and the *ControlDownZone* of a PFC are weighted by the weighting function $f_{weight}(d, c)$. At least one of all devices in one zone has the maximum weighting value. The maximum weighting values are

Table 10.1 Parameters for distributed coordination

Parameter	Value
Crit	100 %
High	95 %
Slope	20
MinSens	0.1

compared between the *ControlUpZone* and the *ControlDownZone*. If the maximum value of the *ControlUpZone* is higher than the value of the *ControlDownZone*, then the PFC is controlled upwards and vice versa.

If the two maximum weighting values are equal or their difference is below a certain threshold (to avoid oscillations depending on the type of PFC) then no control action is carried out. In this case there are at least two overloaded devices with equal weights in the area of influence of the PFC, which could only be compensated through simultaneous, yet contrary control actions. The overloading of one of these devices must be reduced by a second PFC that might be available before the first PFC takes an action.

If both maximum weighting values are zero, then there are no overloaded power system devices in the area of influence of the PFC. In this case a PFC may be controlled back towards its neutral position.

10.3 Comparison Between OPF and Agents Coordination[1]

In this section PST tap positions generated with the multi-agent coordination system are compared with optimal tap positions generated with a commercial SCOPF tool. The test case simulations are carried out based on the topology of the 30-bus New England Test System (NETS) published in [11] with revised controller parameters according to [12]. This test system was chosen to obtain a good compromise between an appropriate size of the test system and the necessary computation time for the simulation of the multiagent system, which increases potentially with rising number of agents. The simulation time should be short enough to be able to carry out N-1 simulations for a variation of scenarios.

10.3.1 PST Devices in the New England Test System

There are not implemented any PFC devices in the original version of the NETS. Therefore, several PSTs have been added to the NETS: One PST is located between the nodes 4 and 14 and another PST is located between the nodes 17 and

[1] All OPF calculations in this chapter have kindly been provided by Tractebel Engineering, which is the coordinator for the FP7 project PEGASE.

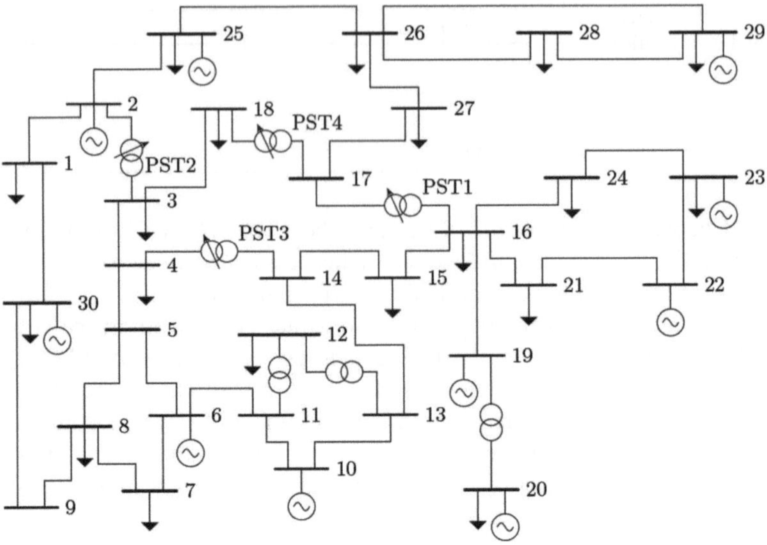

Fig. 10.3 Graphical representation of the NETS topology including the location of PST devices

18 as proposed in [3]. Two further PSTs are located between the nodes 2 and 3 as well as between the nodes 16 and 17 as proposed in [13]. A graphical representation of the NETS topology including all four PST devices is shown in Fig. 10.3. PST devices were chosen instead of FACTS devices to be compatible with the given SCOPF tool. The PST devices are modeled with equal parameters as the 380 kV PST devices installed in Belgium [14].

10.3.2 Test Case Scenarios

When testing a control method it is very important to vary as many boundary conditions as possible, in order to avoid non-general results due to special conditions that might be met coincidentally. In the case of PFC coordination three main test case variations can be identified with significant impact on the coordination result: Variation of the power flow situation, variation of the network topology as well as variation of the PFC locations. Because the number of test case calculations is limited due to computational complexity, the following strategy is pursued:

- Two different generation portfolios according to Table 10.2 are used to vary the power flow situation. These portfolios are determined with OPF calculations that maximize the flows in the NETS under consideration of the constraints of the generators. The loads of the original dataset of the NETS remain unchanged.
- The network topology is varied by performing a full N-1 security calculation with corrective PFC tap positions. That means PFC coordination is carried out for each contingency case.

Table 10.2 Generation portfolio of two different power flow cases

Case	Generation at node # [MW]									
	2	6 (SL)	10	19	20	22	23	25	29	30
1	632.1	652.5	643.6	581.8	585.7	631.4	633.5	622.4	635.5	655.3
2	625.1	554.9	572.4	612.5	618.3	664.7	666.7	636.4	664.6	662

- The PFC location is varied by deactivating the control for different combinations of the four PSTs. In total eight different combinations are analyzed in detail (out of $2^4 = 16$ possible combinations). The selection of these eight combinations is performed with an orthogonal L8 array, as standardized by Taguchi [15]. This orthogonal array is presented in Table 10.3 and it assures that the combinations are chosen in a statistical way to avoid systematic errors.

The total number of test cases is 736, resulting from two generation portfolios multiplied with 46 N − 1 incidents multiplied with eight different combinations of PST locations. Load flow analysis of the total number of 736 test cases shows that 88 test cases are detected with overloaded transmission lines.

10.3.3 Reference Method

As reference method for the evaluation of the multi-agent coordination system the SCOPF method from the PEGASE project [16, 17] is used. The optimization problem is based on the following objective function, in which n_k is the tap position of the PST devices k:

$$f(\underline{x}, \underline{u}) = \sum_k n_k \tag{10.8}$$

The optimization goal is to reduce the deviation of the PST tap positions from the neutral position as much as possible. The vector of decision variables x contains the PST tap positions, as well as the reactive power in-feed of each generator. The reactive power in-feed is included into the decision variables to improve the convergence behavior of the SCOPF. The vector of state variables u consists of the complex voltages V_i.

The boundary conditions of the optimization problem are the equality constraints according to Eqs. (10.9) and (10.10), while all nodal voltage limits and the apparent power limits of all transmission lines are considered as unequality constraints:

$$0.9\,\text{pu} \leq |\underline{V_i}| \leq 1.1\,\text{pu} \tag{10.9}$$

$$|\underline{S_k}| \leq S_{k,\max} \tag{10.10}$$

Table 10.3 Eight experiments with different combinations of activated and deactivated PST control according to Taguchis field L8

Experiment #	PST 1	PST 2	PST 3	PST 4
1	Activated	Activated	Activated	Activated
2	Activated	Activated	Activated	Deactivated
3	Activated	Deactivated	Deactivated	Activated
4	Activated	Deactivated	Deactivated	Deactivated
5	Deactivated	Activated	Deactivated	Activated
6	Deactivated	Activated	Deactivated	Deactivated
7	Deactivated	Deactivated	Activated	Activated
8	Deactivated	Deactivated	Activated	Deactivated

The OPF computation only converges if a solution can be found in which all constraints are fulfilled and in particular if no power system devices are overloaded. Out of the 88 test cases with overloaded transmission lines, the OPF converges in 59 test cases. For the other 29 test cases overloadings could not be resolved only by PST control actions (or modification of the generators reactive power in-feed).

10.3.4 Results Evaluation

In the next step the coordination of the PST tap positions is performed with the multi-agent system for all 59 test cases in which transmission lines are overloaded and the OPF method converges. For each test case the optimal reactive power values of each generator are inserted from the OPF calculation results into the multi-agent simulation to make both calculations comparable. The coordination result of the multi-agent coordination is determined by carrying out a real-time simulation and reading out the PST tap positions after all dynamic oscillations have decayed and a steady state equilibrium is reached.

The comparison between the reference tap positions n_{OPF} of the OPF coordination and the tap positions n_{agents} of the multi-agent coordination is based on the deviation Δn which is calculated as follows:

$$\Delta n_{i,k} = \begin{cases} n_{agents,i,k} - n_{OPF,i,k}, & n_{OPF,i,k} \geq 0 \\ n_{OPF,i,k} - n_{agents,i,k}, & n_{OPF,i,k} < 0 \end{cases} \qquad (10.11)$$

The index i expresses the test case and the index k expresses the PST device. Independent to the control direction Δn gets positive if the multi-agent system performs more control steps than the reference method and Δn gets negative if less control steps are performed.

To obtain a good coordination result that fits to the control needs of the network the coordination system parameters must be tuned accordingly. The development of a detailed parameter tuning procedure is out of scope of this thesis. Therefore,

Table 10.4 Multi-Agent system parameter combination for each PST device

Parameter combination	PST 1			PST 2		
	High (%)	Crit (%)	MinSens	High (%)	Crit (%)	MinSens
1	80	100	0.1	80	100	0.1
2	80	100	0.08	80	100	0.1
3	75	95	0.08	75	95	0.1
4	77	97	0.1	75	98	0.1
Parameter combination	**PST 3**			**PST 4**		
	High (%)	Crit (%)	MinSens	High (%)	Crit (%)	MinSens
1	80	100	0.1	80	100	0.1
2	80	100	0.025	80	100	0.035
3	75	95	0.025	75	95	0.035
4	78	98	0.025	79	99	0.035

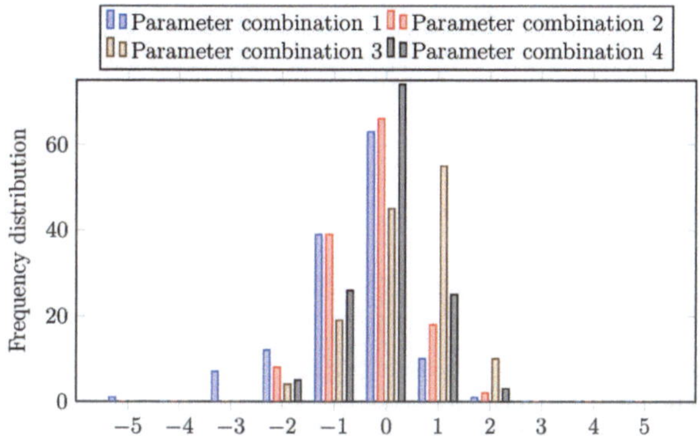

Fig. 10.4 Frequency distribution of the deviation Δn

the parameters are tuned manually. Out of a large number of different simulated parameter combinations, four interesting parameter combinations are summarized in Table 10.4.

For each parameter combination the coordination results of the multi-agent system are compared with the reference results obtained from the OPF calculations. For the evaluation of the coordination results three different statistical evaluations are considered and displayed in the following figures:

- The frequency distribution of the deviation Δn (Fig. 10.4).
- The mean value (Eq. 10.12) and the standard deviation (Eq. 10.13) of the deviation n for each PST (Figs. 10.5 and 10.6). In these equations n_k represents the number of test cases in which PST k is activated.
- The cumulative frequency distribution of the highest line loading per test case (Fig. 10.7).

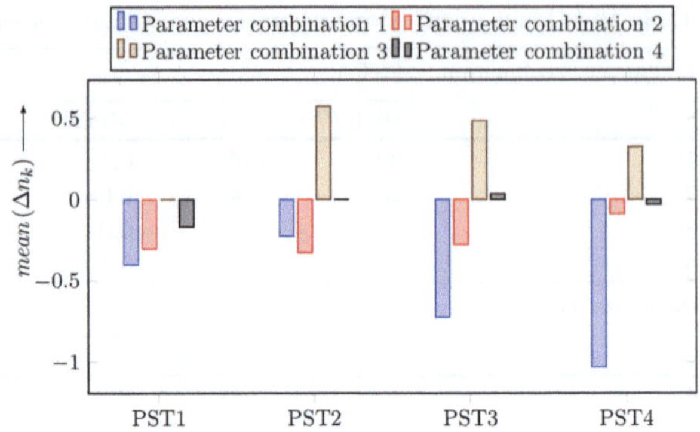

Fig. 10.5 Mean value of the deviation Δn for each PST

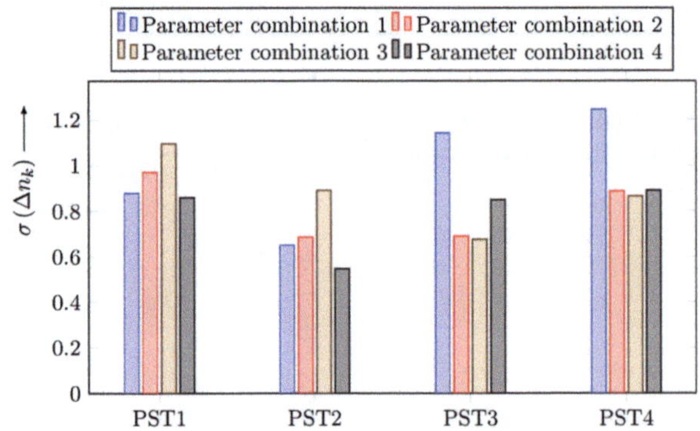

Fig. 10.6 Standard deviation of the deviation Δn for each PST

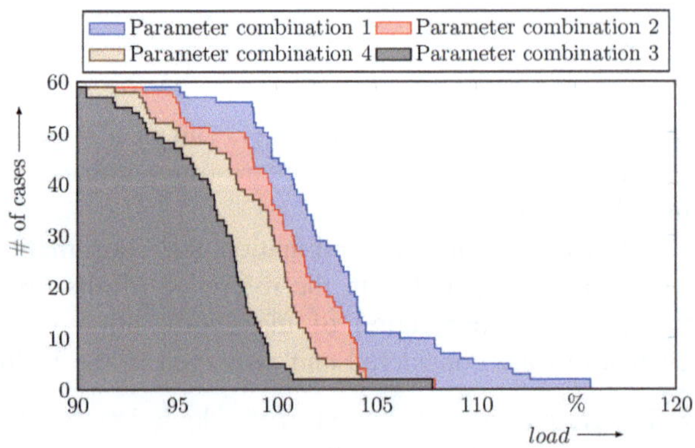

Fig. 10.7 Cumulative frequency distribution of the highest line loading per test case

$$mean(\Delta n_k) = \frac{\sum_{i=1}^{n_k} \Delta n_{i,k}}{n_k} \tag{10.12}$$

$$\sigma^2(\Delta n_k) = \frac{\sum_{i=1}^{n_k} \left(\Delta n_{i,k} - mean(\Delta n_k)\right)^2}{n_k - 1} \tag{10.13}$$

The first parameter combination according to Table 10.4 was used as starting point and corresponds to the parameters proposed in Sect. 10.2. With these parameters the mean value of the deviation Δn is negative for all PSTs. This is caused by the fact that in particular PST3 and PST4 remain at the neutral position in many cases with the multi-agent coordination, while their tap positions are modified in the reference case. As a consequence the frequency distribution of the deviation Δn has several extreme outliers into the negative direction (deviations of up to 5 tap positions).

For these reasons the parameter *minSens* is modified individually for each PST in parameter combination 2. These new parameters cause a significant improvement of the control result, which can be seen in a significant reduction of the standard deviation and a mean value much closer to zero for PST3 and PST4. Also the extreme outliers in the frequency distribution do not occur and the most extreme deviations are two tap positions in both directions. However, there is still a significant tendency that the multi-agent coordination performs less control steps then the OPF coordination. This can be seen in the mean value, which is below zero for all PSTs, as well as in the cumulative frequency distribution, in which a large number of test cases have device loadings over 100 %.

To improve the coordination result further, different values for the parameters *high* and *crit* are applied in parameter combination 3, while the values for *minSens* remain the same as in parameter combination 2. With these modifications the mean value of PST1 is almost zero, while the mean value of the other PSTs is reversed to positive values. In the cumulative frequency distribution the highest line loading in below 100 % for almost all test cases.

In the next step the parameters are further modified individually for each PST in parameter combination 4, in order to bring the mean value of the PSTs closer towards zero while the standard deviation remains at a comparatively low value for each PST.

10.3.5 Conclusions

The control results with parameter combination 4 show that under variation of a large number of boundary conditions, such as the power flow situation, the network topology and the location of PFC devices, the multi-agent coordination system is able to provide a real-time PFC coordination which is close to the reference case. In most cases there is no difference between the multi-agent coordination and the reference case. If there is a deviation, then it is in most cases differing by one tap position and only in a few cases two tap positions. The mean value of the deviation is close to zero for all PSTs. From the cumulative frequency distribution it can be seen that in most of the cases the highest line loading per test case is close to 100 %. Overloading of transmission lines only occurs in very few situations.

There are several reasons why the coordination results of both coordination methods cannot be exactly equal. One of the main differences is that the OPF coordination only performs static computations, while the multi-agent coordination progresses to the coordination result in a dynamic process. For example oscillating power flows caused by tap changing actions might cause some of the differences because the multi-agent coordination directly reacts on them. Another reason for deviations is that both coordination systems are computed in different simulation tools. Although a comparison of the load flow results of the model implementation in both tools has shown that they provide almost equal results, there might be minor differences caused by different modeling details.

10.4 Large Scale Interconnection [19]

The main test case scenario used for verification of the real-time coordination system for PFC is a section of the reduced transmission system model introduced in Chap. 2 reflecting the situation in the CWE region based on the year 2010, including the countries Belgium (B), France (F), Germany (D) and the Netherlands (NL). A geographical representation of the grid model in this region is given in Fig. 10.8. The nodes numbers provide information about their assignment to the different countries according to Table 10.5. The location of PFC devices (as indicated in Fig. 10.8) is determined according to the data collected in [14]. Two different scenarios are assumed: the situation as it is today with conventional Phase Shifting Transformers (PST) and a hypothetical future scenario where the existing PSTs are exchanged with fast PFC devices out of the FACTS family [Thyristor Controlled Phase Angle Regulators (TCPAR) are used in this study] at the same locations as the PSTs are installed today. The technical data of the PST devices is summarized in Table 10.6.

The general objective of the test case simulations is to analyze the control performance of the multi-agent coordination system on a realistic transmission system during an emergency situation under consideration of different scenarios for the available communication latency T_l as well as the available reporting rate F_s of sensor devices of the WAMS.

The selected test case scenario in this paper presents a day with moderate load and low wind in the CWE region. As a consequence the power flow situation in the CWE region is characterized by a high import in NL together with a moderate

Fig. 10.8 Map showing the location of nodes and lines of the ICOEUR network model in the region around Belgium and the Netherlands

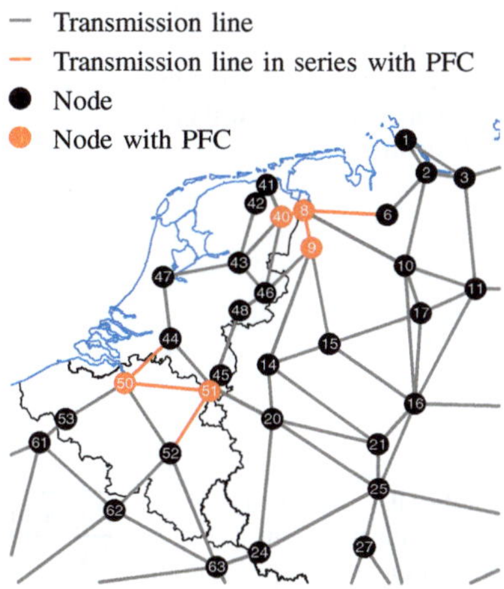

Map based on: © Lutum+Tappert

Table 10.5 Assignment of node numbers to countries

	D	B	NL	F
Node numbers	1–31	40–48	50–53	60–72

Table 10.6 Technical data of PST devices in Belgium, Germany and the Netherlands [14]

Location	n_{neg}/n_{pos}	$\phi_{PST,\min}$	ΔV_{tap} (%)
Zandvliet	−17/17	±25°	2.74
Van Eyck 1	−17/17	±25°	2.74
Van Eyck 2	−17/17	±25°	2.74
Meeden (2x)	−30/30	±30°	1.92
Gronau	−17/17	±12°	1.25
Diele (2x)	−31/31	±24°	1.44

import in D, while B is almost balanced and F is exporting electricity. Very extreme flows arise on the interconnectors in this situation: The flows are from F to B 3403 MW, from B to NL 4202 MW and from D to NL 4548 MW. This situation is not (n − 1) secure. On top of this, to provoke an emergency situation, two events are simulated: At $t = 1$ s one of the two parallel interconnecting lines between nodes 44 and 50 is tripped followed by the tripping of one of the two parallel lines between nodes 45 and 51. The resulting flows on the lines near the border between NL and B are presented in Fig. 10.9. The transmission line between nodes 51 and 52 is almost overloaded at the beginning of the situation. After tripping of one line between nodes 44 and 50, the remaining parallel transmission line gets overloaded as well. Two seconds later, after the tripping of

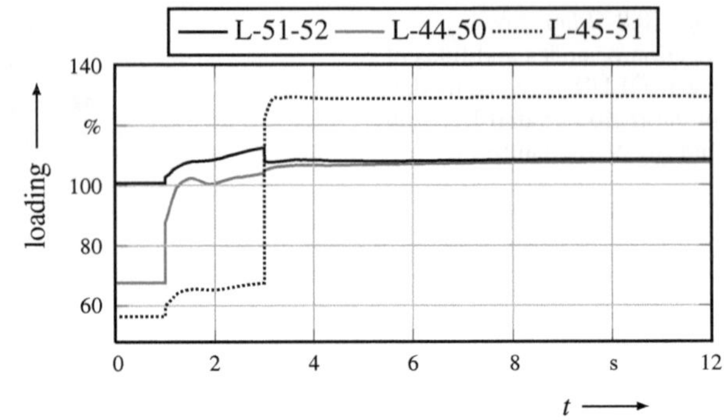

Fig. 10.9 Relative loading of the highest loaded transmission lines during an emergency situation without PFC control

Table 10.7 Necessary bandwidth for agent communication depending on the reporting rate of sensor devices

Reporting rate F_s (Hz)	Bandwidth (Gb/s)
100	1.53
50	1.08
25	0.53
10	0.21

one line between nodes 45 and 51, the remaining parallel transmission line gets significantly overloaded.

10.4.1 Requirements for the Communication System

In order to evaluate the requirements for the communication system, simulations have been performed with varying values for the reporting rate Fs of sensor devices of the WAMS. The situation is simulated with a total number of 161 agents. Table 10.7 presents an approximation of the necessary bandwidth of the communication system, which was estimated by counting the number of messages processed by each agent per second multiplied with the length of a message. The highest value of all agents is the value presented in Table 10.7. As it can be seen, the required bandwidth depends highly on the reporting rate. Nowadays the use of optical ground wires is very common in overhead transmission lines. These ground wires can be used for high speed communication and a bandwidth of up to 10 Gb/s can be achieved [18]. However, since the ground wires are also used for other communication purposes, the reporting rate should be reduced as much as possible, in order to avoid congestions in the communication system.

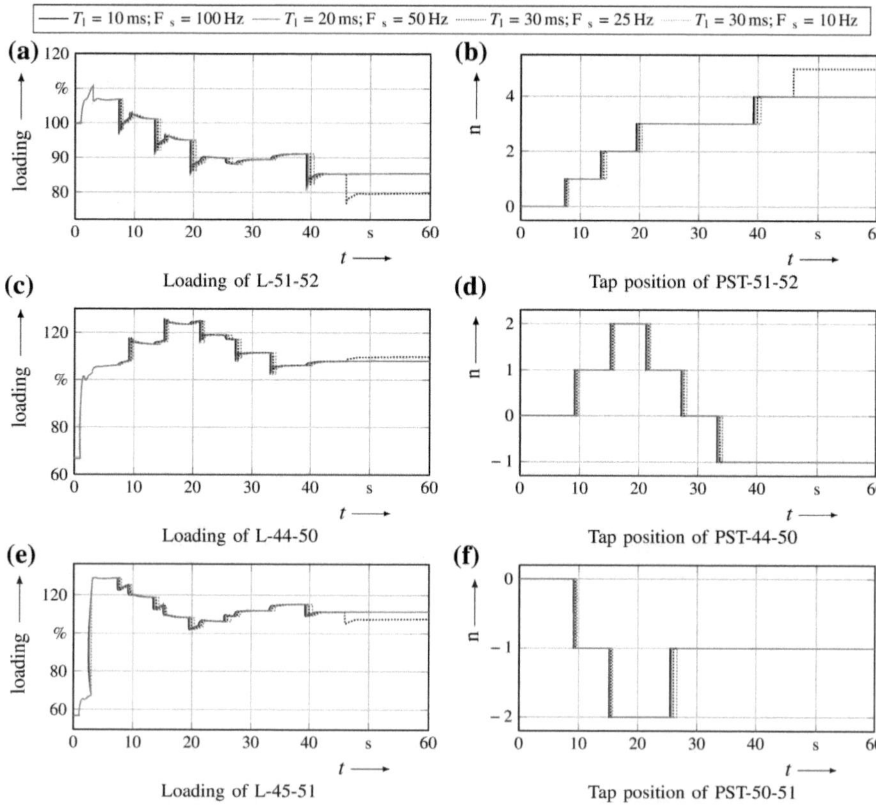

Fig. 10.10 Relative loading of the controlled transmission lines and PST tap positions, controlled by real-time coordination system under an emergency situation

10.4.2 Simulation Scenario with PST

The performance of the real-time coordination system with PST devices is illustrated in Fig. 10.10. Caused by the first transmission line tripping at $t = 1$ s PST-51-52 starts controlling to reduce the loading of L-51-52 and L-45-51 immediately. This first control step can be seen 6 s later, due to the delay of PST tap changes. In the meantime the second transmission line tripping happens at $t = 3$ s and also the devices PST-44-50 and PST-50-51 start changing their tap positions.

Due to the extreme loading of the corridor, overloading cannot be completely avoided on all transmission lines. In such situations the control goal is to equalize the loading of overloaded lines. This goal is reached after $t = 40$ s, at which L-44-50 as well as L-45-51 are not loaded more than 110 %. Compared to the uncontrolled situation (Fig. 10.9) the loading of L-45-51 has been reduced from 130 %

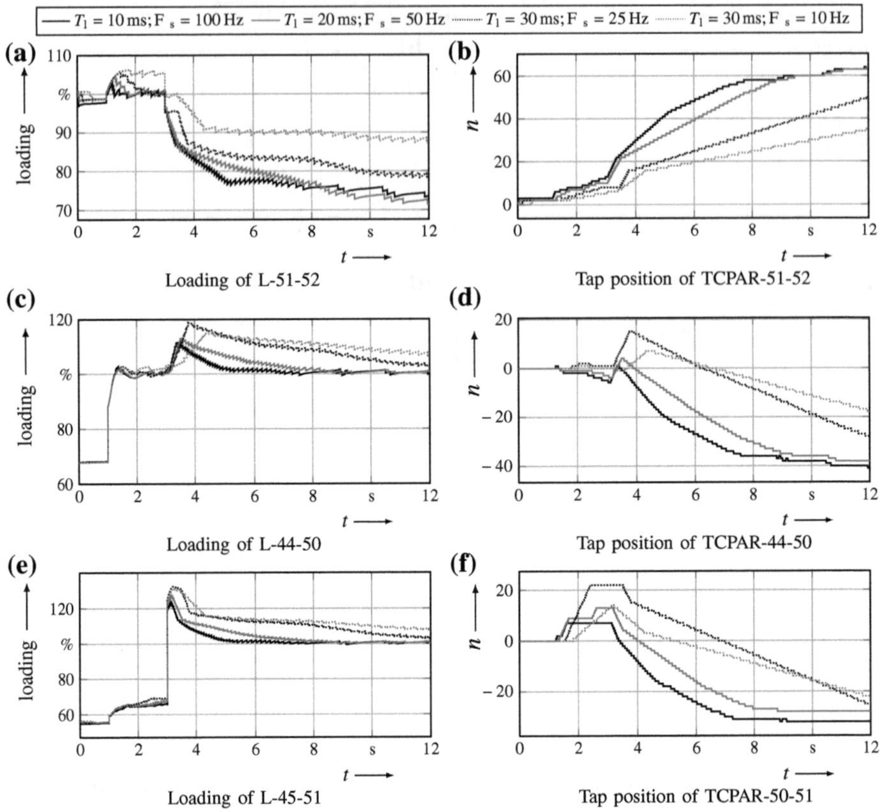

Fig. 10.11 Relative loading of the controlled transmission lines and TCPAR tap positions, controlled by real-time coordination system under an emergency situation

to below 110 % and the overloading of L-51-52 has been resolved while the loading of L-44-50 remains at the same level.

Caused by the low control speed of PFC devices compared to the speed of the communication/sensoring system, the difference between the studied values of communication bandwidth and sensor reporting frequency can be neglected. The performance with $T_l = 30\,\text{ms}$ and $F_s = 10\,\text{Hz}$ proves to be very good and even higher latencies and lower reporting rates might be sufficient.

10.4.3 Simulation Scenario with TCPAR

The performance of the real-time coordination system with fast TCPAR devices is illustrated in Fig. 10.11.

Already before any line tripping happens, TCPAR-51-52 performs minor control actions to reduce the loading of L-51-52. Caused by the first transmission line tripping at $t = 1$ s lines L-51-52 and L-44-50 get overloaded. Since three PFC devices (TCPAR-51-52, TCPAR-44-50 and TCPAR-50-51) have impact on these lines, they start to reduce the overloading in a coordinated way. The second transmission line tripping at $t = 3$ s changes the control situation, since the loading of L-45-51 is significantly increased and the line gets overloaded. TCPAR-44-50 and TCPAR-50-51 change their control priority and reverse their control direction to reduce the overloading of L-45-51. This increases temporarily the loading on L-44-50 until its loading is almost equal to the loading of L-45-51. From that moment on the loading of both lines is reduced to a loading of almost 100 % at $t = 5$ s in the best case. Compared to the uncontrolled situation the overloading of all three devices has been resolved.

In general (also by comparing with other simulation scenarios not presented in this book) it can be stated that the scenario with $T_l = 20$ ms and $F_s = 50$ Hz provides sufficient control performance, together with more compatible bandwidth requirements (compare Sect. 10.4.1). Communication latency as well as required bandwidth of the scenario with $T_l = 10$ ms and $F_s = 100$ Hz is difficult to achieve with today's communication media and protocols of the TSOs, but might be a good scenario for future communication systems [20].

10.5 Summary and Conclusions

To allow for the maximum of adaptability to any possible event, a distributed approach for a real-time PFC coordination system is followed in this chapter. In this multi-agent system each PFC is associated with a controlling agent and each non-controllable power system device (such as a transmission line) is equipped with a non-controlling agent. Controlling agents continuously receive *StateInformMessages* which are originated from the non-controlling agents and contain information about state and loading of the power system devices. Based on these messages the controlling agent performs a topology analysis, followed by a sensitivity analysis to estimate the influence of PFC control actions on each power system device. A weighting function, which uses the loading and the sensitivity information, sorts the power system devices according to their control needs. Based hereon control actions are executed, if necessary, with the PFC device corresponding to the controlling agent. The feedback from the grid communicated through the messages from the non-controlling agents is used to achieve the distributed coordination between several PFC devices.

For verification purposes the PFC set-point values generated with the multi-agent coordination system were compared with optimally coordinated set-points provided by an OPF tool for several test cases with varying power flow situations, grid topologies and PFC locations. Individual tuning of the coordination system parameters was necessary for each PFC to achieve similar coordination results as provided by the OPF tool. With well-chosen agent parameters the deviation between the set-points of both methods is low, although minor deviations occur. As a further verification the dynamic control behavior of the multi-agent coordination system was tested on a reduced representation of the transmission system in the Benelux region with several PFCs with mutual impact on each other. These test cases have shown that the coordination system reacts in an appropriate and robust way on contingency events. The impact of communication delays and different sensor reporting rates on the PFC control is negligible for PST devices, while the control behavior of fast PFC devices is significantly influenced. However, with todays state-of-the-art communication/sensoring technology the coordination speed is sufficiently high.

References

1. T. Aundrup, T. Benz, C. Dörnemann, W. Fischer, C. Gehlen, W. Glaunsinger, H. Hellmuth, J. Kreusel, P. Menke, R. Neumaier, C. Rehtanz, A. Schomberg, J. Schwippe, *Übertragung elektrischer Energie* (VDE, VDE-Positionspapier, 2010)
2. B. Marinescu, J.M. Coulondre, A coordinated phase shifting control and remuneration method for a zonal congestion management scheme, in Power Systems Conference and Exposition. IEEE PES, 2004
3. J. Verboomen, D. Van Hertem, P.H. Schavemaker, W.L. Kling, R. Belmans, Analytical approach to grid operation with phase shifting transformers. IEEE Trans. Power Syst. **23**(1), 41–46 (2008)
4. G. Hug-Glanzmann, G. Andersson, Decentralized optimal power flow control for overlapping areas in power systems. IEEE Trans. Power Syst. **24**(1), 327–336 (2009)
5. A. Marinakis, M. Glavic, T. Van Cutsem, Minimal reduction of unscheduled flows for security restoration: application to phase shifter control. IEEE Trans. Power Syst. **25**(1), 506–515 (2010)
6. J. Verboomen, Optimisation of transmission systems by use of phase shifting transformers. Ph.D. dissertation, Technische Universiteit Delft, 2008
7. S. Lehnhoff, U. Häger, O. Krause, C. Rehtanz, and H. F. Wedde, Towards autonomous distributed coordination of fast power flow controllers in transmission networks, in 4th International IEEE Conference on Liberal and Modernization of Power Systems, Irkutsk, Russia, 2009
8. U. Häger, S. Lehnhoff, C. Rehtanz, H. F. Wedde, Multi-agent system for coordinated control of facts devices, in 15th International Conference on Intelligent System Applications to Power Systems. ISAP'09, 2009
9. U. Häger, A. Seack, C. Rehtanz, S. Lehnhoff, T. Zimmermann, H.F. Wedde, Applicability of coordinated power flow control based on multi-agent systems, in Bulk Power System Dynamics and Control (iREP)—VIII (iREP), iREP Symposium, 2010

10. U. Häger, C. Rehtanz, S. Lehnhoff, Analysis of the robustness of a distributed coordination system for power flow controllers, in 17th international Power Systems Computation Conference (PSCC), Stockholm, Sweden, 2011
11. T.M. Athay, V.R. Sherkat, R. Podmore, S Virmani, C. Puech, in *Systems Engineering for Power: Emergency Operating State Control*, ed. by L. H. Fink and T.A. Trygar. Transient energy stability analysis. Engineering Foundation, U.S. Department of Energy, Sept 1979, pp. IV.122–IV.126
12. G. Rogers, *Power System Oscillations* (Kluwer Academic Publishers, Dordrecht). ISBN: 978-0792377122 (2000)
13. S. Nagalakshmi, N. Kamaraj, Loadability enhancement for pool model with facts devices in transmission system using differential evolution and particle swarm optimization, in India International Conference on Power Electronics (IICPE), 2010. Jan 2011, pp. 1–8
14. D. Van Hertem, The use of power flow controlling devices in the liberalized market. Ph.D. thesis, Katholieke Universiteit Leuven, 2009
15. B. Klein, Versuchsplanung—DoE. Oldenbourg Wissenschaftsverlag GmbH, 2007
16. L. Platbrood, S. Fliscounakis, F. Capitanescu, P. Panciatici, C. Merckx, M.O. Vazquez, D3.2: development of prototype software for steady state optimization of the etn. Pegase Project, Tech. Rep., 2011. [Online]. Available: http://www.fp7-pegase.eu/
17. L. Platbrood, H. Crisciu, F. Capitanescu, L. Wehenkel, Solving very largescale security-constrained optimal power flow problems by combining iterative contingency selection and network compression, in 17th International Power Systems Computation Conference (PSCC), Stockholm, Sweden, 2011
18. M. Ortgies, Neue Infrastruktur für Übertragungsnetzbetreiber. ntz—Fachzeitschrift für Informations—und Kommunikationstechnik, vol 3–4, pp. 22–23, 2011
19. U. Häger, C. Rehtanz, N. I. Voropai, ICOEUR project results on improving observability and flexibility of large scale transmission systems, in IEEE PES General Meeting, San Diego, CA, USA, 2012
20. K.R. Padiyar, *FACTS controllers in Power Transmission and Distribution* (New Age International (P) Ltd, 2007)

Chapter 11
Control of Interconnected Networks

Ettore Bompard and Tao Huang

11.1 Introduction

Interconnections can bring benefits to the involved parties; however, as presented in the previous chapter, they may also contribute to the occurrence of blackouts if sophisticated coordinated control is missing. Each subsystem is jurisdictionally operated by a System Operator (SO); yet, any local decision will instantaneously create changes in the whole system. Under this regime, the overall control of the interconnected system can only be implemented by a set of mandatory coordination rules. In this chapter, a general framework for simulating coordinated control over interconnected power systems is proposed as complex multi-agent system (MAS). Agents are employed to model the decision-making of each SO and simulate the performance of the whole system due to their interaction. Different scenarios are designed to compare the impact of coordination rules, especially the policies for interconnection management and information exchange, on the decision of each SO and on the system feasibility (i.e. maximum line rate, voltage profile, generator capabilities) with reference to individual subsystem and the overall network. The approach is applied to the IEEE-30 bus system to illustrate its application and effectiveness.

11.2 Motivation

An electric power system is a system devoted to make the electricity available for the final use at required locations with appropriate level of quality, in terms of continuity of supply without any interruption and technical performance, such as

E. Bompard (✉) · T. Huang
Politecnico di Torino, Turin, Italy
e-mail: ettore.bompard@polito.it

T. Huang
e-mail: tao.huang@polito.it

U. Häger et al. (eds.), *Monitoring, Control and Protection of Interconnected Power Systems*, Power Systems, DOI: 10.1007/978-3-642-53848-3_11,

voltage, frequency, waveform, etc. In general, a power system includes four subsystems: generation, transmission, distribution and utilization. The first three subsystems are dedicated to assure the quality of electricity to the fourth.

Starting from the 1960s national power grids became interconnected and synchronously operated over the transmission level. The basic idea of the interconnection of two or more power grids involves two main categories of advantages: economy and security. As the increasing recognition of sustainable development in the contemporary society, people have been aware that interconnection could contribute to environmental improvement as well.

The interconnected grid, consisting of multiple interconnected systems, which obviously has a higher total capacity, larger geographic territory, and more various resources for electricity than any individual partner in the interconnection, entitles the partners the following security advantages:

- By sharing the standby and spinning reserves, the unique grid requires less reserve than the sum of requirements for each subsystem.
- A large covering of time zones will remarkably change the daily load diagrams and smooth the total load profiles due to the load shifting.
- With a larger system, impacts from traditional disturbance and unpredictable fluctuation from renewable energy would become smaller; therefore, the system could achieve better frequency consistency.

Besides the security advantages, economic pros can be witnessed as well:

- The interconnection grants possibilities to the interconnected partners to trade energy across their boarders in favoring the acquisition of low cost electric energy in a larger electricity market.
- The interconnected system can accommodate different types of mechanisms between or within connected partners, as well as the overall system. In a long run, the interconnection can facilitate the market integration.
- Different options are available for the economic exploitation of the interconnections. Partners with plenty of energy resources can convert them into electricity and trade with partners which are in poverty of them. As well, it provides enhancement to the security of energy supply for some interconnected partners.
- More flexible choices of economically favorable locations for building new power plants cumulate in a larger geographic sense.

Moreover, interconnections also provide a chance to contribute to the environmental improvement:

- The geographically dependent renewable sources, such as on-shore and off-shore wind farms, spots of photovoltaic, geothermal plants etc., could be available to a demand far away from them.
- Due to the wide adoption of green energy resources in utility scale, total emissions can be diminished with a higher penetration.

Therefore, in practice, interconnections can be witnessed globally. Nowadays the major world interconnected systems are ENTSO-E in Europe, IPS/UPS in the former Soviet Union, Eastern and Western interconnections in America, etc. Those interconnections create the largest and most complicated dynamic systems in history.

However, with the emergence of new technologies, paradigms in power systems are changing towards a more active fashion: power flow is shifting into a bilateral rather than the traditional unilateral direction, owing to the appearance of prosumers who consume and produce electric energy simultaneously. A transfer of focus from a small number of large generators at the transmission level towards an astronomic number of small distributed generators at the distribution level with autonomous behaviors, and the rapid development and massive deployment of measuring and flexible control facilities in transmission networks, such as PST, FACTS, PMU WAMS, etc., prompt a need for the transmission and interconnection of networks to be more "smart" than the old-fashioned passive ones.

Accordingly, more ambitious interconnection plans and proposals are made, considering the changing paradigms and technological developments at rate. A Unified Smart Grid is proposed in the U.S. to setup a national wide high capacity backbone of the transmission network to interconnect local networks having been upgraded to smart grids. The possibilities and other practical issues of interconnecting ENTSO-E and IPS/UPS have been under investigation by some co-funded European projects, such as feasibility studies of UCTE/IPS/UPS with synchronized interconnection, and the ICOEUR project with broader and deeper scopes. Further, a SuperSmart grid, equivalent to the Unified Smart Grid in the U.S., is proposed to unify super grid and smart grid into a comprehensive network which connects ENTSO-E, IPS/UPS with northern Africa, the Middle East, etc.

In general, no matter how large the interconnection would be made, it has to be implemented through synchronous (AC interconnection), asynchronous (DC interconnection) or their hybrid fashion. Each approach has its pros and cons; therefore efforts have been made to figure out general conditions for choosing one of them. Take the AC interconnection for example; it is comparatively simple and cheap if its length is below a certain distance. However, it requires a high compatibility of technical and operational conditions of connected partners. Besides that, the AC interconnection could also create technical difficulties such as large fault currents, increasing risks of blackout due to cascading effects, inter-area oscillations, sub-synchronous oscillations, stability problems in heavily loaded situations, etc.

Some advantages brought by synchronized operation could be cancelled off with the growth of the interconnection in terms of size and distance. Some of them can be solved by using an HVDC interconnection scheme. For example, an HVDC link is able to control the power flow over it and is flexible to adapt to different characteristics of connected partners. It can, as well, effectively increase the system stability, damp inter-area oscillations, diminish fault current, etc. Yet, it is extremely expensive for a short distance connection.

Without interconnection, control in a power system means controlling all sources in terms of generators and compensators in such a way that the combined active and reactive power of the resources match the aggregate power demands plus the losses at the nominal frequency with required voltage profiles, while meet all other constraints imposed by engineering design, regulatory, environmental and economic consideration. This still holds in the interconnected system; further, the interconnection requires coordination among the involved partners for a better global performance and welfare.

The control problems in an interconnected system, as the control of any subsystem, are associated with the operational condition of the systems. Although the boundaries of each state are quite subtle [1–6], the basic control operation in normal conditions is related to the power-frequency control, voltage regulation within each national subsystem and the management of the flows over the interconnection ties, such as ACE, etc. Under emergency conditions, the control target for each system is to keep itself feasible while maximally maintaining its interconnection with the whole network. In either case, we face a situation in which various system operators with full visibility and control over their own subsystems and only partial visibility of their neighboring subsystems provide control actions with the priorities to maximize their own interest. Since the overall system is interconnected, there are cross effects of the control actions. In this perspective, the modeling of various system operators' simultaneously independent decision-making on their own subsystems inside a larger physically interconnected system creates an urgent need for approaches and methods.

In summary, the problem of coordinated control in wide area interconnections, controlled by a variety of different independent system operators, is to determine operational control actions of system operators in steady-state and emergency situations which would result in optimal functioning of the power system interconnection as a whole, respecting individual interests of each partner. Therefore, the problem of developing efficient coordination methods for large interconnections is then to strike a balance between global and local welfares.

11.3 Interconnected Power Grids as a Multiple-Layered System

At present, there are two main reference paradigms for the solution of coordinated control problems in large interconnections. One is called distributed coordination, carried out by issuing a set of unanimously consented rules. In this paradigm, the coordination does not undermine any jurisdictions of each system operator, who still makes its own decisions individually and has full control of its own subsystem. However, additional constraints, issued as coordination institutions, would be imposed and mandatorily compiled by each interconnected partner for achieving a better overall performance of the whole network.

Another type of coordination is based on organizing joint coordination centers, which handle all information from the overall system and issue joint optimal controls on the whole power interconnection. This type of coordination can be referred as centralized coordination.

A traditional overview of a power system usually regards it as a system consisting of different equipment wired together. Under this model, equipment is investigated in full details; and sets of analytical equations are employed to express the physical behavior of the equipment. The component-oriented and phenomenon-oriented modeling successfully solves many practical problems in the power system, but as the grid's size increases, especially after the interconnections of continental areas, the behavior of the interconnected system as a whole becomes drastically complex to analyze and control. The difficulties are not only caused by the increase of size, the basic control structures (centralization or distribution), the jurisdictions and institutions as how to control the interconnected systems also raise other complexities. Due to the jurisdictions, some effective methods under the centralized control may not apply anymore. New perspectives towards interconnected power systems are needed to face the challenge of operation and control of such a large system.

To integratively analyze the power system from the dimension of the decision making process (control) versus the physical equipment to the dimension of a single subsystem versus the unique interconnected grids together, we regard the system as a three-layered system: physical, cyber and decision-making layers which are vertically bounded together, forming a single area (Fig. 11.1). The physical layers, the network (in terms of buses and lines) and physical components (generators, loads, etc.) attached to it are interconnected by tie lines to neighboring subsystems. The cyber layers are virtually connected by the information (measurements, commands) exchange flows, complying with some institutions issued by superior organizations. The decision making layers, usually independent from each other, refer to the process of transferring input measurements of the physical system and quality of data to the output control actions to be implemented on the physical layers.

Besides the complexity of the above-mentioned multiple layers, the operation of the power system shows considerable interaction among physical, technological, economic, institutional and human factors. More precisely, the performance of the physical system definitively depends on the physical laws; for example, the power flows over the network and power injection and withdrawal are governed by power flow equations based on the "Kirchhoff laws" and cannot be overcome. In addition, the power grids need to be operated under strict physical and operational constraints such as the capability of generators, the line flow limits, the static and dynamic stability limits, etc. On the other hand, reliability, security, efficiency and economy, as an expression of the willingness of human, are expected and required when operating the systems. The complex interactions among the social and technical aspects together determine the performance of the power system with very specific features in this environment [7, 8]. This can be referred to as a socio-technical system.

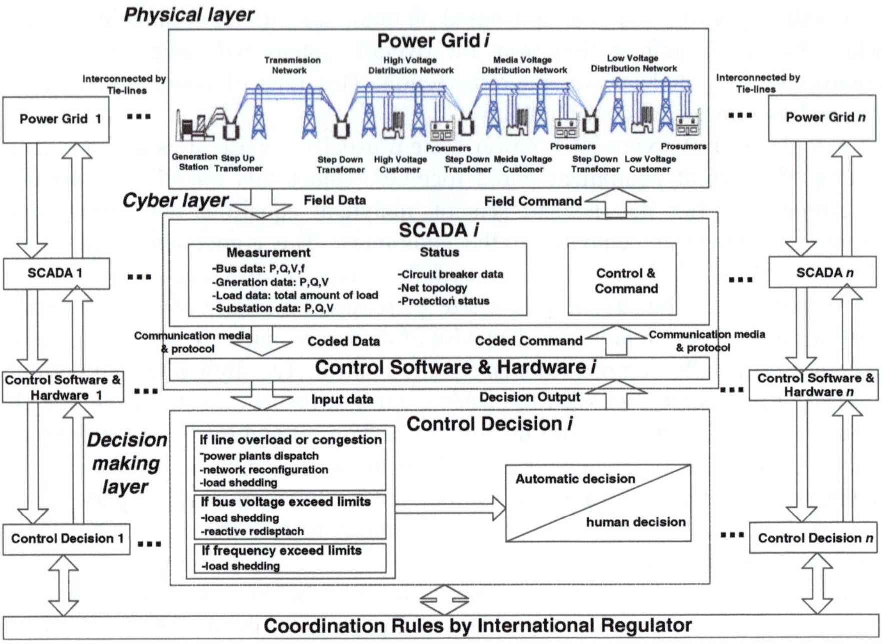

Fig. 11.1 Multiple layers of interconnected power systems

The interconnection of various subsystems into a unique grid, like ENTSO-E/IPS/UPS spanning 13 time zones, outlines the complexity of the socio-technical system. Each system operator has jurisdiction only over a specific part of the interconnected system and pursues individual objectives, usually represented by the local performance of the sub-system while complying with some coordination rules to contribute to the overall performance of the interconnected system. On the institutional level, the cooperation of various SOs is quite loose and rather ambiguous due to the lack of uniquely defined and detailed overall objectives, effective incentives and strong administrative control actions and sanctions.

Due to the interconnected structure of the whole system, any failure within a subsystem will affect simultaneously the whole grid and may eventually evolve to a cascading failure of the whole system. The US-Canada outage and the Italian blackout in 2003, affecting more than 50 million people each, provide very good examples of how a lack of coordination under emergency can result in serious failures of one or more subsystems.

Many studies have been performed on SO's decision making to operate and control the power system under emergency. Xinlei and Chen [9] converts the emergency problem into a feasibility checking problem and solves it based on the Ordered Binary Decision Diagram (OBDD) searching method. Fu et al. [10] uses a security-constrained unit commitment model to obtain the minimum bid-based system operating cost for steady state contingencies. Keqiu et al. [11] takes a measure of the economic equivalent of security and reliability as object function to optimize the coordination of preventive and emergency control using Benders decomposition. A few investigations have taken into consideration the coordination of the decision making of various SOs' under emergency circumstances.

To take all the layers and their interaction collectively into consideration, we propose the general framework to map the complex systems as shown in Fig. 11.2. The subsystems operated by various SOs are connected with each other physically to integrate into a unique grid. An SO operates its corresponding subsystem respectively to maximize its individual objectives, subjected to the physical and operational constraints as "inside" constraints and coordination rules as "outside" obligations. Measurements from the physical layer are sent to the relevant SO as input of the decision making process, while the commands, such as network reconfiguration, load shedding, generation adjustment, etc., which are transferred to the physical layers, are the output of the decision making process. Information from the other SOs might be very valuable to assist the decision making process, hence under the coordination rules subsets of the information concerning the operational data can be exchanged among SOs.

11.4 TSOs as Interacting Agents

The coordinated control problem is complex in its nature due to the interaction of multiple aspects such as system-wide and local performance, social and technical issues, the regulatory/decision making and physical levels, and the ambiguity in the coordination of multiple decision makers both from the theoretical and the computational point of view. The challenge posed by modeling the complexity of the coordination and decision makings of the multi-player and multi-levels as the interconnection of large-scaled dynamic power systems is that no general analytical model can be adapted to represent the performance of the overall system. Multi-agent systems (MAS) are one of the possible theoretical frameworks for modeling the overall performance of the interconnected system, especially with different general structures and control philosophies. It is viewed from the scientific and industrial perspective as one of the most important trends for the next generation of control systems. For examples, Ref. [12] uses MAS to simulate the emergency control of a single SO. Bompard et al. [13] adopts the MAS to search for the system equilibrium under malicious attacks and tries to study the vulnerability of the power system with reference to the dependency on information exchange. Solanki et al. [14] presents a multi-agent approach to power system restoration.

MAS may be regarded as a context in which a population of autonomous agents is interacting with each other through and within an environment is pursuing individual objectives and is able to communicate and interact among themselves and with the environment; the superposition of the individual behavior of the agents determines the status of the system as a whole and its performance. An agent can adapt itself based on the changes occurring in its environment, so that a change in circumstances will still strive for the intended result. The autonomy,

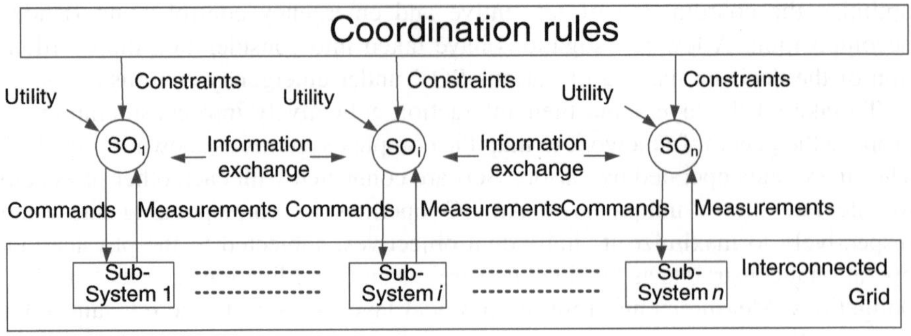

Fig. 11.2 Framework of coordination

social ability and adaptation of an agent provide the way to simulate the complex system which is collectively determined by the interactions of multiple actors and layers. An agent based approach provides a flexible, robust and adaptive mechanism for large-scale distributed systems and this is especially helpful when components of the system are highly heterogeneous.

Each agent is responsible for perceiving the state of the environment, updating its own knowledge, deciding about future actions and finishing the tasks. Therefore, besides the most important feature of an agent—its autonomy, the following features are keys to the functionality of an agent: The ability to communicate with other agents and environment, management of self-organization, as well as coordination and execution. To make the multi-agent based system work, especially aimed at power system coordinated control, many different autonomous components are required, which represent likewise autonomous systems in the sense of subsystems. They interact with each other through and within an environment, pursuing individual objectives while cooperating with each other to collectively achieve the common goal of the system as a whole.

The MAS employed in this framework consists of different layers which correspond to the layers of the power system described in the previous sections, excluding the cyber layer which is in charge of sending measurements and commands in between the physical equipment and decision making units, both human and automation. Therefore, the multi-agent based systems can be depicted as in Fig. 11.3.

The agent here does not necessarily represent a human decision process. It can also be any automatic unit who carries out any decisions, such as the widely used protection devices, smart meters, distributed control systems, etc., in power systems. In this way, the multi-layers and multi-players of the interconnected systems can be simplified and conceptually represented and modeled.

Depending on the nature of a study and the functionalities and details of the models, the design of the MAS used for power system control can be very complex. As a general illustrative example, Fig. 11.4 presents a very natural and simple agent design, i.e. regarding each system operator as a single agent. But

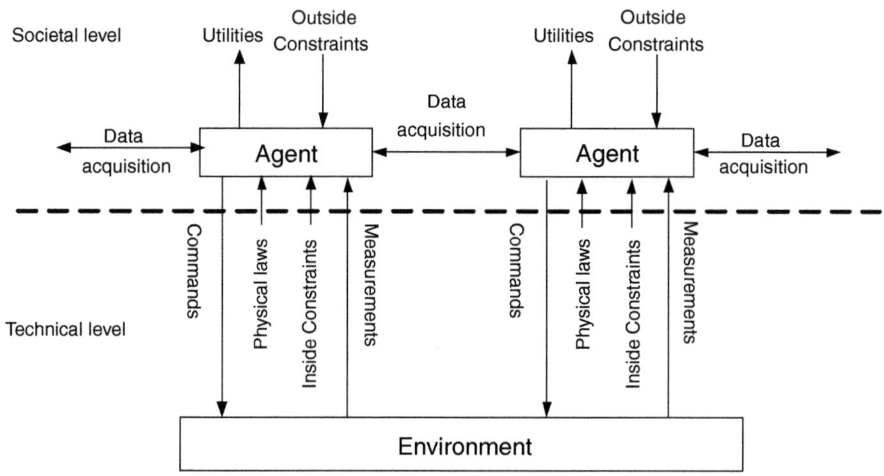

Fig. 11.3 Multi-agent system based decision making

inside each agent, there are a lot of functions that need to be decomposed into more agents who fulfill more specific targets. Each "smaller" agent can also duplicate the structure shown in Fig. 11.3, except for the "environment" which can be any associated part in any layer rather than the measurements from the physical fields.

An agent equivalent to a SO can be modeled as in Fig. 11.4, in which the power flow control and load frequency control are regarded as the main control actions related to interconnections which will directly influence the whole performance of the interconnected grids, while the voltage control is regarded as local control which only affects the local performance of the subsystem. One should notice that only technical aspects are included in this figure; however, when it makes decisions, other aspects will be considered as objectives, constraints, etc.

As mentioned before, the agent in Fig. 11.4 can be decomposed into couples of inferior agents who only deal with one specific task, such as OLTC control, primary control, Automatic Generation Control, load shedding etc.

11.5 Interacting Decision Making in Transmission Systems: An Illustrative Example

As discussed in the previous section, coordinated control in a distributed system is a complex problem. i.e., each system operator makes its own decision to pursue its individual maximal welfare while ignoring the impact on others. The nature of the distributed decision-making process specifies that an "overall control" of the whole interconnected system can only be implemented by coordination rules which eventually mold the performance of the entire system by imposing "outside" constraints to an individual decision maker. In other words, we must respect

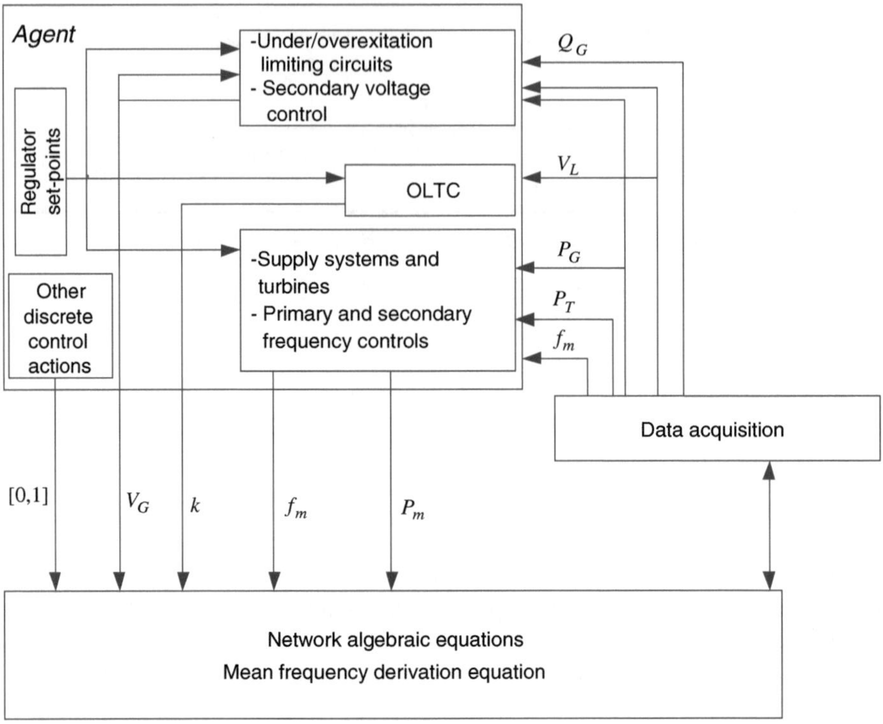

Fig. 11.4 Agent design as a single SO

the fact that the performance of each subsystem and the whole system are provided by independent and combined decisions by a number of autonomous and rational decision makers (i.e. SOs), respectively. The aim is to maximize the performance of their own systems while complying with a set of rules issued by the entity in charge of the coordination of the system (such as ENTSO-E in Europe). How the coordinating rules work and what weaknesses they have are then important to be identified and verified ex-ante.

In this section, with reference to the standard IEEE-30 bus test system (Fig. 11.5) in which 3 SOs are considered, we show how the proposed model can be applied; particularly we use the model to assess the influence of two aspects of the coordination rules: *information set* and *interconnection policy*, on the decision making outputs of each SO and on the system performance. The first refers to the dimension of the information set, in terms basically of the number of buses of the neighboring subsystems that each SO possesses when building the network model for its decision making. The second is related to the constraints on power flow control on the tie-lines under emergency. The simulation is undertaken with the following hypothesizes: steady-state operation, no dynamic considerations.

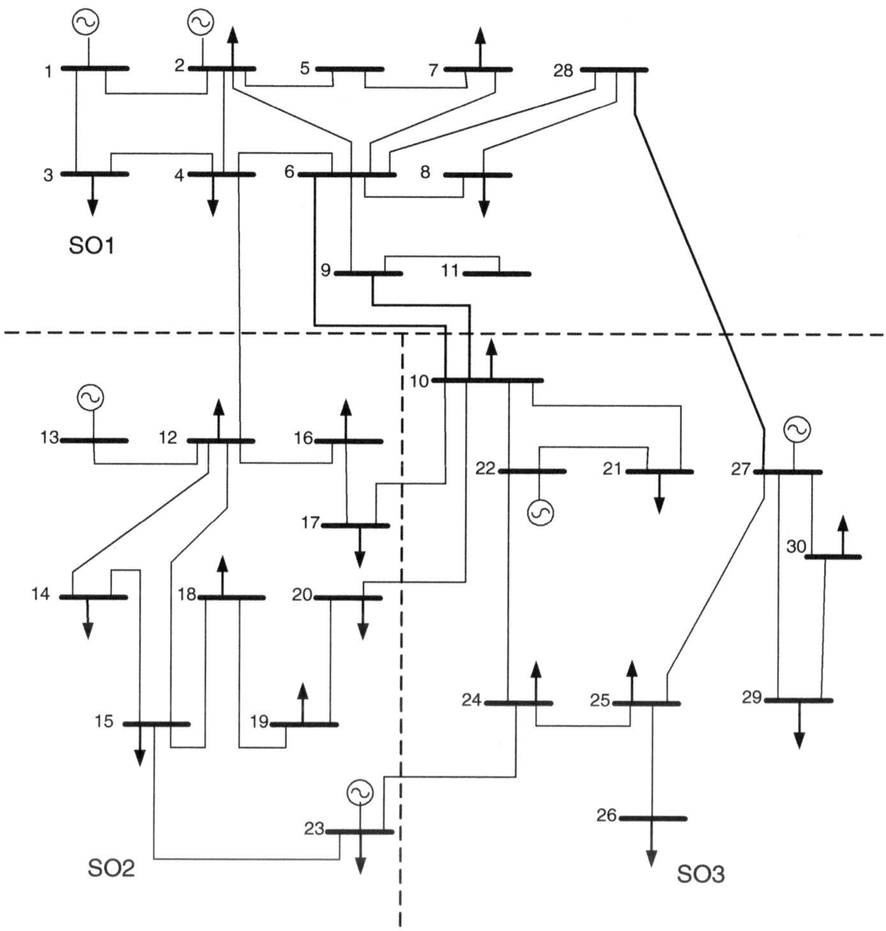

Fig. 11.5 IEEE-30 bus system with area divisions

Under emergency, each SO aims in minimizing the energy un-served to the loads, given the coordination rules he needs to comply with; if that is not technically feasible the SO will curtail loads starting from the less sensitive ones. Let's divide the loads into two sets *interruptible* (D_I) and *superior* (D_S). The interruptible loads are the loads that, also due to special economic incentives or rates, can be curtailed by the SO under emergency control, while the superior loads are the loads that should be supplied and can only be interrupted under severe emergency.

The decision making problem of the SO in this context can be formulated as:

$$\min_{\substack{\Delta P_g, \Delta d_i^p \\ \Delta Q_g, \Delta V_i \\ \Delta \delta_i}} \sum_{i \in D_S} \Delta d_i^p = \sum_{g \in G} \Delta P_g - \sum_{j \in D_I} \Delta d_j^p \tag{11.1}$$

subject to

$$\begin{aligned}\sum_{j=1}^{N} V_i V_j Y_{ij} \cos(\theta_{ij} + \delta_j - \delta_i) - (P_i^0 + \Delta P_i) \\ + (d_i^{p0} - \Delta d_i^p) = 0 \quad i = 1, \cdots, N\end{aligned} \tag{11.2}$$

$$\begin{aligned}\sum_{j=1}^{N} V_i V_j Y_{ij} \sin(\theta_{ij} + \delta_j - \delta_i) - (Q_i^0 + \Delta Q_i) \\ + (d_i^{q0} - \Delta d_i^q(\Delta d_i^p)) = 0 \quad i = 1, \cdots, N\end{aligned} \tag{11.3}$$

$$P_i^m \le P_i^0 + \Delta P_i \le P_i^M \quad i \in G \tag{11.4}$$

$$Q_i^m \le Q_i^0 + \Delta Q_i \le Q_i^M \quad i \in G \tag{11.5}$$

$$d_i^{pm} \le d_i^{p0} + \Delta d_i^p \le d_i^{pM} \quad i \in D \tag{11.6}$$

$$d_i^{qm} \le d_i^{q0} + \Delta d_i^q(\Delta d_i^p) \le d_i^{qM} \quad i \in D \tag{11.7}$$

$$V_i^m \le V_i \le V_i^M \quad i = 1, \cdots, N \tag{11.8}$$

$$\delta_{ji}^m \le \delta_{ji} \le \delta_{ji}^M \quad j, i = 1, \cdots, N \tag{11.9}$$

$$F_{ij}^m \le F_{ij} \le F_{ij}^M \quad j, i = 1, \cdots, N \tag{11.10}$$

$$\frac{\Delta d_i^p}{d_i^{p0}} = \frac{\Delta d_i^q}{d_i^{q0}} \quad i = 1, \cdots, N \tag{11.11}$$

Equations (11.2) and (11.3) represent the power flow equations (real and reactive power balance at each bus); Eqs. (11.4–11.7) are real and reactive power limits of each bus for generators and loads; Eqs. (11.8) and (11.9) take into account the voltage profile and stability limits; Eq. (11.10) incorporates the line flow limits while (11.11) fixes the power factor at each load bus affected by curtailment.

The simulation of the operation of the interconnected power system is based on a continuous sensing of the status of each subsystem by the related SO; if a subsystem has moved to an emergency state, the system operator would be compelled to undertake proper control actions. The objective of the decision making is to minimize the load curtailment after the restorative actions have been undertaken. The decision variables under the control of the SO are, with reference to its system, the generation redispatch, the curtailment of interruptible loads or the

Fig. 11.6 Flow chart of simulation process

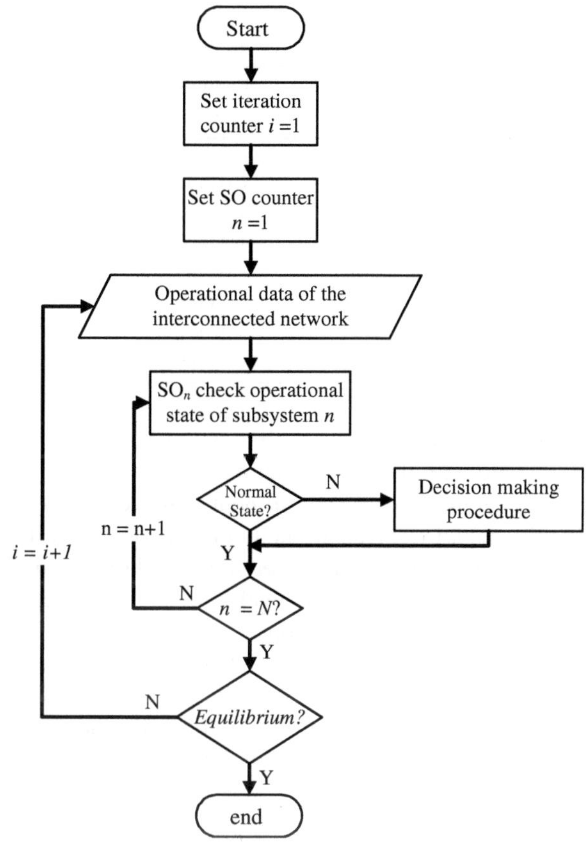

shedding of superior loads; the three actions are undertaken in the order in which they are listed until the normal state is restored. The decision making process is modeled by the optimization problem—. The decision making is performed, as a result of the occurrence of an emergency, by all the SOs at the same time; as a result of the actions the status of each subsystem and of the power system as a whole will change and again the SOs may be called to make decision and perform actions on each subsystem. This process, represented by a sequence of iterations will go on until a Nash equilibrium is reached, if existing, or the system will face a global blackout. The whole process can be illustrated as in Fig. 11.6.

We consider two different coordination scenarios and simulate the outcome of the coordination under two different contingencies.

The coordination scenarios are:

S1—Keeping the power flow on the tie-lines with neighboring systems fixed by each SO, and keeping the amount of exchanged information equal to k (the first k buses closest to the border of each neighbor)
S2—Tripping of the tie lines with neighboring systems.

Table 11.1 Simulation results for contingency *C1* (Post-contingency: $V_8 < V_8^{min}$, $F_{8\text{-}28} > F_{8\text{-}28}^{max}$)

		After restoration			
		Iteration	Load curtailment (%)	Generation adjustment (%)	
				$\sum G^{+}$	$\sum G^{-}$
S1	k = 1	1	−2.002	2.763	−4.196
S1	k = 1	2	−0.006	0.065	−0.072
S1	k = 1	Total	−2.008	2.828	−4.268
S1	k = 2	1	−2.004	4.968	−6.415
S1	k = 2	2	−0.005	3.492	−3.478
S1	k = 2	Total	−2.009	8.460	−9.893
S2		1	−3.193	30.367	−32.747
S2		Total	−3.193	30.367	−32.747

In the scenario *S1,* the tie-line policy imposed by the regulator compels each SO to operate with the goal of maintaining the existing power flow exchange with the other systems even if a curtailment of those flows could alleviate its own problems. In this scenario *k* is the amount of exchanged information of each system. Different values of *k* can represent different rules about the information revealed by each SO to the others as well as the level of detail in the network representation adopted by the SO when taking decisions.

In the scenario *S2* each SO, under emergency, firstly sacrifices the power exchanges with other areas by curtailing the power flow of the tie-lines and, of course, in this case no information about other SOs is needed.

The contingencies, considered as a trigger for the coordinated decision making process, are:

C1. Outage of line 21–22 and line 6–8
C2. Outage of 2 generators, at bus 2 and at bus 23, located in 2 different faulted subsystems.

Contingency *C1* refers to a case in which simultaneously two heavily loaded transmission lines in two different systems have a failure while contingency *C2* refers to the case in which two generators in two different subsystems are tripped simultaneously due to failures and this leads to a lack of power supply.

The results of the simulation are summarized, for both scenarios, in Table 11.1 and Figs. 11.7, 11.8 for contingency *C1* and in Table 11.2 and Figs. 11.9, 11.10 for contingency *C2*.

From Table 11.1 we can see that for the emergency state caused by contingency *C1*, the impact of the tie-line policy is more profound than the size of the information set. In other words, during the emergency of transmission lines tripping, keeping the integration of the whole system is much more important than exchanging more information since the support from other subsystems could facilitate its recovering to a normal state.

The percentage of load curtailment in the scenario *S2* is larger than that for *S1*, and the same applies for the adjustment of generation. For *S2* the adjustment

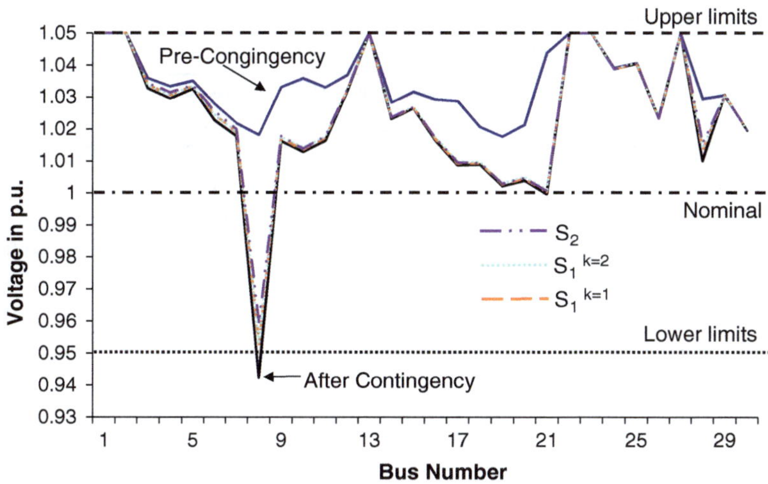

Fig. 11.7 Voltage profiles for emergency C1

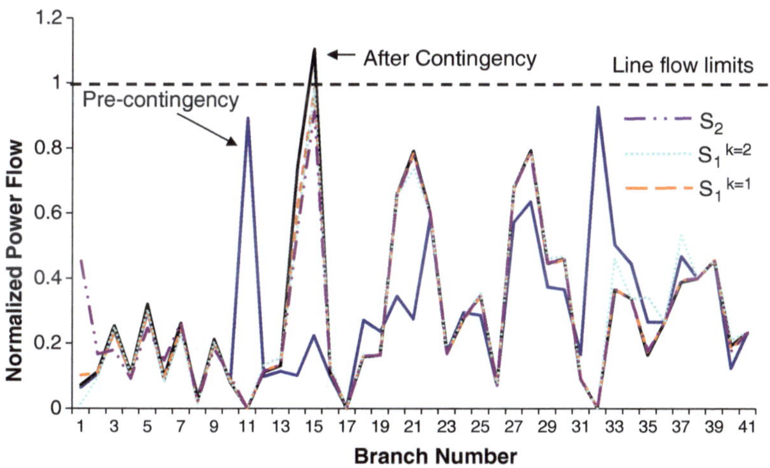

Fig. 11.8 Normalized power flows for emergency C1

percentage of generation is 7–10 times larger than for *S1* because the actions of curtailing the tie-line power under the emergency situation immediately induces a lack of backup between the neighboring areas. Each SO needs to redispach its own subsystem independently, which increases the generation adjustment of each subsystem and eventually increases the total generation adjustment of the interconnected system.

The percentage of load curtailment in *S1* is almost equal for both $k = 1$ and $k = 2$; on the other hand, the information affects the decisions of SOs in terms of

Table 11.2 Simulation results for emergency *C2* (post-contingency: $P_{g1} > P_{g1}^{max}$)

		After restoration			
		Iteration	Load curtailment (%)	Generation adjustment (%)	
				$\sum G^{+}$	$\sum G^{a}$
S1	k = 1	1	−11.74	31.77	−43.06
	k = 1	Total	−11.74	31.77	−43.06
	k = 2	1	−23.71	19.53	−43.06
	k = 2	Total	−23.71	19.53	−43.06
S2		1	−9.627	33.77	−43.06
		Total	−9.627	33.77	−43.06

[a] The generation decrease includes the active power loss from the triggered out units

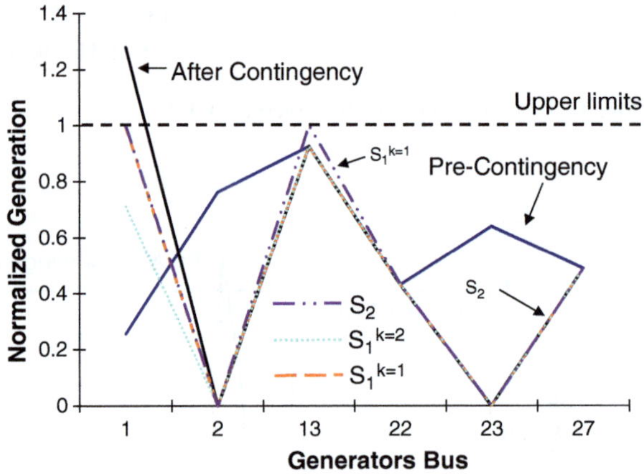

Fig. 11.9 Generation adjustment for emergency C2

generators' set-points. For both $k = 1$ and $k = 2$, the sum of the increase and decrease of generation is almost equal, but it is differently allocated among the units increase and decrease.

The status of the system after the contingency is assessed with reference to the voltage profile and the flows on the system lines as presented in Figs. 11.7 and 11.8.

After contingency C1, both voltage and line flows constraints are violated. In this case the decision making of the SOs is able within the first 2 iteration to bring the system back from the emergency state to a normal state that corresponds to equilibrium. Table 11.2 lists the simulation results of the 2 scenarios under emergency C2.

In the case of contingency *C2* (Table 11.2) the impact of the size of the information set are more important than those related to the policy for tie-line management; the decremental adjustment, which is equal to the lost power of the generators at bus 2 and 23, is identical for all the scenarios. At scenario *S2* the load curtailment is smallest due to the independent and local redispatch of generators in

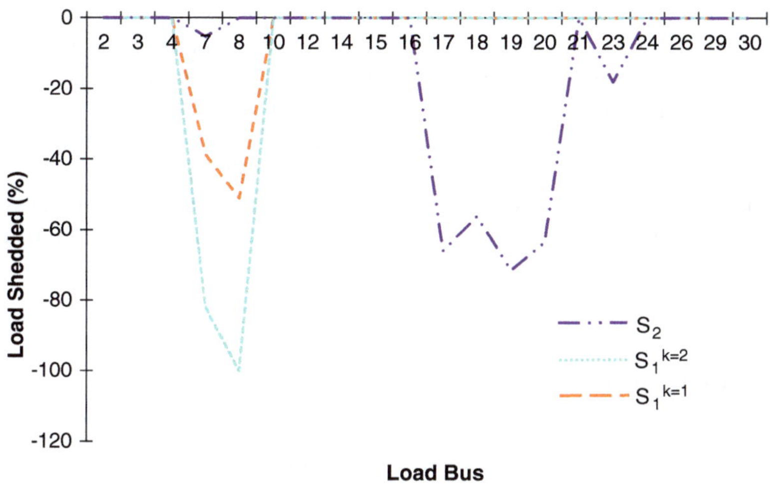

Fig. 11.10 Load shedding for emergency C2

such a way that each SO can decrease the cross-border power transmission which slightly decrease the system losses. Thus, fewer loads are tripped compared to *S1* with $k = 1$.

For *S1* with $k = 2$, the demand of operational control of more buses drastically increases the difficulties for the SO who needs to make decisions to minimize its load curtailment while complying with the coordination rules. When a SO takes a decision, more information about the buses of neighboring subsystems implies more consideration of buses beyond the control of the SO. That the SO does not possess the jurisdiction to control those buses prohibits the use of the resources at those buses to bring the system back to normal state. Mathematically speaking, it adds more constraints to the optimization problem and may result in the diminishment of the feasible region or even make the problem infeasible. In this case, more loads must be tripped to obtain feasibility of the system.

Figures 11.9 and 11.10 show the details of generation adjustment and load curtailment for emergency *C2* for all scenarios. After the failure of the generators at bus 2 and 23, the generator at bus 1 skyrockets from 26 % of its capacity to 128 % of that because the generator at bus 1 is selected as slack generator, which is responsible for keeping the balance between generation and load in the entire system. Except for the generator at bus 1, the rest of the generators produce almost the same power for all scenarios. It is also worth noting that the adjustment curves (Fig. 11.9) for *S2* and *S1* with $k = 1$ almost overlap, implying that the total load curtailment is equal if the tiny difference in system losses is neglected (Table 11.2).

Although the sum of the load curtailed for *S2* and *S1* with $k = 1$ are identical, their allocation among the loads is different. For *S2*, the curtailed loads are distributed among 6 buses, 5 of which are located within subsystem 2 (Figs. 11.10

and 11.5). For *S1* with $k = 1$, the loads that have been curtailed are located at bus 7 and 8 within subsystem 1, while SO2 does not interrupt any load since SO1 is exporting energy to SO2 for *S1*, so that this SO curtails its own load to comply with the coordination rules.

Besides obeying the coordination rules, the impacts of more information as we pointed out before raises the load curtailment under *S1* with $k = 2$.

In addition, for all scenarios, we find the Nash equilibrium at which the systems with reference to the subsystems and interconnected grid are restored back to normal state in the first iteration.

11.6 Conclusions

In the interconnected power systems which exhibit interactions and interdependencies between subsystems, decision-makers act to maximize its own utilities under the compliance with both technical and social constraints. A challenge is to formulate the social aspects which have not been studied deeply and are still formidably defined.

This chapter makes a different study about coordination and decision making of interconnected power systems by viewing the huge complex system as a socio-technical infrastructure. We propose a framework to capture the influences from the societal level on the technical level. Expectations and constraints from both levels collectively determine the decisions from each decision maker and thus shape the performance of the system. The study also provides an approach to investigate the impact of technical constraints posed to the design and operation of huge infrastructures.

The coordinated recovery of interconnected power transmission grids under emergency can be modeled base on this framework. Under the framework, MAS are appealed to simulate the decision making and interactions. As the simulation has shown, under different emergencies, the information set and the policy for tie-line management can be of different importance.

The framework and approach used in this chapter can be easily modified and extended in the future according to various requirements such as different coordination rules concerning frequency regulation and/or different context such as market environment, etc.

References

1. K.S. Swarup, P.B. Corthis, ANN approach assesses system security. IEEE Comput. Appl. Power **15**(3), 32–38 (2002)
2. H. Yan, J.C. Chow, M. Kam, C. Sepich, R. Fischl, Design of a binary neural network for security classification in power system operation, in (IEEE, 1991), vol. 1122, pp. 1121–1124

3. O.I. Elgerd, *Electric Energy Systems Theory: An Introduction*, 2nd edn. (McGraw-Hill, New York City, 1983)
4. PJM manual 13: emergency operations. http://www.pjm.com
5. NERC, *Standard: capacity and energy emergencies*, vol. EOP-002-2 (2007)
6. ENTSO-E, Continental Europe Operation Handbook: policy 5: emergency operations. https://www.entsoe.eu/publications/system-operations-reports/operation-handbook/
7. Z. Lukszo, P. Heijnen, Better design and operation of infrastructures through bi-level decision making, in (IEEE, 2007), pp. 181–186
8. E. Bompard, T. Huang, R. Napoli, Emergency coordination and decision making over interconnected power systems, in (IEEE, 2009), pp. 3545–3551
9. W. Xinlei, S. Chen, A new scheme for power system emergency control based on the OBDD searching method, in (IEEE, 2006), pp. 1–5
10. Y. Fu, M. Shahidehpour, Z. Li, AC contingency dispatch based on security-constrained unit commitment. IEEE Trans. Power Syst. **21**(2), 897–908 (2006)
11. W. Keqiu, W. Liyong, G. Zhen, Z. Baohui, Z. Guochang, Assessment of preventive control and emergency control coordinating economic benefit, in (IEEE, 2006), pp. 1–6
12. D. Panasetsky, P. Etingov, P, Voropai N Multi-agent approach to emergency control of power system. in (IEEE, 2008), pp. 2157–2161
13. E. Bompard, R. Napoli, F. Xue, Vulnerability of interconnected power systems to malicious attacks under limited information. Eur. Trans. Electri. Power **18**(8), 820–834 (2008)
14. J.M. Solanki, S. Khushalani, N.N. Schulz, A multi-agent solution to distribution systems restoration. IEEE Trans. Power Syst. **22**(3), 1026–1034 (2007)

Chapter 12
Distributed Optimization of Interconnections

Oleg Voitov, Lembit Krumm and Oleg Soukhanov

12.1 Introduction

This section presents a short overview of the main stages in development and application of decomposition principles for solving the problems of current flow optimization in electrical networks divided into several subsystems. The main goal of decomposition is to decrease the laboriousness for the solution of large-dimensional optimization problems in electric power systems by reducing them to problems of smaller dimension. Consideration is given to two approaches. In the first approach each subsystem is represented by an equivalent fuel cost function (equivalent characteristic) and a feasibility region characteristic and in the second approach a functional characteristic is defined for each subsystem that defines the relation between the values of Lagrange multipliers and currents in the interconnecting branches at a solution point. In order to simplify the identification of essential properties of the optimization procedure in terms of decomposition this section considers, firstly, a linear electrical network in which variables are nodal currents and currents in branches, and, secondly, the problem of determining a one-criterion optimal current flow.

To this end the necessary terminology and notations are introduced. Then a general algorithm for determining optimal current flow in terms of decomposition is described for two forms of representation of external characteristics—equivalent

O. Voitov (✉)
Energy Systems Institute, Irkutsk, Russia
e-mail: sdo@isem.sei.irk.ru

L. Krumm
TUT, Tallinn, Estonia
e-mail: lembitkrumm@gmail.com

O. Soukhanov
VEI, Moscow, Russia
e-mail: soukh@enersys.ru

U. Häger et al. (eds.), *Monitoring, Control and Protection of Interconnected Power Systems*, Power Systems, DOI: 10.1007/978-3-642-53848-3_12,

and functional. A detailed description of the algorithms is presented. The main statements have been checked by numerical calculations made in a MATLAB environment [1].

12.2 Overview

General issues of decomposition for solving optimization problems were considered in [2, 3]. As applied to the electric power industry the research presented in [4–6] laid the foundations for studies on methods intended to solve and adjust the process of solving multi-stage hierarchical sub-problems of complex optimization that were considered in space, time and situation aspects of control and adjustment of expansion. The authors of [4–6] propose procedures for building multi-stage hierarchical subproblems for complex optimization of control over operation and expansion of Interconnected Power Systems (IPS) under consideration of aspects such as time, space as well as state (i.e. normal and disturbed conditions). For the first time the equations of the physical-technical state of power systems were considered in the optimization problem together with requirements to quality and reliability of operation in the form of inequality constraints on parameters of the physical-technical systems. In doing so, first of all, consideration was given to only one optimization criterion, i.e. the minimization of total costs. The problems of complex optimization in a general form were considered as hierarchical and multi-stage control problems in time, space and state. For solving these problems and sub-problems a generalized reduced gradient method (GRGM) was developed together with respective rational mathematical apparatus depending on specific properties of the problems and sub-problems to be solved. These developments rely on the theory of complex optimization and control with application of methods for reference modeling. Different modifications of the generalized Newton type method play an important role in GRGM [4–7].

Currently multi-criteria optimization problems have become very relevant. In particular they can be reduced to one-criterion problems, for example, under certain agreements between the partners about their profit distribution. The agreements can also be related to the direct improvement of the equilibrium states between the partners, which can be established by any market mechanism if the solution is not located in the Pareto region. Here GRGM was generalized for solving such problems of Pareto-optimal correction as well.

However, in the case of a coalition of partners, say different power interconnections, which have considerably different interests and contradictions, these approaches should be developed in the framework of game theory. Antagonistic games should be considered only in extreme cases. At best the correction is carried out in terms of different narrower interests.

These problems are discussed more in detail in the survey papers [8, 9]. These papers present six main principles of solving the problem of complex optimization in the most general form and give the main methodological approach to the analysis and optimization of IPS operation and expansion reliability at inter-country levels. The methods developed by the authors of [2–4, 8, 9] provide the following features:

1. The sub-problems are related to the principle of optimality.
2. The market mechanisms of corresponding subsystems that have to provide feasibility of steady states are taken into account if necessary.
3. Optimal values of the objective functions and the boundaries of the feasibility region are approximated in their hierarchical structure. To this end we use a rational combination of the mathematical apparatus of GRGM and the neural network method.
4. The boundaries of controllability regions are approximated in the space of independent boundary variables, which characterize the distribution of different reserves, in particular, energy and power. This makes it possible to study controllability regions for the condition of large random disturbances as well.
5. In a general case it is necessary to take into account incompleteness and fuzziness of the initial information starting from the lower levels of the optimization sub-problems.

In [7] a decomposition methodology is presented for solving the steady state equations of electric power systems with the use of the implicit functions theory which is also applied in GRGM. The system is divided into natural and artificial subsystems. The main goal of this study was to decrease the laboriousness for the solution of large-dimensional optimization problems in electric power systems by reducing it to problems of smaller dimension. In [8, 9] the authors suggest a similar methodology for the decomposition of steady state equations of electric power systems on the basis of the theory of implicit functions as well as the creation of optimal conditions on the basis of a generalized Lagrange method. This methodology is called functional modeling [10]. Functions of dependent boundary variables of individual subsystems of lower level are called by functional characteristics. According to this methodology different conditions of modeling are considered with the use of a distributed network of coordinating centers with computers of the lower hierarchy level. The above mentioned methodologies [7–9] have advantages and disadvantages. In the first methodology (application of mathematical apparatus of GRGM) the number of linear equations at each iteration is considerably lower than in the functional modeling method. Furthermore, with a large number of inequality constraints, particularly at optimal correction of operating conditions in case of large disturbances, GRGM can converge much faster and more accurately than the Lagrange method in the second approach. It is clear that in the general case it is necessary to study the possibilities for a rational combination of these methods. To this end, we consider a common calculation algorithm that can be represented by a two-level procedure. At the lower (second) level the first level problem of optimal current flow is solved for each subsystem at specified values of variables of the problem of the upper level. As a result we obtain an approximation of functional characteristics. This approximation analytically defines a dependency of variables in the lower level problem on the variables of the upper level. Then the problem of the upper level, which is often called a coordinating problem, is solved, and the optimal value of variables in the problem of this level is determined. Algorithms based on decomposition usually cause problems due to the division into subsystems and the consideration for nonlinearity and inequality constraints. In the following we consider and compare:

- Methods intended to solve and correct the solution of multi-stage hierarchical sub-problems of a complex optimization that are considers time, space and state aspects with the use of GRGM.
- Functional modeling methods based on the Lagrange method.

12.3 Definitions and Terms

The main propositions and notions that provide the basis for the decomposition method as well as the implemented algorithms are formulated as follows.

An electrical network consists of a set of components, which can be assigned to nodes and branches. Each component of the network is described by a set of network parameters and state variables. Network parameters are a system of connections and electrical parameters of network components. State variables are nodal voltages, transformation factors and currents along branches. The electrical network is represented by a set of specified subsystems that are connected by interconnecting branches. Network parameters and state variables of the network components that belong only to one subsystem are called internal and those belonging to two subsystems are called external. Below we describe principles of division or decomposition of electrical networks and a two-level hierarchy of problems that arise when the current flow is optimized with decomposition methods.

The current flow optimization on the second, lower, level is solved independently on each level. The solutions for each subsystem are optimal values of the internal parameters at specified values of the external parameters, and a characteristic or functional relationship (linear or nonlinear) between the values of internal and external variables.

12.4 Formulation of the Current Flow Optimization Problem with Decomposition in Electrical Networks

Let an electrical network have N nodes, L branches and N_S subsystems. The current flow is formulated as a one-criterion statement as follows:

$$\min_{J_G} \sum_{i=1}^{N} F_{Gi}(j_{Gi}) \tag{12.1}$$

$$W(Z) = 0 \tag{12.2}$$

$$Z_{\min} \leq Z \leq Z_{\max} \tag{12.3}$$

where Z is the vector of the state variables of the electrical network. The components of this vector are the nodal current injections j_{Gi} for each node $i = 1, N$ and the currents on the branches I_{Fj}, for each branch $j = 1, L$. $F_{Gi}(j_{Gi}) = a_{1i}j_{Gi}^2 + a_{2i}j_{Gi} + a_{3i}$ is the function of fuel cost for the generation at node i. $W(Z)$ is a system of equality constraints of the nodal currents. $Z_{\min}$, $Z_{\max}$ are vectors of minimum and maximum values of the vector of the state variables Z. The vector Z is divided into sub-vectors of internal Z_i^{kk} and external Z^{nn} variables for all $k, n = 1, N_S$. The set of vector components of Z^{kk} is different to that of Z^{nn}. The determination of an optimal current flow with decomposition techniques can be represented as a two-level procedure. On the first, higher level, the initial optimal current flow problem (12.1)–(12.3) is reduced to solve the following set of second level problems.

The problems of the second, lower level, at which the optimal current flow is determined for each subsystem $k = 1, N_S$, have the following form:

$$\min_{Z^{kk}} F^k(Z^{kk}, Z^{k0}) \tag{12.4}$$

$$W^k(Z^{kk}, Z^{k0}) = 0 \tag{12.5}$$

$$Z^{kk}_{\min} \leq Z^{kk} \leq Z^{kk}_{\max} \tag{12.6}$$

$$Z^{k0} = const \tag{12.7}$$

where $W^k(Z)$ is a set of balance equations of nodal currents for subsystem k. $Z^{kk}_{\min}$, $Z^{kk}_{\max}$ are the vectors of minimum and maximum values of the internal state variables Z^{kk} of subsystem k.

By solving the problems (12.4)–(12.7) we form a functional characteristic $F_X^k(Z^{kk}, Z^{k0})$ and a characteristic of the feasibility region for each subsystem k.

At the first, upper level we find a solution to the coordinating problem of the form

$$\min_{Z^0} \sum_{k=1}^{N_S} F_X^k(Z^{kk}, Z^{k0}) \tag{12.8}$$

$$W^0(Z^{kk}, Z^{k0}) = 0 \tag{12.9}$$

$$Z^0_{\min} \leq Z^0 \leq Z^0_{\max} \tag{12.10}$$

$$Z^{kk} = const, \quad \forall k = 1, N_S \tag{12.11}$$

where $Z^0_{\min}$, $Z^0_{\max}$ are the vectors of the minimum and maximum values of the external state variables Z^0. By solving the problems (12.8)–(12.11) we determine an approximation of the optimal value of the vector of external state variables Z^0.

12.5 Algorithm for Optimization of Operating Conditions in the Electrical Network with Decomposition Methods

The main idea lies in the choice of state variables that provide an effective decomposition in the problem solution process. According to [5] such variables can represent flows or currents in the interconnected branches of the electrical network. A formalized description of the algorithm consists of the following steps:

1. Division into subsystems. Setting up a list of internal nodes, cutsets and a list of external lines entering into the subsystem. This procedure can hardly be formalized.
2. Obtaining initial approximations for internal and external variables. This is necessary to use the results of previous calculations.
3. Solving the optimization problem (12.4)–(12.7) for each subsystem. The method of Lagrange multipliers makes it possible to obtain a direct algorithm by solving a set of linear equations. With inequality constraints the algorithm becomes iterative and the use of GRGM becomes effective.
4. Forming functional or equivalent characteristics and characteristics of the feasibility region. In decomposition each subsystem is represented by a characteristic which is a functional linear or nonlinear relationship between optimal values of internal and external variables. With inequality constraints the feasibility region characteristic supplements a functional characteristic that contains limiting values of external variables which correspond to feasible current flows in the subsystem. The feasibility region characteristic is thus a system of inequality constraints imposed on changes in the external variables.
5. Solving the coordination problem (12.8)–(12.11) and specifying the values of the external variables. When only inequality constraints and functional characteristics are taken into account, the algorithm is iterative. When equivalent characteristics and the method of Lagrange multipliers are used, the algorithm is direct. In this case the algorithms for solving the problems coincide for all subsystems. When inequality constraints are introduced the algorithms become iterative.
6. Checking the termination criterion. If the criterion is fulfilled, terminate the algorithm. Otherwise go to step 3.

12.6 Description of Algorithms for Solving the Problem of Current Flow Optimization on the Basis of Decomposition

Let the network consist of N nodes and L branches and be divided into N_S subsystems. Each subsystem $k = 1, N_S$ includes N^k nodes, L^k branches and N_G^k generators. The subsystem of upper level contains L^0 interconnecting branches that connect the subsystems. For simplification assume that the subsystems have no common branches. The slack node belongs to subsystem N_S and is situated at a boundary node, i.e. a node which is one of the ends of an interconnecting branch. The matrix notation '/n' shows that the row corresponding to the slack node is not included in the matrix. The list of electrical network branches is sorted in such a way that the first incidence matrix $M_{/n}$ of the size $(N-1 \times L)$ can be represented in a block form as $M_{/n} = \left(M_{/n}^1 \ldots M_{/n}^{N_S} M_{/n}^0\right)$, where $M_{/n}^k$ is a matrix block of the size $\left(N-1 \times L^k\right)$ which is made up for subsystem k and $M_{/n}^0$ is a matrix block of size $\left(N-1 \times L^0\right)$ containing the common elements of the subsystems. For simplification of the description the matrix $M_{/n}^{0k}$ of the size $\left(N^k-1 \times L^0\right)$ is introduced. This matrix is obtained from the matrix $M_{/n}^0$ by deleting all rows corresponding to the nodes that do not belong to subsystem k. Let us represent the row of the first incidence matrix m_n of the size $(1 \times L)$, made up for the slack node n, also in a block form as $m_n = \left(m_n^{/0} \quad m_n^0\right)$, where $m_n^{/0} = \left(m_n^1 \ldots m_n^k \ldots m_n^{N_S}\right)$ contains the submatrices m_n^k of the size $\left(1 \times L^k\right)$, made up for subsystems $k = 1, N_S$. The matrix notation '/0' shows that the matrix is made up for branches that are not connected with the slack node n. The matrix of branch currents $I_F^T = \left(I_F^{/0^T} \; I_F^{0^T}\right)$ has a similar structure, where $\left(I_F^{/0}\right)^T = \left(I_F^{1^T} \ldots I_F^{k^T} \ldots I_F^{N_S^T}\right)$ contains the submatrices I_F^k of the size $\left(L^k \times 1\right)$, that are made up for each subsystem, $k = 1, N_S$. Matrices of load currents $J_{L/n}^T = \left(J_L^{1^T} \ldots J_L^{k^T} \ldots J_L^{N_S^T}\right)$ and matrices of generation currents $J_{G/n}^T = \left(J_G^{1^T} \ldots J_G^{k^T} \ldots J_G^{N_S^T}\right)$ are of the size $\left(N-1 \times 1\right)$, that are made up for each subsystem, $k = 1, N_S$.

12.6.1 Formulation of the Current Flow Optimization Problem

Below we describe the algorithm for solving the problem of current flow optimization. This problem includes the balance equations of currents at all nodes of the electrical network and additionally the currents of the interconnecting branches I_F. The formulation of the optimal current flow problem has the following form:

$$\min_{J_G} F = \sum_{k=1}^{N_S} \sum_{i=1}^{N_G^k} F_i^k (j_{Gi}^k) \tag{12.12}$$

$$\sum_{k=1}^{N_S} \left(M_{/n}^k I_F^k + J_L^k - J_G^k \right) + M_{/n}^0 I_F^0 = 0 \tag{12.13}$$

$$m_n^{/0} I_F^{/0} + j_{Ln} - j_{Gn} + m_n^0 I_F^0 = 0 \tag{12.14}$$

The Lagrange function of the problem (12.12)–(12.14) has the form

$$\begin{aligned} \min_{J_G, I_F} L = & \sum_{k=1}^{N_S} \sum_{i=1}^{N_G^k} F_i^k (j_{Gi}^k) + \sum_{k=1}^{N_S} \sum_{i=1}^{N^k} \lambda_{Wi}^k (j_{Li}^k - j_{Gi}^k + m_i^k I_F^k) \\ & + \lambda_{Wn} (j_{Ln} - j_{Gn} + m_n I_F) \end{aligned} \tag{12.15}$$

where m_i^k is the row i of the matrix $M_{/n}^k$ of the size $(1 \times L^k)$, λ_{Wi}^k is the component i of the vector of the Lagrange multipliers $\Lambda_{W/n}$ of the size $(N-1)$ and λ_{Wn} is the Lagrange multiplier corresponding to the balance equation of currents for the slack node n. In a matrix form (12.15) becomes:

$$\begin{aligned} \min L(J_G, I_F, \Lambda_{W/n}, \lambda_{Wn}) = & F + \Lambda_{W/n}^T (J_{L/n} - J_{G/n} + M_{/n} I_F) \\ & + \lambda_{Wn} (j_{Ln} - j_{Gn} + m_n I_F) \end{aligned} \tag{12.16}$$

Let us decompose the problem. We assume that the currents in the interconnecting branches I_F^0 are fixed at a value $\widehat{I}_F^0$. Hereinafter the symbol '^' shows the fixation of the variable. In each subsystem $k = 1, N_S$ a slack node is chosen and denote it by symbol n. According to the decomposition principles the following problems are set up.

The lower level problems are created by fixing $\widehat{I}_F^0$ for each subsystem $k = 1, N_S$:

$$\begin{aligned} \min L\left(J_G^k, I_F^k, \Lambda_{W/n}^k, \lambda_{Wn}^k\right) = & F^k + \left(\Lambda_{W/n}^k\right)^T \left(J_{L/n}^k - J_{G/n}^k + M_{/n}^{0k} \widehat{I}_F^0 + M_{/n}^k I_F^k\right) \\ & + \lambda_{Wn}^k \left(j_{Ln}^k - j_{Gn}^k + m_n^{k0} \widehat{I}_F^0 + m_n^{k/0} I_F^k\right) \end{aligned} \tag{12.17}$$

By solving the optimal current flow problem for each subsystem $k = 1, N_S$ we determine functional relationships between the optimal values of internal and external variables (currents of interconnecting branches) $J_{G/n}^k(\widehat{I}_F^0)$, $I_F^k(\widehat{I}_F^0)$, $\Lambda_{W/n}^k(\widehat{I}_F^0)$ and $\lambda_{Wn}^k(\widehat{I}_F^0)$.

The upper level problem is created by fixing the state variables I_F^k of each subsystem $k = 1, N_S$ at the values obtained by solving the problems of the lower level:

$$\min_{I_F^0} L(I_F^0) = \widehat{F} + \widehat{\Lambda}_{W/n}^T \left(J_{L/n} - \widehat{J}_{G/n} + M_{/n}^{/0} \widehat{I}_F^{/0} + M_{/n}^0 I_F^0 \right) + \widehat{\Lambda}_{W/n} \left(j_{Ln} - \widehat{j}_{Gn} + m_n^{/0} \widehat{I}_F^{/0} + m_n^0 I_F^0 \right) \tag{12.18}$$

By grouping the terms with variable and fixed values in (12.18) we obtain the expression

$$\min_{I_F^0} L\left(\widehat{J}_G, I_F^0, \widehat{\Lambda}_{W/n}, \widehat{\lambda}_{W/n}\right) = \widehat{\Lambda}_{W/n}^T M_{/n}^0 I_F^0 + \widehat{\lambda}_{Wn} m_n^0 I_F^0 + Const \tag{12.19}$$

where $Const = \widehat{F} + \widehat{\Lambda}_{W/n}^T \left(J_{L/n} - \widehat{J}_{G/n} + M_{/n}^{/0} \widehat{I}_F^{/0} \right) + \widehat{\lambda}_{Wn} \left(j_{Ln} - \widehat{j}_{Gn} + m_n^{/0} \widehat{I}_F^{/0} \right)$. The solution of the problem (12.19) is used to determine the values of currents on the interconnecting branches I_F^0.

12.6.2 Decomposition Algorithm with Functional Characteristics Only Considering Equality Constraints

Let us consider more in detail the algorithm for solving the problems of each level. For subsystem k the functional characteristics have the following form

$$\Lambda_{W/n}^k = 2A_{/n}^k \left(J_{L/n}^k + M_{/n}^k \widehat{I}_F^k + M_{/n}^{0k} I_F^0 \right) + A_{2/n}^k \tag{12.20}$$

$$\lambda_{Wn}^k = 2a_{1n}^{xk} \left(m_n^k \widehat{I}_F^k + m_n^{0k} I_F^0 + j_{Ln}^k \right) + a_{2n}^{xk} \tag{12.21}$$

where $M_{/n}^{0k}$ is the block of matrix $M_{/n}^0$ which only contains rows and columns corresponding to the nodes and branches subsystem k, m_n^{0k} is the row of matrix $M_{/n}^{0k}$ which corresponds to the slack node of subsystem k and contains only columns corresponding to interconnecting branches. Note that the index n indicates that these values are related to a slack node of either the entire network, which is marked by the superscript 0, or the subsystem which is also marked by an additional superscript k. Let us consider more in detail the algorithm for solving the problem of the upper level by writing the required optimality conditions for the branch currents in function (12.19):

$$\frac{\partial L}{\partial I_F^0} = \left(M_{/n}^0 \right)^T \widehat{\Lambda}_{Wn} + \left(m_n^0 \right)^T \widehat{\lambda}_{Wn} = 0 \tag{12.22}$$

By substituting the functional characteristics of each subsystem (12.20)–(12.21) into (12.22) we obtain the following system of linear equations:

$$\begin{aligned} \frac{\partial L}{\partial I_F^0} = {} & \left(M_{/n}^0\right)^T \left(2A_{1/n}\left(J_{L/n} + M_{/n}^{/0}\widehat{I}_F^{/0} + M_{/n}^0 I_F^0\right) + A_{2/n}\right) \\ & + \left(m_n^0\right)^T \left(2a_{1n}^{xN_S}\left(m_n^{/0}\widehat{I}_F^{/0} + m_n^0 I_F^0 + j_{Ln}^{N_S}\right) + a_{2n}^{xN_S}\right) = 0 \end{aligned} \tag{12.23}$$

Expression (12.23) can be converted into the following form

$$CI_F^0 = B \tag{12.24}$$

with $C = 2(M_{/n}^0)^T A_{1/n} M_{/n}^0 + 2(m_n^0)^T a_{1n}^{xN_S} m_n^0$ and $B = -2\left(M_{/n}^0\right)^T (A_{1/n}(J_{L/n} + M_{/n}^{/0}\widehat{I}_F^{/0}) + A_{2/n}) - 2\left(m_n^0\right)^T \left(a_{1n}^{xN_S}\left(m_n^{/0}\widehat{I}_F^{/0} + j_{Ln}^{N_S}\right) + a_{2n}^{xN_S}\right)$.

The matrix C has the size $(L^0 \times L^0)$ and the matrix B has the size $(L^0 \times 1)$.

The decomposition algorithm with functional characteristics and consideration of equality constraints has the following steps.

1. Set the iteration index $m = 0$. Initial approximation of currents in the inter-system branches $I_{F\,m}^0$.
2. Calculation of variable values for each subsystem and formation of functional characteristics by use of (12.20) and (12.21).
3. Calculation of the currents on the interconnecting branches $I_{F\,m}^0$ by solving the system of linear equations (12.24). Increase the iteration index $m = m + 1$.
4. Check of termination criterion. If $\left\|I_{F\,m}^0 - I_{F\,m+1}^0\right\| > \xi$, go to step 2. Otherwise terminate the algorithm. $\xi$is a specified accuracy.

12.6.3 Decomposition Algorithm with Equivalent Characteristics Considering Only Equality Constraints

Let us consider an algorithm for solving the decomposition problem is not based on functional characteristics but on a function of equivalent fuel costs of each subsystem. Note that for solving the problem for subsystem k the variable parameters are state variables, i.e. generation currents J_G^k and currents in the interconnecting branches I_F^0. Let us consider the lower level problems at which each subsystem k consists of N_T problems of the form:

$$\min_{J_G^k} F^k = \sum_{i=1}^{N^k} F_i^k(j_{Gi}^k) \tag{12.25}$$

$$M^k_{/n} I^k_F + J^k_L - J^k_G + M^0_{/n} I^0_{F\,t} = 0 \tag{12.26}$$

$$m^{/0}_n I^{/0}_F + j_{Ln} - j_{Gn} + m^0_n I^0_{F\,t} = 0 \tag{12.27}$$

where $I^0_{F\,t} = I^0_{F\,0} + t \cdot \Delta I^0_F$, $t = 1, N_T$, $I^0_{F\,0}$ is a matrix consisting of the initial value of the of currents on the interconnecting branches and ΔI^0_F is a specified increment of the elements of the matrix I^0_F. Each problem solution is represented by a pair of numbers: the total fuel cost function $F^k_{opt\,t} = \sum_{i=1}^{N^k} F^k_i\left(j^k_{G\,i\,opt}(I^0_{F\,t})\right)$ and the sum of optimal values of the generator currents $G^k_t = \sum_{i=1}^{N^k} j^k_{G\,i\,opt}(I^0_{F\,t})$. In total there are N_T pairs of values of the form $\left\{F^k_{opt\,t}, G^k_t\right\}$. The least squares is used to find the quadratic approximation of the equivalent fuel cost function of subsystem k:

$$F^k_{eq}(G^k) = a^k_{1\,eq}(G^k)^2 + a^k_{2\,eq} G^k + a^k_{3\,eq} \tag{12.28}$$

On the upper level each subsystem k is represented by the total load $L^k = \sum_{i=1}^{N^k} j^k_{L\,i}$, the total generation G^k and the equivalent fuel cost function $F^k_{eq}(G^k)$.

The upper level problem is composed as follows:

$$\min_G F = \sum_{k=1}^{N_S} F^k_{eq}(G^k) \tag{12.29}$$

$$M^0_{/n} I^0_F + J^0_L - J^0_G = 0 \tag{12.30}$$

$$j^0_{Ln} - j^0_{Gn} + m^0_n I^0_F = 0 \tag{12.31}$$

where $J^0_L$is a matrix of the size $(N_S - 1 \times 1)$ containing the total load currents in the subsystems, J^0_G is a matrix of size $(N_S - 1 \times 1)$ containing the total generator currents in the subsystems.

The decomposition algorithm with equivalent characteristics only taking into account equality constraints consists of the following steps.

1. Construction of equivalent characteristics for each subsystem $k = 1, N_S$ by solving the current flow optimization problem (12.25)–(12.27).
2. Determination of optimal values of the total generation G^k for each subsystem $k = 1, N_S$ and the currents on the interconnecting branches I^0_F by solving the problem (12.29)–(12.31).
3. Calculation of optimal values for the generation currents J^k_G of each subsystem $k = 1, N_S$ for the currents I^0_F. Obtained in the previous step.

This algorithm needs only one iteration.

12.6.4 Decomposition Algorithm with Functional Characteristics that Considers Equality and Inequality Constraints

The objective function of this optimization problem is formulated in the following way:

$$\min_{J_G} F = \sum_{k=1}^{N_S} \sum_{i=1}^{N_G^k} F_i^k(j_{Gi}^k) \tag{12.32}$$

with the following constraints for each subsystem $k = 1, N_S$

$$M_{/n}^k I_F^k + J_L^k - J_G^k + M_{/n}^0 I_{Ft}^0 = 0 \tag{12.33}$$

$$m_n^{/0} I_F^{/0} + j_{Ln} - j_{Gn} + m_n^0 I_{Ft}^0 = 0 \tag{12.34}$$

$$I_{F\min}^0 \le I_F^0 \le I_{F\max}^0 \tag{12.35}$$

where $I_{F\min}^0$, $I_{F\max}^0$ are matrices containing the minimum and maximum values of the currents on the interconnecting branches I_F^0.

The Lagrange function for the problem (12.32)–(12.35) is

$$\begin{aligned}\min_{J_G, I_F} L = &\sum_{k=1}^{N_S} \sum_{i=1}^{N_G^k} F_i^k(j_{Gi}^k) + \sum_{k=1}^{N_S} \sum_{i=1}^{N^k} \lambda_{Wi}^k(j_{Li}^k - j_{Gi}^k + m_i^k I_F^k) \\ &+ \lambda_{Wn}(j_{Ln} - j_{Gn} + m_n I_F) + \sum_{j=1}^{N_F} \mu_{Fj}(I_{Fj} - I_{F\max j})\end{aligned} \tag{12.36}$$

where m_i^k is row i of matrix $M_{/n}^k$ and has L^k elements, λ_{Wi}^k is component i of the vector of the Lagrange multipliers $\Lambda_{W/n}$ and has $(N-1)$ elements, λ_{Wn} is the Lagrange multiplier corresponding to the balance equation of currents at the slack node n and μ_{Fj} is the component j of the vector of the Lagrange multipliers Ξ_F and has L^0 elements.

The Lagrange function (12.36) can be converted into a matrix form

$$\begin{aligned}\min L(J_G, I_F, \Lambda_{W/n}, \lambda_{Wn}) = &F + \Lambda_{W/n}^T(J_{L/n} - J_{G/n} + M_{/n} I_F) \\ &+ \lambda_{Wn}(j_{Ln} - j_{Gn} + m_n I_F) + \Xi_F^T(I_F - I_{F\max})\end{aligned} \tag{12.37}$$

where $J_{G/n}$ is the vector of generation currents consisting of $\sum_{k=1}^{N_S} N_G^k$ elements.

The optimality conditions of function (12.37) coincides with the conditions (12.22)–(12.24), written for function (12.19), while the optimality condition for currents on interconnecting lines different to condition (12.22) and can be written as follows:

$$\frac{\partial L}{\partial I_F} = M_{/n}^T \Lambda_{w/n} + m_n^T \lambda_{Wn} + \Xi_F = 0 \tag{12.38}$$

With consideration of the active inequality constraint for the current on the interconnecting branch $i - j$ condition (12.38) can be written as $\lambda_i - \lambda_j + \mu_{i-j} = 0$, where μ_{i-j} is the Lagrange multiplier corresponding to the considered active inequality constraint. If the constraint is not considered, condition (12.38) can be written as $\lambda_i - \lambda_j = 0$.

Since the set of considered active inequality constraints imposed on interconnecting branches is not known in advance a solution can be obtained by use of the problem

$$\min\left(\Lambda_w^T \Lambda_w\right) \mid I_{F\min}^0 \leq I_F^0 \leq I_{F\max}^0 \tag{12.39}$$

where Λ_w is the expanded matrix of the Lagrange multipliers of the size N_S, which consists of the sub-matrices $\Lambda_{w/n}$ and $\lambda_{w/n}$, each of them containing functional characteristic of the form (12.20) and (12.21).

A decomposition algorithm with functional characteristics that takes into account equality and inequality constraints can be split up into the following steps.

1. Set the iteration index $m = 0$. Initial approximation of currents in the interconnecting branches $I_{F\,m}^0$.
2. Calculation of variable values for each subsystem and formation of functional characteristics by use of (12.20) and (12.21).
3. Calculation of the currents on the interconnecting branches $I_{F\,m}^0$ by solving the optimization problem (12.39). Increase the iteration index $m = m + 1$.
4. Check of termination criterion. If $\left\| I_{F\,m}^0 - I_{F\,m+1}^0 \right\| > \xi$, go to step 2. Otherwise terminate the algorithm. ξ is a specified accuracy.

12.6.5 Decomposition Algorithm with Equivalent Characteristics that Considers Equality and Inequality Constraints

This decomposition algorithm has a two-level structure. At the lower level we form an equivalent fuel cost function for each subsystem $k = 1, N_S$ and a feasibility region characteristic. In order to construct the equivalent characteristic of subsystem k, first of all, we find a range of change in the total generation current $\left(\tilde{I}_{G\min}^k \div \tilde{I}_{G\max}^k\right)$, which will further be called the characteristic of the feasibility region of subsystem k. Without considering the current constraints in the branches of subsystem k the range is determined as

$$\tilde{I}^k_{G\min} = \min\left(I^0_{F\min}, \sum_{i=1}^{N_k} J^k_{Gi\min}\right), \tilde{I}^k_{G\max} = \min\left(\sum_j I^0_{Fj\max}, \sum_{i=1}^{N_k} J^k_{Gi\max}\right) \quad (12.40)$$

with $I^0_{F\,t} = I^0_{F\,0} + t \cdot \Delta I^0_F$ and $t = 1, N_T$. $I^0_{F\,0}$ is a specified initial value of the current matrix of the interconnecting branches and ΔI^0_F is the increment of the elements of the matrix I^0_F, which is determined as $\Delta I^0_F = (\tilde{I}^0_{F\max} - \tilde{I}^0_{F\min})/N_T$.

To construct the equivalent characteristic of subsystem k we solve N_T problems of the form

$$\min_{J^k_G} F^k = \sum_{i=1}^{N^k} F^k_i(j^k_{G\,i}) \quad (12.41)$$

$$M^k_{/n} I^k_F + J^k_L - J^k_G + M^0_{/n} I^0_{F\,t} = 0 \quad (12.42)$$

$$m^{/0}_n I^{/0}_F + j_{Ln} - j_{Gn} + m^0_n I^0_{F\,t} = 0 \quad (12.43)$$

$$J^k_{G\min} \le J^k_G \le J^k_{G\max} \quad (12.44)$$

$$I^0_{F\min} \le I^0_F \le I^0_{F\max} \quad (12.45)$$

where $J^k_{G\min}, J^k_{G\max}$ are matrices containing the minimum and maximum values of the generation currents J^k_G of subsystem k and $I^0_{F\min}, I^0_{F\max}$ are matrices containing the minimum and maximum values of the currents on the interconnecting branches I^0_F.

Each solution of the problem (12.41)–(12.45) is represented by a pair of numbers constituting the sum of the fuel cost function $F^k_{opt} = \sum_{i=1}^{N^k} F^k_i(j^k_{G\,i\,opt}(I^0_{F\,t}))$ and the sum of the optimal values of the generation currents $G^k = \sum_{i=1}^{N^k} j^k_{G\,i\,opt}(I^0_{F\,t})$. In total there are N_T pairs of values of the form $\left\{F^k_{opt}, G^k\right\}$. Furthermore, we will use the least square method to find a quadratic approximation of the equivalent fuel cost function of subsystem k of the form

$$F^k_{eq}(G^k) = a^k_{1\,eq}(G^k)^2 + a^k_{2\,eq} G^k + a^k_{3\,eq} \quad (12.46)$$

On the upper level each subsystem k is represented by the total load $L^k = \sum_{i=1}^{N^k} j^k_{L\,i}$, the equivalent fuel cost function $F^k_{eq}(G^k)$ and the characteristic of the feasible region $\left(\tilde{I}^k_{G\min} \div \tilde{I}^k_{G\max}\right)$.

The problem of the following type is solved on the upper level

$$\min_G F = \sum_{i=1}^{N^k} F^k_{eq}(G^k) \quad (12.47)$$

$$M^0_{/n} I^0_F + J^0_L - J^0_G = 0 \tag{12.48}$$

$$j^0_{Ln} - j^0_{Gn} + m^0_n I^0_F = 0 \tag{12.49}$$

$$\tilde{I}^k_{G\min} \le G^k \le \tilde{I}^k_{G\max}, \quad k = 1, N_S \tag{12.50}$$

$$I^0_{F\min} \le I^0_F \le I^0_{F\max} \tag{12.51}$$

The matrix of currents I^0_F the interconnecting branches is determined by solving problem (12.47)–(12.51). The Decomposition Algorithm with Equivalent Characteristics that Considers Equality and Inequality Constraints algorithm consists of the following steps:

1. Construction of the equivalent characteristics and the characteristics of the feasibility region for each subsystem $k = 1, N_S$ based on the solution of the problem (12.41)–(12.45).
2. Determination of the currents I^0_F on the interconnecting branches by solving the problem (12.47)–(12.51).
3. Calculation of state variables for each subsystem $k = 1, N_S$ based on the currents I^0_F obtained in the previous step.

12.7 Formulation of the Short-Term Optimization Problem

Approaches that are based on application of functional and equivalent characteristics are used to solve short-term optimization problems [11, 12]. For their illustration we introduce the following notation and terms: N_{int} is the number of time intervals, into which the control cycle or period is divided; N_T—is the number of thermal power plants (TPP); N_G—is the number of hydro power plants (HPP); P^t_{Ti} and P^t_{Gj}—are active power injections of the TPP i and HPP j in during the time interval t; P^t_H—is the load in the time interval t; $B^t_i(P^t_{Ti}) = a^t_{1i} {P^t_{Ti}}^2 + a^t_{2i} P^t_{Ti} + a^t_{3i}$—is the fuel cost function and $\varepsilon^t_{Ti} = \frac{dB^t_i}{dP^t_{Ti}} = 2a^t_{1i} P^t_{Ti} + a^t_{2i}$—is the characteristic of relative incremental costs for TPP i in during the interval t; $Q^t_i(P^t_{Gi}) = b^t_{1i} P^{t2}_{Gi} + b^t_{2i} P^t_{Gi} + b^t_{3i}$—is the fuel cost function; $\varepsilon^t_{Gi} = \frac{dQ^t_i}{dP^t_{Gi}} = 2b^t_{1i} P^t_{Gi} + b^t_{2i}$—is the characteristic of the relative incremental costs for HPP i in during the time interval t; and ΔV_j—is the specified water inflow of HPP j.

The short-term optimization problem is formulated as follows:

$$\min_{P_T} \sum_{t=1}^{N_{int}} \sum_{i=1}^{N_T} B^t_i(P^t_{Ti}) \tag{12.52}$$

$$P_H^t - n_T^T P_T^t - n_G^T P_G^t = 0, \quad t = 1, N_{int} \tag{12.53}$$

$$\Delta V_j - \sum_{t=1}^{N_{int}} n_G^t Q_{Gj}^t \left(P_{Gj}^t \right), \quad j = 1, N_G \tag{12.54}$$

where n_T and n_G—are the column vectors with unit components, whose dimensions are equal to N_T and N_G, respectively.

12.7.1 Algorithm for Solving the Short-Term Optimization Problem Without Decomposition

The optimality conditions required for the Lagrange function of the problem (12.52)–(12.54) have the following form for $t = 1, N_{int}$:

$$\frac{\partial L}{\partial P_{Ti}^t} = \varepsilon_{Ti}^t - \lambda_P^t = 0, \quad i = 1, N_T \tag{12.55}$$

$$\frac{\partial L}{\partial P_{Gj}^t} = -\lambda_P^t - \lambda_G^t \varepsilon_{Gj}^t = 0, \quad j = 1, N_G \tag{12.56}$$

$$\frac{\partial L}{\partial \lambda_P^t} = P_H^t - n_T^T P_T^t - n_G^T P_G^t = 0 \tag{12.57}$$

$$\frac{\partial L}{\partial \lambda_G^t} = \Delta V_j - \sum_{t=1}^{N_{int}} n_G^T Q_{Gj}^t \left(P_{Gj}^t \right) = 0, \quad j = 1, N_G \tag{12.58}$$

where λ_P^t and λ_G^t—are Lagrange multipliers related to the equality constraints on the active power balance in time interval t and the water flow rate balance for the control cycle of HPP j in time interval t, respectively.

These conditions can be written as a system of linear equations with a square matrix of the degree $(N_T * N_{int} + N_{int} + N_G)$. The solution of this system is used to determine active power injections of power plants and values of the Lagrange multipliers at the optimality point.

12.7.2 Algorithm for Solving the Short-Term Optimization Problem with Decomposition

Based on the application of the group relaxation principle [12] and the assumption that the initial values of active power injections P_{Gi}^t of HPP are satisfying the constraints (12.54), the initial problem of short-term optimization can be divided into N_{int} problems, which can be solved independently of each other. Hence, this approach can be treated as two level decomposition of the initial problem. On the first level the number of subsystems is equal to the number of intervals N_{int}. Either the equivalent characteristic $B^t_{eq}\left(P^t_H - \sum_{j=1}^{N_G} P^t_{Gj}\right)$ or the characteristic of relative incremental costs $\lambda^t_{eq}\left(P^t_H - \sum_{j=1}^{N_G} P^t_{Gj}\right)$ is formed for each subsystem as a function of the active power injection of the HPPs. At the second level the coordinating problem is solved and its solution is used to improve the values of HPP active power injections P^t_{Gi}.

Let us consider the algorithm for solving the short-term optimization problem more in detail. The equivalent fuel cost functions $B^t_{eq}(P_H)$ or the equivalent characteristics of relative incremental costs $\varepsilon^t_{eq}(P_H)$ is constructed for each interval $t = 1, N_{int}$. The algorithm of constructing such characteristics consists of in the following way. Let $\Delta P_{II} = \frac{P_{Hmax} - P_{Hmin}}{N_{points}}$, where N_{points}—is the number of points for calculating the equivalent characteristic values, and the number of iterations is initiated with $k = 0$.

1.
$$P^k_H = P_{Hmin} + (k+1) * \Delta P_H$$

2. Determine the optimal problem solution for each point $k = 1, N_{\text{points}}$

$$\min_{P_T}\left(P_T^T A_1 P_T + A_2^T P_T + A_3\right) \tag{12.59}$$

$$P^k_H - n_T^T P_T = 0 \tag{12.60}$$

where A_1—is the diagonal matrix of the degree N_T, for which the i-th diagonal element is equal to the coefficient a_{1i} of the fuel cost function for TPP i; A_1 and A_2—are column matrices of the degree N_T, for which the elements in row i are equal to the coefficients a_{2i} and a_{3i} of the fuel cost function for TPP i.

The necessary optimality conditions for the Lagrange function of the problem (12.59)–(12.60) can be represented as a system of linear equations:

$$\begin{pmatrix} 2A_1 & -n_T \\ -n_T^T & 0 \end{pmatrix}\begin{pmatrix} P_T \\ \lambda \end{pmatrix} = \begin{pmatrix} -A_2 \\ -P^k_H \end{pmatrix} \tag{12.61}$$

The optimal values of active power generation P_{Topt} of the power plants and the Lagrange multiplier λ_{opt} are obtained at the solution point of system (12.61). To construct the equivalent characteristic a pair of values either (P_{opt}, P^k_H) or (λ_{opt}, P^k_H) that determine point k characteristic $B^t_{eq}(P_H)$ or characteristic $\varepsilon^t_{eq}(P_H)$ are preserved.

The analytic description of the equivalent characteristics is obtained by using the least squares method. The equivalent characteristics are represented by a quadratic polynomial of the form

$$B^t_{\mathrm{eq}}\left(P^t_{Hred}\right) = a^t_{1\mathrm{eq}}\left(P^t_{Hred}\right)^2 + a^t_{2\mathrm{eq}}P^t_{Hred} + a^t_{3\mathrm{eq}} \tag{12.62}$$

Where P^t_{Hred}—is the reduced load in the interval t that is equal to $P^t_H - \sum_{j=1}^{N_G} P^t_{Gj}$. Properties of the equivalent characteristics $B^t_{eq}(P_H)$ and $\varepsilon^t_{eq}(P_H)$ are described in the following. These characteristics have the following relation:

$$\frac{dB_{eq}(P_H)}{dP_H} = \varepsilon_{eq}(P_H) \tag{12.63}$$

It results first, from the method for constructing the characteristics and second, from interpretation of the Lagrange multiplier value.

If the load P_H is known, then the corresponding optimal value of the Lagrange multiplier $\lambda_{opt} = \varepsilon_{eq}(P_H)$ is also known. As noted above, the following relation is met at the solution point for each generator i

$$\varepsilon_{Ti} = \frac{dB_i}{dP_{Ti}} = 2a_{1i}P_{Ti} + a_{2i} = \lambda_{\mathrm{opt}} \tag{12.64}$$

The optimal values of active power generation $\mathrm{P_{Ti}, i = 1, N_T}$ are determined on the basis of this relation.

The formulation of the upper level coordination problem with equivalent characteristics has the following form:

$$\min_{P_G} \sum_{t=1}^{N_{int}} B^t_{eq}\left(P^t_H - \sum_{j=1}^{N_G} P^t_{Gj}\right) \tag{12.65}$$

$$\sum_{t=1}^{N_{int}}\left(n^T_G P^t_{Gj} - \Delta V_j\right) = 0, \quad j = 1, N_G \tag{12.66}$$

The conditions for the Lagrange function of the problem (12.65)–(12.66) are represented as a system of linear equations of the form:

$$\begin{pmatrix} 2A_{1G} & -N \\ -N^T & 0 \end{pmatrix}\begin{pmatrix} P_G \\ \Lambda_G \end{pmatrix} = \begin{pmatrix} A_{2G} \\ -\Delta V_G \end{pmatrix}, \tag{12.67}$$

where P_G—is a column matrix of the degree $N_{int}N_G$ that contains the sequential values of active power injections in each interval for each HPP $\left\{P^1_{G1}, \ldots, P^{N_{int}}_{G1}, \ldots, P^1_{GN_G}, \ldots, P^{N_{int}}_{GN_G}\right\}$; Λ_G—is a column matrix of the degree N_G that contains the values of the Lagrange multipliers corresponding to the constraints (12.66); A_{1G}—is a square matrix of the degree $N_G N_{int}$, whose rows correspond to the derivative $\frac{d\sum B^t_{eq}}{dP^t_{Gj}}$ has and contain non-zero values. The columns correspond to the variables P^t_{Gj} for $j = 1, N_G$; N—is a matrix with $N_G N_{int}$ rows and N_G columns and each column j in it has unities in rows from $(j-1) * N_G + 1$ to

$j * N_G$ and zeros in the rest of the rows; A_{2G}—is a column matrix of the degree $N_G N_{int}$, whose rows correspond to the derivative $\frac{d\sum B^t_{eq}}{dP^t_{Gj}}$ which is equal to $2a^t_{1eq}P^t_H + a^t_{2eq}$; ΔV_G—is a column matrix of the degree N_G, and the element of row j is equal to the value of the water inflow ΔV_j for of HPP j.

The active power generation P^t_{Gj} of a HPP with $j = 1, N_G$ and $t = 1, N_{int}$ and as well as the Lagrange multipliers $\Lambda_G = \{\lambda_{Gj}\}, \quad j = 1, N_G$ will be determined from the solution of system (12.67).

The first derivative is determined for each interval for the calculated values of the HPP active power generation P^t_G with respect to the equivalent characteristic $\frac{dB^t_{eq}}{d\sum_{j=1}^{N_G} P^t_{Gj}}$, whose value is equal to the optimal value of the Lagrange multiplier λ^t_{opt}, is determined for each interval for the calculated values of HPP active powers. In accordance with (12.64) the optimal values of the TPP active power generation can be determined in each interval by the relation

$$P^t_{Ti} = \frac{\lambda^t_{opt} - a^t_{21}}{2a^t_{1i}}, \quad i = 1, N_T \tag{12.68}$$

The short-term optimization problem can be solved by the above presented algorithms without any iterations and with or without decomposition of the initial problem.

12.7.3 After-Effect Function for Correction of the Short-Term Optimization

Planning and optimization of short-term operating conditions is solved with a lead time between one day and a week. Deviations between the planned operation and the real-time operation are inevitable. An approach for correction of the scheduled values is suggested in [13]. Let the water inflow ΔV for HPPs, the fuel cost functions $B^t_i(P_T)$ for TPPs and the average load curve P^t_H, where with $t = 1, N_{int}$, be given. For each interval the equivalent characteristic $B^t_{eq}(P_G)$ is formed. Then the following problem is solved

$$\min_{P_G} \sum_{t=1}^{N_{int}} B^t_{eq}(P^t_G) \tag{12.69}$$

$$\sum_{t=1}^{N_{int}} n^T_G Q^t_G(P^t_G) - \Delta V = 0 \tag{12.70}$$

The solution of this problem determines estimates of the optimal water inflow of HPPs for each interval.

If more accurate information is available for the current interval t^i, for example about the consumption P^t_H, the water inflow ΔV, then the optimal values of P^t_G, P^t_T that were obtained earlier during the planning can be corrected. One of the approaches to solve the correction problem is based on the construction of after-effect functions which are linear equivalent functions. In this case the control cycle or period is divided into two intervals. The first interval t^i_1 is the present interval t^i and the second interval t^i_2 continues from the end of the first interval to the end of the period $t^i_2 = (N_{int} - t^i)$. In this case the after-effect function $B^{t^i_2}_{ae}\left(P^{t^i}_G\right)$ depends on the optimal total fuel consumption in interval t^i_2 based on the HPP active power $P^{t^i}_G t^i_1$.

Thus, the after-effect function consist of the derivative

$$B^{t^i_2}_{ae} = \frac{dB^{t^i_2}_{opt}}{dP^{t^i_1}_G} \tag{12.71}$$

that shows how the total optimal fuel consumption will change in the second interval of the length $(N_{int} - t^i)$ at caused by changing power generation of the HPPs in the first or present interval.

At first a family of problems is solved to construct the after-effect function. This is done by solving the following problem for each fixed value of P^k_G the following problem is solved:

$$B^{t^i_2}_{opt}(P^k_G) = \min_{P_G} \sum_{t=t^i+1}^{N_{int}} B^t_{eq}(P^t_G) \tag{12.72}$$

$$\sum_{t=t^i+1}^{N_{int}} P^t_G - \left(\Delta V - \sum_{t=1}^{t^i-1} P^t_G - P^{t^i}_G\right) = 0 \tag{12.73}$$

After determination of the analytic form of the after-effect function, for example by use of the least squares method, the following planning problem is solved for the interval i:

$$\min_{P^{t^i}_G}\left(B^{t^i}_{opt}\left(P^{t^i}_G\right) + B^{t^i_2}_{a\_e}\left(P^{t^i}_G\right)\right) \tag{12.74}$$

Note that the specification of information requires an adjustment of the optimal characteristic in the current interval.

12.8 Numerical Results

In this section the optimal current flow problem is solved with and without consideration of inequality constraints on the example of the test network scheme, whose graph is shown in Fig. 12.1 an parameters in Table 12.1.

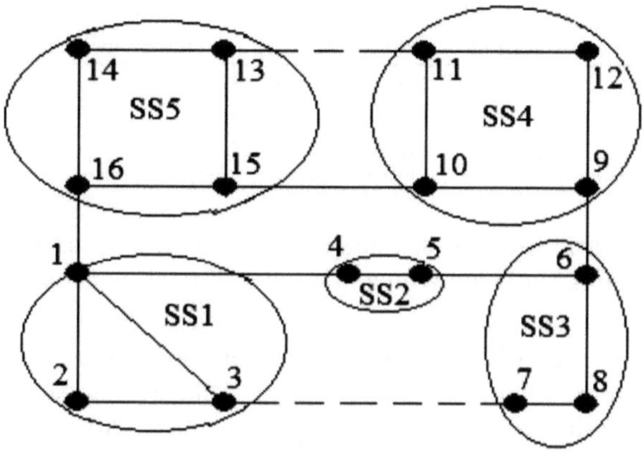

Fig. 12.1 Scheme of the test network

Table 12.1 Electric network parameters

S/s: k	Nodes	$F_i^k(j_{Gi}^k) = a_{1i}^{xk}(j_{Gi}^k)^2 + a_{2i}^{xk} j_{Gi}^k + a_{3i}^{xk}$	Load : j_{Li}^k, [A]
1	1	$0.05j_{G1}^2 + 2j_{G1}$	20
2	2	$0.1j_{G2}^2 + j_{G2}$	20
1	3	$0.05j_{G3}^2 + 2j_{G3}$	20
2	4	$0.02j_{G4}^2 + 0.5j_{G4}$	20
3	5	$0.015j_{G5}^2 + j_{G5}$	20
3	6	$0.1j_{G6}^2 + 0.5j_{G6}$	20
3	7	$0.5j_{G7}^2 + 2j_{G7}$	20
4	8	$0.25j_{G8}^2 + 0.5j_{G8}$	20
4	9	$0.05j_{G9}^2 + j_{G9}$	20
1	10	$0.05j_{G10}^2 + 2j_{G10}$	20
2	11	$0.02j_{G11}^2 + 0.5j_{G11}$	20
3	12	$0.015j_{G12}^2 + j_{G12}$	20
3	13	$0.1j_{G13}^2 + 0.5j_{G13}$	20
3	14	$0.5j_{G14}^2 + 2j_{G14}$	20
4	15	$0.25j_{G15}^2 + 0.5j_{G15}$	20
4	16	$0.05j_{G16}^2 + j_{G16}$	20

The network scheme consists of 16 nodes and 11 branches out of which 5 branches are interconnectors between subsystems. In order to be able to model different types of electric networks (open-loop, meshed) two interconnecting branches indicated by a dashed line in Fig. 12.1 are assumed to be switched branches. The resistance of all branches is identical and equal to 1 Ω. The electric network is divided into 5 subsystems (SS). At each node i of subsystem k load currents j_{Li}^k and generator currents j_{Gi}^k are defined. The fuel cost function $F_i^k(j_{Gi}^k)$ and the convex characteristic of relative incremental costs (the first derivative with respect to the fuel cost function) $E_i^k(j_{Gi}^k)$ are given for each generator at node i of subsystem k. For simplification of the calculations and to consider losses in the

Table 12.2 Results of current flow optimization with different algorithms

No.	Algorithm	Decomp.	F_{onm}	Iterations on lower level	Iterations on upper level
1	Lagrange	No	533.23		1
2	DFE	Func.	533.23	1	20
3	DEE	Equiv.	533.24	3	1

electric network the study is performed for currents. The goal is to determine an optimal solution with a minimum of total fuel costsF_{opt}. The problem variables are the injections $j_i^k = j_{Li}^k - j_{Gi}^k$ at each node i and the currents i_{Fl}^k in each branch l of subsystem k. The solutions obtained by using each of the above presented algorithms are presented in Table 12.1 and are compared with the solutions obtained as a result of the current flow optimization without decomposition (row No. 1) (by help of a routine program for solving nonlinear programming problems in terms of equality and inequality constraints).

For the DEE (Decomposition with Equivalent characteristics and Equality constraints) algorithm the equivalent fuel cost functions (row No.3) are obtained by a preliminary calculation of four optimal operating conditions for each subsystem. The calculation of the coordination problem requires a single iteration. It should be underlined that after the solution of the coordination problem there is no need to recalculate the equivalent characteristics. In the DFE (Decomposition algorithm with Functional characteristics and Equality constraints) algorithm (row No. 2) each subsystem is calculated in a single iteration. However, about 20 iterations are needed to determine the optimal values of the external variables and the current on the intersystem branches (Table 12.2).

The considered decomposition examples with the simplified representation of the network show that different external characteristics (functional or equivalent) allow for obtaining a solution of the initial control problem. The reason of the relatively slow convergence of the functional characteristics method in comparison with the equivalent characteristics method requires additional studies. Apparently this is explained by application of first order minimization methods such as the gradient.

12.9 Conclusion

Algorithms are suggested to determine the optimal current and load flow in electric networks by using decomposition, in which the initial problem is reduced to a two-level procedure. At the first level the initial network is divided into small-dimension subsystems and the optimal solution is determined for each of them. At the second level the overall coordination problem is solved based on the solutions obtained for each subsystem.

The properties of two decomposition algorithms were studied: he algorithm with construction of functional characteristics of the subsystems and the algorithm with generation of equivalent characteristics of the. It is revealed that the functional characteristics can be obtained by differentiation of equivalent characteristics. For the equivalent characteristics the mathematical description of the coordination problem and the optimization problems correspond to each other.

The convergence characteristics of the algorithms have been studied numerically for the equivalent and the functional characteristics of the calculated electric network schemes, in which loads were represented as current sources, and inequality constraints were not taken into consideration. Testing of the algorithms has shown that the functional characteristics method is more laborious. The convergence characteristic depends on the interconnections between the subsystems. Testing of the algorithms has shown that if an electric network does not contain loops, the solution requires a fewer number of iterations. Consideration is given to the algorithms for solving the short-term optimization problem, in which the control cycle is divided into several intervals. It is indicated that the application of the equivalent fuel cost function in each interval is efficient to solve the optimal load flow problem. On the whole, the analysis has shown the possibility of using decomposition of an electric network to solve the optimal current and load flow problems in an electric network.

References

1. N. Martynov, *Introduction to MATLAB 6: Manual* (KUDITs-OBRAZ, Moscow, 2002) (in Russian)
2. L. Lesdon, *D1.2.2 Optimization of Large Systems*. (Nauka, Moscow, 1975) (in Russian)
3. V.I. Tsurkov, *Decomposition in High-Dimension Problems* (Nauka, Moscow, 1981) (in Russian)
4. L. Krumm, *The reduced gradient methods in electric power system control* (Nauka, Novosibirsk, 1977) (in Russian)
5. L. Krumm, *The Optimization Methods in the Control of Power Systems* (Nauka, Novosibirsk, 1981) (in Russian)
6. V. Balanovski, A. Gamm, Y. Grumbkov, L. Krumm, O. Voitov et al., *The Theory and Methods of the Analysis and Control of Steady States in Power Systems* (Nauka, Novosibirsk, 1987) (in Russian)
7. A. Gamm, L. Krumm, I. Sher, Two algorithms for calculating steady states of complex electric power systems with division into subsystems. Izvestiya AN SSSR Energetika i transport **1**, 51–53 (1966) (in Russian)
8. L. Krumm, Analysis and optimization of control and reliability of the interconnected power system operation and development at the interstate level in free market conditions. Collected papers: methodological problems in study on large energy system reliability. Issue 55. Methodological and applied problems of energy system reliability, (ISEM SO RAN, Irkutsk, 2005) (in Russian)
9. L. Krumm, Development of the theory and methods for complex optimization of control and reliability of interconnected power system operation at the interstate level in free market conditions. Collected papers: methodological problems in study on large energy systems. Issue 56. Reliability problems in the restructured energy systems and methods for their solving. (ISEM SO RAN, Irkutsk, 2006) (in Russian)

10. O. Sukhanov, Y. Sharov, *Hierarchical Models in the Analysis and Control of Power System Operating Conditions*. (Izdatelsky dom MEI, Moscow, 2007) (in Russian)
11. V. Gornstein, *The Most Expedient Operating Conditions of Hydro Power Plants in Electric Power Systems* (GEI, Moscow, 1959) (in Russian)
12. L. Krumm, N. Murashko, Complex calculation of the feasible optimal short-term operating condition of electric power systems by using the reduced gradient method. Izvestiya AN SSSR. Energetika i transport **1**, 3–15 (in Russian)
13. E. Tsvetkov, T. Alyabysheva, L. Parfenov, *Optimal Operating Conditions of Hydro Power Plants in Electric Power Systems*. (Energoatomizdat, Moscow, 1984) (in Russian)

Part V
Stability and Protection Techniques in Interconnected Power Systems

Chapter 13
Protection Technologies

Anna Arestova and Andrey Grobovoy

13.1 Introduction

The general goals of system protection can be outlined as follows: maintaining the ability to deliver electric power, protection of equipment, keeping power system integrity and power quality, etc.

Hence, the objective of protective relaying is to selectively and reliably isolate faulty power system components in the shortest time possible. The clearing time depends on the kind of dynamic phenomena (compare Chap. 3) which the protective relay deals with. Usually protection devices are divided into two categories: local protection and wide-area protection systems. Local protection deals with local faulty conditions, while wide-area protection needs information from the overall system to determine whether a network situation is allowable or intolerable. Table 13.1 presents an overview over the most important protections devices and classifies them into one of the two categories.

Most of modern protection devices based on microprocessors combine the following functions: (a) protection, (b) monitoring, (c) reclosing, (d) regulating, (e) synchronization. The main types of digital and microprocessor based relaying functions are surveyed in the present chapter. The chapter consists of two parts. In the first part the local protection systems are observed. The second part examines the general direction and principles of wide area protection.

A. Arestova (✉) · A. Grobovoy
Power System Emergency Control Laboratory, Novosibirsk, Russia
e-mail: Ann.arestova@gmail.com

A. Grobovoy
e-mail: grobovoy@ieee.org

U. Häger et al. (eds.), *Monitoring, Control and Protection of Interconnected Power Systems*, Power Systems, DOI: 10.1007/978-3-642-53848-3_13,

Table 13.1 General components of protection system

Local protection relaying	Wide area protection system
• Overcurrent • Overvoltage • Undervoltage • Distance • Differential • Pilot • Gas-accumulator • Back-up • Loss of synchronism	• Fault clearance • Wide-area out-of-step • Wide-area oscillation damping • Adaptive protection • Adaptive self-healing

13.2 Local Protection Systems

The fundamental types of transmission device protection are **overcurrent, overvoltage, and undervoltage relays**. These protection types require a current or voltage transformer which simply measures the current or voltage in a circuit. Overload protection, for instance, is divided into two types: instantaneous overcurrent and time delayed overcurrent. Instantaneous overcurrent requires that the current exceeds a pre-determined level for the circuit breaker to operate. Time delayed overcurrent protection operates based on a current versus time curve.

Another type is **distance protection** that observes both voltage and current. A fault on a circuit will generally create a sag in the voltage level. If the ratio of voltage to current measured at the relay terminals is within a pre-determined level the circuit breaker will operate. If the relay setting is determined to be below the apparent impedance it is determined that the fault is within the zone of protection.

Differential relays can have a variety of different designs, depending on the equipment they protect. It might be a transmission line, a generator, a transformer, or a portion of a bus, etc. The main equipment needed are current transformers (CT) placed at both ends of the protected zone. The differential relay current is proportional to the vector difference between the currents entering and leaving the protected circuit; and, if the differential current exceeds the relays pickup value, the relay will operate.

Pilot relaying is an adaptation of the differential relaying principles for the protection of transmission line sections. Differential relaying is not appropriate for transmission line protection because the terminals of a line are highly separated to interconnect the CT secondaries. There is an interconnecting channel between the ends of the transmission line. Three different types of such a channel are presently in use: wire pilot, carrier-current pilot, and microwave pilot.

Gas-accumulator relays are applicable for oil-filled transformer. The principle of the protection is based on a gas-collecting chamber. There is a certain amount of gas which should be collected for a contact closing that usually sounds an alarm.

Back-up protection is assigned to protect the power system in case of a failure of primary protection. The primary protection has to remove only the affected items to clear the fault while the back-up protection might remove both the affected and unaffected items.

The **loss of synchronism relay** protects a generator against out-of-step conditions. Distance relays are often used to provide an out-of-step protection function, whereby it is called to provide blocking or tripping signals upon detection of an out-of-step condition.

13.3 Wide-Area Protection Systems

There are several conventional protection systems which are installed in a distributed way. The following types of protection systems belong to this group: under frequency load shedding, under voltage load shedding, remote load shedding, HVDC fast power change, automatic shunt switching, dynamic breaking, area isolation, etc.

Up-to-date satellite-based time-keeping systems provide new approaches for protection and control of electric power system with wide-area applications (compare Chaps. 4 and 5). The application of synchrophasor measurements in a protective relay results in increased power system reliability and provides easier disturbance analysis, protection, and control capabilities than do approaches with other information sources. For example the use of wide-area measurements allows for minimization of the error of line distance protection phase angles [1].

There is a large potential for advanced wide-area protection and control systems, based on powerful, flexible and reliable system protection terminals, high-speed communication, and GPS synchronization in conjunction with careful and skilled engineering by power system analysts and protection engineers.

According to [2] the major application areas of synchronized measurement technologies for wide area system protection are:

1. Real-time visualization of power systems
2. Design of an advanced warning system
3. Analysis of the causes of a total or partial blackout
4. Benchmarking, validation, and fine-tuning of system models
5. Enhancement in state estimation
6. Real-time congestion management
7. Real-time angular and voltage stability analysis and enhancement
8. Improved damping of inter-area oscillations
9. Design of an adaptive protection system.

All these protection systems have in common that they are based on the use of both local measurements and remote information, which allows categorizing them as System Integrity Protection Schemes (SIPSs). The concept of SIPSs is usually associated with centralized as well as local (decentralized) emergency control systems. Modern computing technologies, communication systems and measurement systems are required to create decentralized SIPSs systems [3]. This subsection is divided into five segments. Firstly, the fault clearance relay principles are described. The second part is devoted to modern out-of-step protection ideas. The third part observes principles and tools for wide area oscillation damping. The adaptive protection in modern power system is considered in the fourth segment. At last, adaptive self-healing is the subject of the fifth part.

13.3.1 Fault Clearance Relays

The idea of wide-area current differential backup protection employing broadband communications and time transfer systems is discussed in [4]. The basic difference compared to conventional backup protection is that wide-area current differential backup protection uses the PMU measurements of several power system devices communicated over a broadband network instead of comparing the currents only between the two ends of a single device. This provides better selectivity for the tripping of fault currents, smaller outage areas and faster operation times than conventional backup protection. In [5], the results of an experimental examination of a wide area backup protection system are presented. Several different approaches for a wide area backup protection system are available in the literature and discussed in the following.

A new concept for the design of a wide-area backup protection system is presented in [6]. The proposed system comprises a master system and several local units. A similar concept is also examined in [7]. This system is arranged in three layers. Connected via fiber-optic communication links, the devices can process intelligent algorithms based on locally collected data. This adds an intelligent analyzing and controlling function to the key process of power system backup relay protection.

A primary and backup cooperative protection system based on wide area agents is investigated in [8], in which agents are used to give each protection relay control capacity as well as the ability to communicate with other agents of relays. This method might become a new philosophy for primary and backup protection.

In [9] it is shown that wide-area measurements offer a possibility for preventing the remote back-up relays from tripping if load swings are interpreted as a fault by the relay. A concept of the wide area backup protection techniques for power transmission grids using PMUs is evolved in [10]. The relay described in this chapter represents a new state-of-the-art in the field of interconnected grid protection for the following reasons:

1. The relay is based on shared data from all areas
2. One relay is used instead of several standalone relays with different coordination complexity
3. The relay can provide unit protection by identification of the faulted zone
4. Only one tripping decision is issued from the protection center.

Another application of WAMS technology with reference to back-up relaying was described recently. In this case the fault components are suggested to form a wide area backup protection algorithm. In [11], the algorithm is based on the voltage distribution of the fault component. A steady state fault component-based method for precise localization of the fault branch is proposed in [12].

Today distance protection relays include fault-location capabilities. In the following the use of wide area measurements will be discussed to improve the precision and speed of this function. One of the possible concepts is presented in [13]. This advanced high-speed fault location concept is based on the use of voltage and current samples that are synchronously taken at both ends of the transmission line and can be implemented either as a simple add-on to the digital fault records or as a stand-alone new relaying function. This technique can be very fast, selective and accurate, providing fault analysis performance exceeding other known techniques [14]. It is important to take into account the equipment, especially PMUs, which is required for the fault location. Considering the installation cost of PMUs, it is important to investigate the placement scheme of the PMUs to allow for a minimal number of locations in the network. An algorithm for the determination of the place and the minimum number of PMUs allowing for detection of the accurate place of any fault in the power system is presented in [15].

Another algorithm for fault location on transmission lines using WAMS technologies is presented in [16]. It is based on linear state estimation where fault location and voltage at the fault point are added as a new state variable. The accuracy of the fault location strongly depends on the data procession. A tool to improve the data visualization of the fault location with traditional SCADA system data is presented in [17].

Another fault location method based on principal component analysis (PCA) is proposed in [18]. The principal components fault location is based on data provided by the WAMS technologies. The PCA method is an efficient mathematic tool to reduce dimension of PMU data matrix with huge information. Therefore, the state matrix, consisting of raw synchrophasor information at multi-points and various times, is compressed effectively. In general, the fault component can be determined exactly with the use of this method.

13.3.2 Wide-Area Out-of-Step Protection

Distance relays are often used to provide an out-of-step protection function by sending blocking or tripping signals upon the detection of out-of-step conditions. This section deals with possible improvements of precision and speed of out-of-step relays by use of wide area measurements.

One of the first realizations of out-of-step protection devices using measurements of the phase angle differences between substations is discussed in [19]. In [20] a coordinated out-of-step interface separation system is presented based on

WAMS for providing more comprehensive and synchronized data for the out-of-step protection. The global oscillation state in the power system can be observed from the systematic layer, so that the decision to separate the power system interface is more coordinated and accurate when the out-of-step situation happens in the interconnected system. If a detected oscillation center is located in the protected zone, the system gets split into two areas, avoiding a blackout in the entire interconnected power system.

An attractive solution is to adapt the security-dependability balance in response to changing system conditions based on real-time measurements from PMUs. Two types of reliability have been designated as "security" and "dependability" by protection engineers. The existing protection systems with their multiple zones of protection and redundant systems are biased towards dependability, i.e., a fault is always cleared by a certain relay. The result is a system that virtually always clears the fault but as a consequence permits larger numbers of false trips. High dependability is recognized as being a desirable protection principle when the power system is in a normal "healthy" state, and high-speed fault clearing is highly desirable in order to avoid instabilities in the network. The consequent price paid in occasional false trip is an acceptable risk under "normal system" conditions. In the system suggested in [9] the main principle of out-of-step detection is based on the evaluation of the system state by the use of WAMS technology.

Out-of-step protection is usually linked with islanding of parts of the system. A difficult task after islanding has happened is to determine the precise borders of the islands. To overcome this problem, an intelligent Instability Detection for Islanding Prediction, based on rotor angles, voltage magnitudes, current magnitudes, current angles and voltage angles, is proposed in [21]. The main philosophy of islanding detection in the proposed method is to use off-line performed decision trees and neural network data mining algorithms to determine the PMU locations, detection parameters, and their triggering values for islanding detection. With information from accurate system models PMUs can be used online to predict system islanding with high reliability.

Islanding can either be triggered by use of conventional out-of-step relays or by use of synchronized phasor measurements. The advantages and disadvantages of wide area measurements in this area of power system protection are discussed in the following.

When a power system is disturbed, the electric torque output and mechanical torque input of generators is in imbalance. If the disturbance is strong, a contingency involving the loss of several important transmission lines may trigger cascading events and possibly lead to a blackout. Controlled islanding is the last line of defense to stabilize the whole system. Reference [22] presents a decision tree assisted scheme to determine the timing of controlled islanding in real-time by use of phasor measurements. One can see that this method is an effective tool for resolving this problem.

The necessity and importance for the use of PMUs for monitoring and capturing the event of re-synchronization is shown in [23]. A synchroscope can be used to connect the islanded parts.

Based on WAMS an adaptive framework for out-of-step splitting is proposed in [24] which makes use of remote communication to obtain quick and selected disconnection actions. An adaptive out-of-step scheme for system separation is performed by employing PMU based WAMS data. The new system splitting strategy proposed in this chapter is based on the identification of generator coherency groups with the use of voltage angle differences.

After the occurrence of an islanding situation WAMS data can be used to improve the power system restoration process. In [25] a multi-agent approach for power system restoration is described. The proposed system consists of a number of bus agents (BAGs) and a single facilitator agent (FAG). The BAG is developed to propose a suboptimal target configuration after a fault occurrence, by interacting with other BAGs based on only locally available information. The FAG is used to act as a manager in the decision process. The power system restoration model assumes the execution of several rules that might be realized by use of WAMS technology.

GPS-synchronization technology in combination with WAMS technologies allows for more efficient power system state estimation (compare Chaps. 6 and 3). This is the most common and reliable method to achieve coordination with respect to system restoration [26]. The synchronized phasor measurement system provides a clear picture of most of the significant events that lead to a blackout and are further on aspects of the system restoration. The data recorded by the phasor measurement prototype, which is presented in [27], is proved to be a valuable tool for clarifying and adding new information to the data obtained in a conventional way.

For a more detailed discussion of this subject refer to Chap. 16.

13.3.3 Wide-Area Oscillation Damping

One very important task in power system operation is the ability to damp low frequency oscillations (compare Chap. 3). The most efficient way for that consists in the use of FACTS devices including HVDC systems. It is known that wide area measurements can be used both for emergency control and damping of low frequency oscillations in power systems. Two applications for HVDC technologies in Greece and Italy are described in [28]. As noted in this chapter there are several areas of application of HVDC systems such as:

1. Special dynamics control
2. Network frequency control
3. Small signal stabilization
4. Large signal stabilization
5. Damping of sub-synchronous resonance
6. Voltage regulation and compensation.

Wide-area coordinated HVDC modulation controllers to damp out inter-area oscillations for China Southern Power Grid of China are presented in [29]. Such inter-area oscillations can arise in so-called hybrid transmission systems i.e. a mixed transmission system comprising both AC and DC components. The way to perform parameter optimization for the HVDC stabilizers in this project is presented in [30].

A three-layer-architecture for system wide area control based on coordinated control of network controllers with optimal power flow and damping control with embedded HVDC is introduced in [31]. Wide Area Control System (WACS) technology is used for that. The basic damping behavior of a HVDC scheme utilizing PMU measurement values can be a very efficient tool for power oscillation damping. It can be seen that the prerequisite for a proper functioning of damping control is a high quality of data delivered by the WACS. The main idea is the use of a hierarchical scheme for the optimization. This scheme consists of a measurement layer, an economic layer and a network layer.

A comparison between local and remote signals in different control strategies for power oscillations damping in the control of a Voltage Source Converter HVDC (VSC-HVDC) is done in [32]. The use of VSC-HVDC is the most potential way for improving power system security. The location of VSC-HVDC and the selection of the local feedback signal for the damping is purposely determined so that a high residue magnitude of the inter-area mode is achieved. This allows local signals to provide as much damping as remote signals. However, the efficiency of the damping with local signals depends on the location of the VSC-HVDC.

The problem of communication delays in wide-area damping control to prevent power oscillations in HVDC/AC interconnected power systems is examined in [33]. It is shown that the wide-area damping controller can effectively damp inter-area oscillations in the parallel HVDC/AC interconnected system, even if there exists a certain time delay in the wide-area feedback control signals, the designed controller can still achieve good oscillation damping performance, and prevent the inter-area oscillations. Controllers which take into account time delay control might be useful in the near future.

One more power oscillation damping controller for VSC-HVDC systems is suggested in [34]. It is shown that for power oscillation damping the multi input single output modal linear quadratic Gaussian controller is very promising. This might allow assessment of the effectiveness of HVDC to damp inter-area oscillations within highly meshed networks.

The design of WAMS-based multiple HVDC damping control systems with reference to Central China power grid is examined in [35]. A control signal is generated from a damping controller and added to the reference signal of HVDC to regulate the active power transferred by the HVDC link during oscillations. Results obtained in a wide variety of operating conditions indicate that the wide-area HVDC damping controller designed in this chapter is effective and robust.

A new HVDC control concept which effectively copes with disturbances in weak systems is represented in [36]. This concept is termed as Combined and Coordinated Control Method (CCCM). According to this method the rectifier directly controls the DC power, and the inverter controls the equivalent resistance. The studies show that the operating point does not return directly to the pre-disturbance operating point when the conventional control concept is applied. However, direct movement can be observed when the new control is used. Such a concept might be efficient in the frame of modern trends of HVDC implementation.

13.3.4 Adaptive Protection

A further area for implementation of WAMS technologies is adaptive protection. The authors of [37] define adaptive protection as (1) "automatically altering, on-line, the protection system settings, functions and characteristics. This suggestion implies the need for computing and transmitting new operating parameters from a central site. Digital relays make on-line changes possible. An expanded communications system is required for realization", and (2) "finding optimal, condition dependent relay settings. i.e. settings which will result in the most desirable operation instead of a conventional trade-off between conflicting requirements, such as dependability and security, or for different load conditions".

A further definition of adaptive protection is presented in [38]: "a protection philosophy which permits and seeks to make adjustments automatically in various protection functions in order to make them more attuned to prevailing system conditions".

Many research works have shown the applicability of adaptive protection principles among others to transmission line relaying, out-of-step relaying, loss-of-field relaying and load-and-generation-tripping relaying. Adaptive relaying diminishes the likelihood of false tripping during system disturbances and reduces the probability that such disturbances spread into catastrophic failures.

The concept of adaptive protection as a tool for preventive and emergency control is discussed in [37], which also includes a survey since the start period of adaptive protection development. The next steps of the adaptive protection concept development from 1988 to 1999 may be outlined as follows:

1. 1988–1990:
   - basic concepts of adaptive protection,
   - rapid coordination calculation experiments,
   - subsystem and localization studies.
2. 1991–1993:
   - digital relaying systems,
   - specific adaptive relaying examples,
   - use of synchronized phasor measurements,
   - surveys of industrial practice and wishes.

3. 1994–1996:
   - identify relays as major blackouts contributors,
   - special purpose adaptive applications,
   - identify concept of hidden failures,
   - discussion of on-line coordination.
4. 1997–1999:
   - Additional applications: automatic reclosure, transformer,
   - additional techniques: agents, decision trees,
   - experiments with wide-area concepts.

The methodology for implementation a security/dependability adaptive protection scheme with the use of data mining is tested in [39]. The methodology to implement the proposed security/dependability adaptive voting scheme is concerned with the following:

1. Determination of the PMU placement
2. Defining an objective function to classify the system state into "stressed" or "safe"
3. Identifying which measurements are most suitable for classifying the system state
4. Defining the decision logic to alter the security/dependability bias of the adaptive protection scheme.

Thus, WAMS technologies open a new area in power system adaptive protective relaying.

13.3.5 Adaptive Self-Healing

The new concept of power system protection against catastrophic failures on the basis of the adaptive self-healing has become a popular choice in protective system relaying. Using vulnerability assessment techniques (that will be explained more in detail in Chap. 14), the severity of any given contingency can be assessed. The role of wide area measurements for this concept is described in the following.

The underlying concept for the self-healing, distributed control of an electric power system involves treating individual components as independent intelligent agents, competing and cooperating to achieve global optimization in the context of the whole system's environment with the use of PMU and WAMS technology [40]. An example for a self-healing system for power flow control was presented in Chap. 10 of the present book.

In [41], the envisioned framework requires deployment of autonomous intelligent agents across an interconnection-wide system to support a self-healing power grid. This would enable the system to be adaptable to varying operating conditions of the power system for analyzing and maintaining its reliability in real-time. The interaction between the intelligent components of the infrastructure would be orchestrated through a set of execution cycles, tailored to the physical phenomena and operating concerns in the power system. Such an approach can become the fundamental technology for organization of adaptive self-healing power systems.

The paper [42] presents a general methodology and scalable framework for business case analyses to assess the costs and benefits of implementing a high performance IT infrastructure for a self-healing grid. The infrastructure calls for a scalable and distributed architecture as well as geographically and temporally coordinated autonomous intelligent control to address power system operating concerns and realize major improvements in reliability. The framework facilitates the evaluation of the financial feasibility of this far reaching IT infrastructure. The methodology exploits published industry statistics regarding physical and financial attributes and can be adapted. This technology can allow for realizing the adaptive self-healing concept in large scale power grids.

13.4 Conclusions

The wide penetration of WAMS technologies into protective relaying should be approved as the determinative for improving of power system reliability. The future developments in this area should give even more benefits for power system dependability. The wide area protection systems enlarge functional facility of traditional protective relaying. The short overview of state-of-the-art protection technologies indicates that the role of PMUs is changing. They are no longer simply monitoring the grid and being used for post-mortem analysis. They have stepped into a truly operational role of helping to identify and manage a major event on the grid. And with this, PMUs have become a piece of equipment which is no longer optional but vital to grid reliability.

References

1. G. Benmouyal, E.O. Schweitzer, A. Guzman, Synchronized phasor measurement in protective relays for protection, control, and analysis of electric protection, control, and analysis of electric power systems (2004) (Printed USA 20020927)
2. S. Chakrabarti, E. Kyriakides, T. Bi, D. Cai, V. Terzija, Measurements get together. IEEE Power Energ. Mag. (January–February 2009)
3. M. Begovic, D. Novosel, D. Karlsson et al., in *Wide-Area Protection and Emergency Control*. Proceedings of the IEEE, **93**(5), 876–891 (May 2005)
4. U. Serizawa, M. Myoujin, K. Kitamura et al., Wide-area differential backup protection employing broadband communications and time transfer systems. IEEE Trans. Power Deliv. **13**(4) (October 1998)

5. Y. Serizawa, H. Imamura, N. Sugaya, in *Experimental Examination of Wide-Area Current Differential Backup Protection Employing Broadband Communications and Time Transfer Systems* (1999) (0-7803-5569-5/99)
6. R. Giovanini, D.V. Coury, K.M. Hopkinson, J.S. Thorp, in *A Primary and Backup Cooperative Protection System Based on Wide Area Agents*. Transmission and Distribution Conference and Exposition: Latin America. IEEE/PES, pp. 428–434, doi: 10.1109/TDC.2004.1432418 (2004)
7. D. Wang, S. Miao, X. Lin et al., in *Design of a Novel Wide-Area Backup Protection System*. Transmission and Distribution Conference and Exhibition: Asia and Pacific. IEEE/PES, pp. 1–6, doi: 10.1109/TDC.2001546766 (2005)
8. R. Giovanini, D.V. Coury, K.M. Hopkinson, J.S. Thorp, A primary and backup cooperative protection system based on wide area agents. IEEE Trans. Power Deliv. **21**(3) (July 2006)
9. J. De La Ree, V. Centeno, J.S. Thorp, A.G. Phadke, Synchronized phasor measurement applications in power systems. IEEE Trans. Smart Grid, **1**(1) (June 2010)
10. M.M. Essia, M.E. Masoud, M.M. Elanwar, A novel back up wide area protection technique for power transmission grids using phasor measurement unit. IEEE Trans. Power Deliv. **25**(1), 270–278 (January 2010)
11. Z. He, Z. Zhang, W. Chen, O.P. Malik, X. Yin, Wide-area backup protection algorithm based on fault component voltage distribution. IEEE Trans. Power Deliv. **26**(4) (October 2011)
12. J. Ma, J. Li, J.S. Thorp, A.J. Arana, Q. Yang, A.G. Phadke, Fault steady state component-based wide area backup protection algorithm. IEEE Trans. Smart Grid **2**(3) (September 2011)
13. M. Kezunovic, B. Perunicic, Synchronized sampling improves fault location. IEEE Comp. Appl. Power **8**(2), 30 (1995). ISSN O8950156/95
14. M. Kezunovic, B. Perunicic, Automated transmission line fault analysis using synchronized sampling at two ends. IEEE Trans. Power Syst. **11**(1) (February 1996)
15. K. Mazlumi, A. Abyaneh, S.H.H. Sadeghi, in *Determination of Optimal PMU Placement for Fault-Location Observability, DRPT2008*, Nanjing (April 2008)
16. M. Shiroei, S. Daniar, M. Akhbari, *A New Algorithm for Fault Location on Transmission Lines*. (IEEE, New York, 2009). 978-1-4244-4241-6/09
17. C. Zheng, Y. Dong, O. Gonen, M. Kezunovic, in *Data Integration Used in New Applications and Control Center Visualization Tool*, 2010 (978-1-4244-6551-410)
18. Z. Wang, Y. Zhang, J. Zhang, *Principal Components Fault Location Based on WAMS-PMU Measure System*. (IEEE, New York, 2011) 978-1-4577-1002-511
19. Y. Ohura, M. Suzuki et al., A predictive out-of-step protection system based on observation of the phase difference between substations. IEEE Trans. Power Deliv. **5**(4), 1695–1704 (November 1990)
20. Y. Wang, Y. Yin, J. Hou, *Coordinated Out-of-Step Protection System Based on WAMS*. (IEEE, New York, 2005) 0-7803-9114-4/05
21. Z. Pakdel, *Intelligent Instability Detection for Islanding Prediction* (Virginia Polytehnic Institute, USA, 2011)
22. R. Diao, V. Vittal, K. Sun et al., *Decision Tree Assisted Controlled Islanding for Preventing Cascading Events* (IEEE, New york 2009) 978-1-4244-3811-209
23. F. Galvan, S. Mandal, M. Thomas, *Phasor Measurement Units (PMU) Instrumental in Detecting and Managing the Electrical Island Created in the Aftermath of Hurrican* (IEEE, New York, 2009) 978-1-4244-3811-2/09
24. L. Li, Y. Liu, *Out-of-Step Splitting Framework Based on Adaptive Separation Detecting Criterion*. (IEEE T&D Asia 2009)
25. T. Nagata, H. Sasaki, A multi-agent approach to power system restoration. IEEE Trans. Power Syst. **17**(2), 457–462 (May 2002)
26. CIGRE Task Force C2.02.24, *Defense Plan Against Extreme Contingencies*, Technical Brochure 316, CIGRE, Paris (April 2007)
27. I.C. Decker, M.N. Agostini, A.S. Silva, D. Dotta, in *Monitoring of a Large Scale Event in the Brazilian Power System by WAMS*. IREP Symposium—Bulk Power System Dynamics and Control-VIII (IREP), Buzios (August 2010)

28. S. Corsi, A. Danelli, M. Pozzi, Emergency—Stability Controls Through HVDC Links. Power Engineering Society Summer Meeting, IEEE **2**, 774–779 (25 July 2002)
29. P. Li, X. Wu, Y. Zhang, *Analysis of Modulation Controllers of Multi-Infeed HVDC for CSG in 2008* (IEEE, New York, 2006) 1-4244-0111-906
30. X-M. Mao, Y. Zhang, L. Guan, X-C. Wu, *Coordinated Control of Inter Area Oscillation in the China Southern Power Grid* (IEEE, New York, 2006) 0885-8950
31. D. Westermann, H. Sauvain, *Experiences with Wide Area Coordinated Control of Facts Devices and HVDC in a Real Time Environment* (IEEE, New York, 2007) 978-1-4244-2
32. H.F. Latorre, M. Ghandhari, L.S. Eoder, *Use of Local and Remote Information in POD Control of a VSC-HVDC* (IEEE, New York, 2009) 978-1-4244-2235-7/09
33. Y. Li, C. Rehtanz, D.C. Yang, K. Gomer et al., in *Wide-Area Time-Delay Damping Control to Prevent Power Oscillations in HVDC/AC Interconnected Power Systems*. 2010 International Conference on Power System Technology (2010)
34. R. Preece, A.M. Almutairi, O. Marjanovic, J.V. Milanovic, *Damping of Inter-area Oscillations by VSC-HVDC Active Power Modulation with Supplementary WAMS Based Modal LQG Controller*. (IEEE, New York, 2011) 978-1-4577-1002-5/11
35. W. Juanjuan, F. Chuang, Z. Yao, Design of WAMS-based multiple HVDC damping control system. IEEE Trans. Smart Grid **2**(2),363–374 (June 2011)
36. D. Povh, P. Thepparat, D. Westermann, *Further Development of HVDC Control* (IEEE, New York, 2011) 978-1-4244-8417-111
37. M.J. Damborg, M. Kim, J. Huang, S.S. Venkata, A.G. Phadke, in *Adaptive Protection as Preventive and Emergency Control*, 2000 (0-7803-6420-1)
38. S.H. Horowitz, D. Novosel, V. Madani, M. Adamiak, System-wide protection. IEEE Power Energ. Mag. (September–October 2008)
39. E.E. Bernabeu, J.S. Thorp, V. Centeno, Methodology for a security-dependability adaptive protection scheme based on data mining. IEEE Trans. Power (2011)
40. S.M. Amin, B.F. Wollenberg, Toward a smart grid. IEEE Power Energ. Mag. (September–October 2005)
41. K. Moslehi, R. Kumar, Vision for a self healing. ABB Rev. (April 2006)
42. H. Moslehi, A.B. Ranjit Kumar, P. Hirsch, in *Feasibility of a Self-Healing Grid—Part I Methodology and Cost Models* (IEEE, New York, 2006) 1-4244-0493-206

Chapter 14
Dynamic Security Assessment and Risk Estimation

Antans Sauhats, Evgenijs Kucajevs, Dmitrijs Antonovs and Romans Petrichenko

14.1 Introduction

In consequence of numerous random and uncertain parameters, the operating condition in the power system constantly changes. Protection technologies, as introduced in the previous chapter, are used to switch of power system devices if a risk of damages occurs. As a result, sometimes system blackouts can take place, which lead to enormous economic, social and environmental losses. The wish to diminish them is only natural. To ensure this task, the concept of risk, which reflects a combination of the probability of failure states and the level of their consequences, is used. To minimize risk, it is possible to use two main approaches: either on probabilistic basis or on deterministic basis. If the probabilistic approach is used for estimating the risk level, it is necessary to know the loss function as well as the distribution laws of numerous influencing parameters. When estimating the risk level, the Monte-Carlo method has to be used.

For determining the loss function, experience in designing power units and selection of substation's parameters can be used. To manage the risk, the anti-emergency measures (energy flow and reserves operation, under-frequency load shedding, power system separation) are applied.

A. Sauhats (✉) · E. Kucajevs · D. Antonovs · R. Petrichenko
Riga Technical University, Riga, Latvia
e-mail: sauhatas@eef.rtu.lv

E. Kucajevs
e-mail: kucajevs@sep.lv

D. Antonovs
e-mail: dantonov@eef.rtu.lv

R. Petrichenko
e-mail: romoff@inbox.lv

U. Häger et al. (eds.), *Monitoring, Control and Protection of Interconnected Power Systems*, Power Systems, DOI: 10.1007/978-3-642-53848-3_14,

14.2 Risks in Power Systems, Sources and Consequences

Securing the reliability and efficiency of electric power supply for consumers is the main task to be solved in the course of controlling the operation of a power system. By solving this task, energy companies are striving to gain profit. In consequence of numerous random and uncertain parameters, the operating condition in the power system changes continuously. Power system elements such as loads, generators, transformers, transmission lines, are switched on and off, causing changes of power flows. Some of these changes are initiated by the power system operator due to the need to satisfy technical limitations, ensuring power supply quality and efficiency. Another part is caused by the influence of numerous external factors and can be treated as uncertain and random changes. An essential part of control impacts are generated by protective relaying and power system automation devices. The elements of a power system are subject to failures. Unintended disconnection of any technical element of the power system is possible. Wrong actions both of automation devices and operating personnel are possible, resulting in violation of technical limitations and use of unforeseen dangerous modes. Changes in the modes may be caused by random events or by the actions of the operators of the neighboring power systems.

The most severe losses (in this chapter the term "losses" accumulates and reflects all kinds of possible economic, social and environmental negative consequences) occur in the case of blackouts [1–5]. Usually, a blackout starts with the disconnection of one main element, causing a dangerous overload of the transmission grid, which is followed by a chain of fast-developing cascade-wise events: multiple line tripping, voltage instability and collapse, loss of stability, chaotic division of power systems into parts, frequency instability and so on.

As a result, massive-scale power plant outages and system blackouts take place. In such cases, electricity supply to consumers is completely disrupted in large territories, so it is important to ensure that the stoppage of equipment proceeds without damage. Since this is not achievable in all cases, expensive equipment may be damaged, requiring long-lasting repair afterwards. To restore the operation of such a power system, the electric plants have to be energized, for which it is necessary to supply the plant with electricity for its process (auxiliary) equipment to operate. For this purpose, up to 5–8 % of the plant's installed power could be required. This power is supplied by the "islands" that have been favorably separated from the power system, with their plants functioning. Step by step, the energy is delivered from one plant to another, until all of them are operating. Depending on the specific operation features of the power plant and the amount of shed load, the restoration of electricity supply may last from several hours to several days. Nowadays, under urbanization conditions, a collapse of a city power system causes serious trouble and leads to enormous losses [5]. The wish to diminish them is only natural. It is important to notice and emphasize that losses depend on random and uncertain factors and parameters and are uncertain as a result. By controlling the structure and parameters of a power system, the following two tasks can be solved simultaneously:

- Diminish the probability of unfavorable events or situations.
- Diminish the level of losses emerging in the case of unfavorable events.

To ensure the conditions for solving both of the above-mentioned tasks, the concept of risk is used, which reflects a combination of the probability of failure states and the level of their consequences.

There are many possible risk indices for different purposes. These risk indices reflect various factors, including component capacities and outages, load profiles and forecast uncertainties, system configurations and operating conditions, and so on. It is extremely inconvenient to use a number of indices for describing risk with the aim of taking a responsible decision [6]. At the same time, the complexity of modern power systems, their large dimensions, the random and uncertain character of many parameters predetermines the difficulties in describing and calculating the risks by using one integral index. The problems that arise can be divided into three types:

1. Computational problems caused by the need of modeling an extremely large and complex system.
2. Methodological problems caused by the impossibility of describing the losses by using a single measure, for example, in monetary way.
3. Methodological problems caused by the uncertain nature of the initial information and the possible losses.

The problems of the first type are solved in pace with the development of computing methods and aids. The problems of the second type are explained by the following: as a result of major energy emergencies, various losses occur, which practically cannot be represented in terms of money, for example, loss of human life, considerable social consequences, environment pollution. The losses of such types are not dealt with in this chapter.

The essence of the difficulties, which arise when using losses of uncertain nature as the measure of risk, can be explained by a simple example shown on Fig. 14.1.

Let us suppose that the power system operator can choose one of the two modes. In the first case (Fig. 14.1), losses may occur, which are distributed within the range from A to B (the distribution law is not known). In the second case, losses are distributed within the range from C to D. Which mode is preferable? There is no unambiguous answer. Furthermore, the statistical decision-making theory confirms and proves that the decision depends on the subjective preferences of the person making the decision [6]. An optimist is likely to prefer the distribution A–B. In this case, the minimum possible losses are lower. A pessimist, who is oriented towards the minimization of the maximum possible losses, will prefer the distribution C–D. For solving tasks that are similar to the above-mentioned one, it is possible to use game theory criteria: Bayes, Laplace, Hurwitz, Minimal

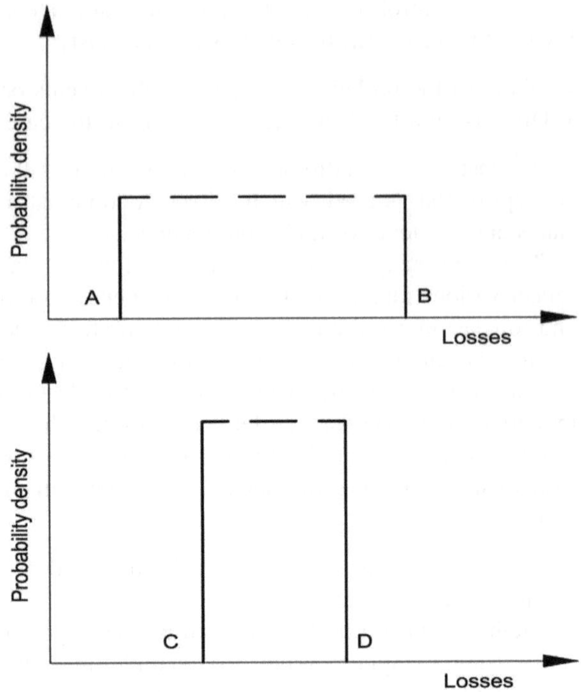

Fig. 14.1 Selection of mode and the possible losses

risk. Parts of these are based on setting a subjective function of losses distribution [6]. The setting of a sufficiently substantiated subjective function of losses distribution makes it possible to use the average losses or other numerical distribution characteristics as the measure of risk. The present chapter is confined to this approach.

14.2.1 Control of a Power System in Terms of Its Security

According to analyses of the operating conditions of systems, power system risk assessment can be divided into two basic aspects: system adequacy and system security [7]. Adequacy relates to the existence of sufficient facilities within the system to satisfy the load demand of the consumers and the operational constraints of the system. Adequacy is therefore associated with static conditions that do not include the dynamic and transient processes taking place in the system. Security relates to the ability of the system to respond to dynamic and transient disturbances arising within the system.

Security is therefore associated with the response of the system to whatever perturbations it is subject to. Normally, security evaluation requires the analysis of dynamic, transient, or voltage stability in the system.

From the security point of view, the operating conditions of a power system can be classified as [8, 9]:

- normal;
- pre-emergency (alert);
- emergency or extreme emergency (if the disturbance is very severe);
- post-emergency (restorative).

In the normal operating condition, all the system variables are within the normal range and no equipment is being overloaded. The system operates in a secure manner and is able to withstand a contingency without violating any of the constraints [7]. For normal operating conditions, mandatory security standards exist. Commonly [10, 11], this concerns the observance of the single-contingency (N − 1) criterion, which means that in case of emergency tripping of one important element of a power system, it is kept in normal operation, thus preventing the development of a cascading emergency. A power system operator performs non-stop monitoring of security criteria, ensuring the power reserves needed for frequency and voltage control; a definite reserve is provided to secure the allowable power exchange over the transmission lines.

A pre-emergency condition sets in when during normal running an element of a power system is tripped, which disturbs its operation. In this case, the emergency condition has not yet begun but the security criteria are no longer observed, and an emergency process may start developing at any time if the severity of the operating condition increases. Still, the situation can be normalized by mobilizing the capacities and reserves [7, 10].

In the cases when under a pre-emergency state of a power system countermeasures are not applied or prove inefficient, emergency processes develop, which can be of local or cascade-wise character. Local emergencies are eliminated by protective means that trip the faulty element, its functions being taken over by other elements of the system. A cascading emergency is followed by other events that still more aggravate its development; as a result, a full collapse of the power system can be expected, with cascading outages of power plants, making it necessary to restore their operation. Once the cascading emergency condition has set in, the process is developing rapidly. The personnel are unable to control the situation any more, and a blackout can only be prevented by fast-acting automatic protection means. Control actions, such as load shedding and controlled system separation, are aimed at preventing a widespread blackout [12, 13].

In the post-emergency condition, the control actions are taken to reconnect all the facilities and to restore system load [13].

To manage the above-mentioned situation, the anti-emergency measures shown on Fig. 14.2 are applied [13].

Organizational measures are associated with the staff actions that are to be taken in compliance with written or electronic instructions.

These measures should correspond to the human abilities to act and make decisions promptly. To provide a higher level of security, the execution of instructions can be made partly automatic. At normal operation, these are applied

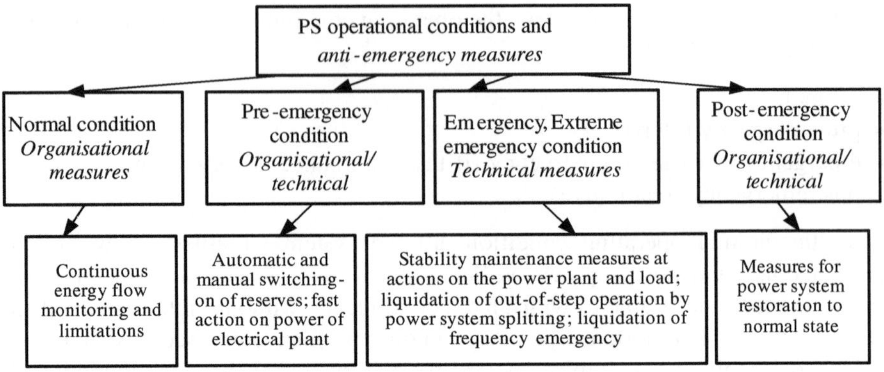

Fig. 14.2 Classification of anti-emergency measures by the operating conditions of a power system

in the usual course; however, in a pre-emergency condition, special programs are required that can provide instructions for urgent actions in a changed situation.

Technical measures are indispensable in the management of emergencies and extreme emergencies, when the events develop too fast for human action and a blackout can only be prevented with fast-acting technical means. In these cases, the staff should be involved at the post-emergency stage.

Power market deregulation and economic benefits have resulted in more stressed operating conditions and more vulnerable networks. When the transmission system is most heavily loaded, the power flows through some of the tie-lines are approaching their limits. A slight increase in these flows along with a critical event (a credible contingency) might lead to the development of a cascading emergency and a collapse of the power system. As a consequence, the need to identify the operational security level of the transmission system increases.

14.2.2 Losses During Emergency Situations

There are some difficulties during estimation of economic losses caused by emergency situations in the power system. For example CIGRE Task Force 38.02.21 [14] identified it as a difficult problem in probabilistic security assessment tasks. It is possible to use several different approaches [7]:

- Estimation of losses as undelivered energy to the customer and not received income to the power company. In this case level of losses is understated because many parameters are not taken into consideration (losses of consumers caused by sudden interruption of power supply, negative social aspect, loss of companies image and so on).

- The use of penalty system. In this case a power company pays penalties with coefficient exceeded many times the cost of undelivered energy (the coefficient can reach a level of 100 and more).
- Measurement of losses by the value of not produced national product.
- Measurement of losses by the use of partial coefficients.

The cost of energy interruption consists of the losses raised during power supply interruption. Complete costs associated with power supply interruption are:

- Additional expenses of industrial enterprises associated with no produced goods, spoilage of production, damage of equipment and so on.
- Expenses of some consumers for spoiled food products, interruption of heating system and so on.
- Expenses which are difficult to measure in monetary way—losses related to comfort and life threat problems.

Researchers of many countries try to find methods for determination of losses for different power consumers—agricultural, offices, factories and so on. Energy interruption expenses depend on duration of interruption. There are interesting peculiarities associated with the mentioned phenomenon:

- Results of research changes in a wide diapason;
- Results are different for different countries;
- Results depend on the type of consumers, industrial branch or agricultural and so on;
- Very often cost of expenses is a nonlinear function of interruption duration.

In many countries the relative interruption cost value is used for calculation of energy interruption cost. This is attitude of all expenses caused by energy interruption towards non delivered electric energy.

Results of research determine the value of relative interruption costs depending on the type of consumers as 0.5–75 €/kW h [7, 14, 15]. For instance the Latvian Ministry cabinet's law determines that during energy interruption caused by a supplier the undelivered energy should be compensated to the consumer by paying triple energy tariff and for very important consumers—penalties are six times the regular tariff. In case that the interruption time is larger than the standard limit the supplier should cover consumers the losses caused by reduction of good's production (in case that the losses level is larger than the preset multiple tariff level). When reserve power supply is activated, additional penalties are not used.

14.2.3 Probability of Failures and Disturbances in the Power System

In the world, blackouts of power systems occur regularly. For example [1] shows the data on blackouts occurrences in the North-American power systems, during which (a) over 50 thousand consumers were deprived of electricity, and (b) the

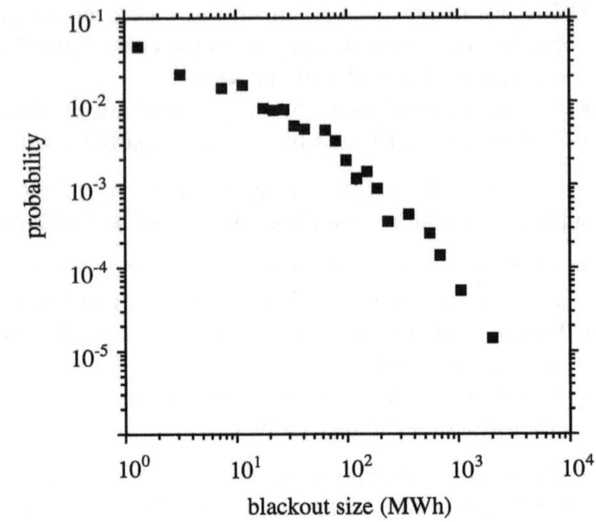

Fig. 14.3 North American blackout size probability distribution

Table 14.1 Failures' parameters and repair time of 330 kV network

No.	Equipment	Frequency (λ), 1/100a	Time (r), h
1	Gas-blast circuit breaker	$\lambda_j = 0.44$	$r_j = 16$
2	Bus system	$\lambda_k = 0.26$	$r_k = 8$
3	Isolators	$\lambda_a = 0.25$	$r_a = 10$
4	Arresters	$\lambda_{pn} = 0.36$	$r_{pn} = 8$
5	Voltage transformers	$\lambda_{sp} = 0.07$	$r_{sp} = 8$
6	Current transformers	$\lambda_{st} = 0.14$	$r_{st} = 8$
7	Autotransformers	$\lambda_{at} = 5$	$r_{at} = 50$
8	Overhead transmission (100 km)	$\lambda_{LN} = 25$	$r_{LN} = 10$
9	Autotransformer protective relays	$\lambda_{AT(RZ)} = 0.05$	$r_{AT(RZ)} = 4$
10	Overhead transmission protective relays	$\lambda_{LN(RZ)} = 0.1$	$r_{LN(RZ)} = 4$

Data of Canadian CEA

tripped power exceeded 100 MW. The situation is also similar in power systems of West-European countries, Russia, and many others [16, 17].

Data [1, 7] shown in the Fig. 14.3 and Table 14.1 confirm that the probability of a main equipment failure (per 100 year) is low, consequently probability of blackout development is quite small. In the same time collapse of a power system causes serious trouble and leads to enormous losses. The low probability of troubles and blackouts essentially complicates calculations of risk indices.

14.3 Mathematical Determination (Definition) of Risk

When describing risks, two main approaches are possible: Either on probabilistic basis or on deterministic basis [18, 19].

The probabilistic approach is more common, however, using it requires essential expenses for the collection of information and calculations. The use of the deterministic approach leads to loss of accuracy in describing the risks and a decrease of efficiency in managing them.

14.3.1 Probabilistic Approach

Let us assume that the objective of an electrical utility is to gain the profit Pr f. This profit depends on the set of many random and uncertain parameters $X(x_1, x_2, \ldots, x_n)$. In the case of large power systems there could be thousands of such parameters.

The profit Pr f of each electrical utility depends on the topology T chosen by this utility and the parameters Π of the power system controlled by this electrical utility. The freedom in choosing the topologies and parameters of the electrical utilities is limited by technical and legislative factors and regulations that are defined, in particular, by the Regulator's institution. Let us assume $s = \{T, \Pi\}$ as chosen combination of the topology and parameters and $F(s, X)$ as subjective distribution functions [6], which can be assigned to the uncertain factors X. The existence of distribution functions makes it possible not to distinguish between random and uncertain parameters in the future.

Hence, we can affirm that profit is any function Φ:

$$\mathrm{Pr\,f}_i = \Phi(s, X) \tag{14.1}$$

It naturally follows that in the accepted conditions Pr f is also a stochastic variable and can be characterized by:

- In a most complete way—by the distribution function or probability density function.
- Numerical parameters of the distribution, for example, expectation and the standard deviation or by the minimal and maximal value of the profit. In the majority of the applications in practice only numerical characteristics Pr f are used.

In statistical theory of the decision making function of risk estimation for parameters computed for a random variable X is determined as mathematical expectation of the losses [1, 6, 7]:

$$E(L, s) = \int_X L(\mathrm{Pr\,f})dF(s, X) \tag{14.2}$$

The functions L can be chosen depending on how one measure the distance between the estimate Pr f* and the unknown profit Pr. For example:

$$L(\Pr f) = \begin{cases} -a(\Pr f - \Pr f^*), & \text{for } \Pr f - \Pr f^* > 0 \\ -b(\Pr f - \Pr f^*), & \text{for } \Pr f - \Pr f^* < 0 \end{cases}, \tag{14.3}$$

where Prf* is an income level below which the critical losses appear.

This level is determined by the decision maker [6]. The selection of the coefficient b, as it will be shown below, can be realized using experience of the power units' design.

Unfortunately for real power systems (s, X) are vectors of high dimension and it is difficult to determine the function L and the distribution function $F(s, X)$ with sufficient proof.

So Eq. (14.2) is difficult to apply for the solution of power engineering problems.

The cost measure of risk, the so called Value of Risk (VaR) [15], is often used for the solution of complex financial problems. This is expressed in monetary unit estimation of losses which are computed with a given confidence level.

For banking business the confidence level usually is assumed equal to 0.99. For the solution of power engineering problems seldom events should be considered therefore such confidence level is insufficient [6]. Hence, very complex problem is raised for estimation of a sufficient confidence level. Such an approach for determination of the risk measure is not suitable for a power engineering problem. As shown in [6] for practical application of the VaR concept in power systems asset management, modifications and simplifications are needed.

Essential simplification of the problem for estimation of the risk level can be applying the technical risk *TR* concept, which is the sum:

$$TR = \sum p_i(X, s)\, L_i(X, s) \tag{14.4}$$

where p_i is the probability of one undesirable event, L_i are losses (for example proportional) due to undelivered power to the consumer caused by one undesirable event and can be expressed in monetary units. The use of Eq. (14.4) allows to simplify significantly the analysis in comparison with application of (14.2).

The optimization task became single-criterion using (14.4) for determination of technical risk's measure. In this case the optimization task whose target is the choice of an optimal topology T_{opt} and parameters Π_{opt} can be expressed as:

$$\{T_{opt}; \Pi_{opt}\} = Arg \min \sum p_i(X, s) L_i(X, s). \tag{14.5}$$

For practical implementation even the utilization of Eq. (14.5) can be very complicated. Cardinal simplification can be done using additional assumptions:

- An undesirable event in the power system should be considered as large blackout i.e. the negligence of stability in the power system, frequency actuated emergency situation, voltage avalanche or preceded cascade type disconnection of the overloaded transmission lines.
- Damage rate caused by the mentioned emergency situation is considered as inadmissible.

- The initial event of such an emergency is considered as failure of one power system element.

Acceptance of the above mentioned assumptions results in consideration the most popular and in the power systems wide-used (N − 1) criterion.

14.3.2 Deterministic Approach (N − 1 Criterion and Its Lacks)

The determination of security levels, for given operating conditions, traditionally has been done using deterministic criteria, according to [7, 9, 13, 20, 21]:

- An operating condition is identified as secure if it can withstand the effects of each and every contingency in a pre-specified contingency set ("Withstanding the effects" means that the given contingencies will not violate loading or voltage criteria or make the system unstable).
- The mentioned consideration results in an acceptable level of risk.
- In case when a failure of any element can cause an emergency situation the risk level is very high. In such a case control actions should be taken to move the system into the secure region. To realize these control actions specific time limits should be foreseen.

Use of the N − 1 criterion determines the level of risk during power system control in most cases by losses L_{n-1}, having only two values. The first one corresponds to fulfillment of the considered criterion ($L_{n-1} = 0$) and the second—to non-fulfillment (losses level is inadmissible). So verification of the parameter L_{n-1} is reduced to the verification of the N − 1 criterion. Hence, the problem, having a probabilistically character, is transformed into a deterministic solvable problem using simplified methods. However these simplifications are achieved at the expense of substantially decrease of risk estimation accuracy. In particular, it is not possible to formulate the risk optimization task at the conditions when the N − 1 criterion is fulfilled. It is easy to recognize a decision-making problem in the above described process; the decision is whether to take actions and if so, what kind and how much. The deterministic method provides a simple rule for use in making this decision: optimize economy within hard constraints of the secure operational region. It is this simplicity that has made the deterministic method so attractive, and so useful, in the past. Today, however, with the industry's emphasis on economic competition, and with the associated increased network vulnerability, there is a growing recognition that this simplicity also carries with it significant subjectivity, and this can result in constraints that are not uniform with respect to the security level. This suggests that the ultimate decisions that are made may not be the best ones [7, 9].

Many critical publications are devoted to the description of the N − 1 criterion drawbacks. Most of them describe the development of methods for calculation of probabilities of undesirable events [7–9, 13]. The risk's measure is presented by the set of events and probabilities. If there is need for solution of an optimization problem multi-objective optimization should be applied which is a rather complicated approach.

14.4 Ways of Risks Management

Action impacts O, which are used for minimization of technical risks and choice of optimal topology T_{opt} and parameters Π_{opt}, can be separated in two parts:

- Action O_1 which specifies configuration of the power system, methods and means of its control for long-term planning (construction of transmission lines, generating units, implementation of emergency automation devices and so on).
- Action O_2 which specifies a state of the power system during on-line control (control of power flows, introduction of reserves, regulation of voltage and consumption, preventive measures for maintenance of efficiency and detection of faulty equipment, fault detectors and so on).

Both action impacts can be used for minimizing the probability of undesirable events and for minimizing losses, caused by such events. In the following we present consideration of impact of emergency automation devices during on-line control for minimization of risks, caused by possible important emergency situations with the loss of system stability and under-frequency situation.

14.4.1 Automatic Control Actions

The currently applied protective automatics (incl. special protective schemes) mainly consist of the following components [7, 9, 13, 22, 23]:

- relay protection that trips a faulty element;
- automatic re-closing to make this element operational;
- automatic connection of reserve;
- automatic emergency control of active power by means of fulfilling the functions of: stability loss avoidance, out-of-step running liquidation, under-frequency limitation, over frequency limitation, equipment load relief [24, 25];
- devices for automatic emergency control of voltage [26].

The different kinds of protective automatics and widely adopted control actions are shown in Fig. 14.4.

For all above mentioned cases controlling actions significantly minimize the negative consequences.

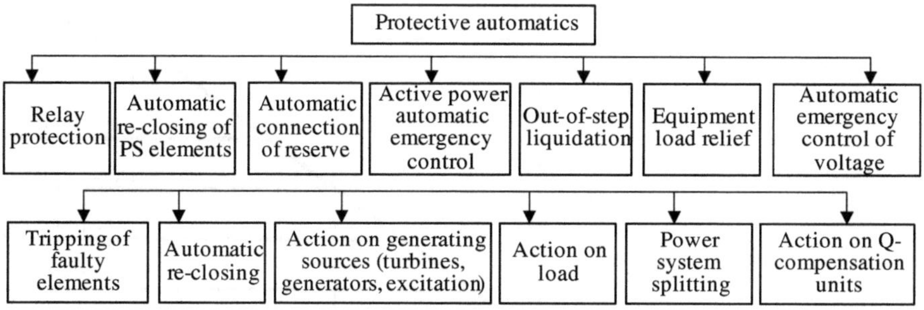

Fig. 14.4 Different kinds of protective automatics and control actions

14.4.2 Reserves for Power Balance Maintenance

Security measures should contain the requirements on reserve maintenance and control. The placement of reserves should be dictated by economic reasons. A reserved unit should not be fully loaded, which, however, is undesirable from the viewpoint of its competitiveness. At the same time, low-economical equipment often has no sufficient power to keep a reserve. In many power systems the requirements on reserve capacities and their placement are standardized, and the task of reservation is in these cases economically justified.

The active power balance should be provided in real time, using reserves divided into two parts: rotating and stationary power reserves. To provide reserve for major power systems definite capacity proportions should be territorially distributed, involving power plants that participate in frequency control. The corresponding capacities should have rotating power reserves, which are associated with the dependence of consumption on the weather conditions and forecasting accuracy. The reserves can be divided by the speed of their mobilization into three groups as follows.

The primary reserves are mobilized in seconds, with their amounts depending on the system's power. As shown in Fig. 14.5, the reserve mobilization time for the major European power system (ENTSO-E CE) under normal conditions is 5–30 s, and under emergency conditions—up to 120 s.

The secondary reserves are mobilized within the time from several tens of seconds to minutes. The recommended reserve capacities in major power systems depend on their loads; for the mentioned system this dependence is shown in Fig. 14.6.

For primary and secondary reserves not fully loaded capacities are employed, with a power that would be sufficient both for additional load and for load relief during control.

The tertiary reserves are mobilized in the time up to 15 min. To these reserves belong: stand-by hydro- and pumped storage power plants as well as gas turbines, which, being started, can assume load within 15 min.

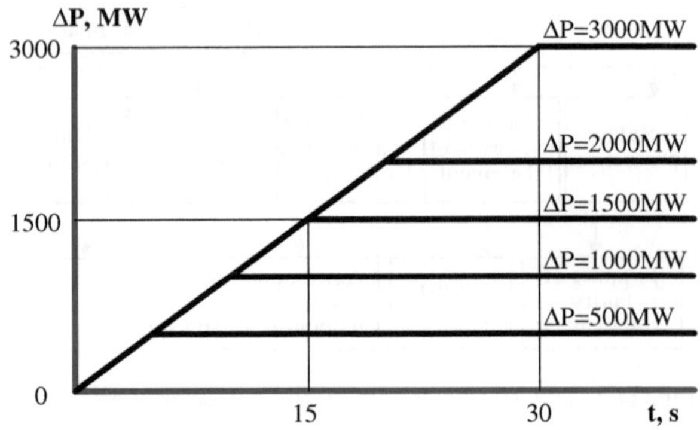

Fig. 14.5 The minimum time for deployment of primary reserves

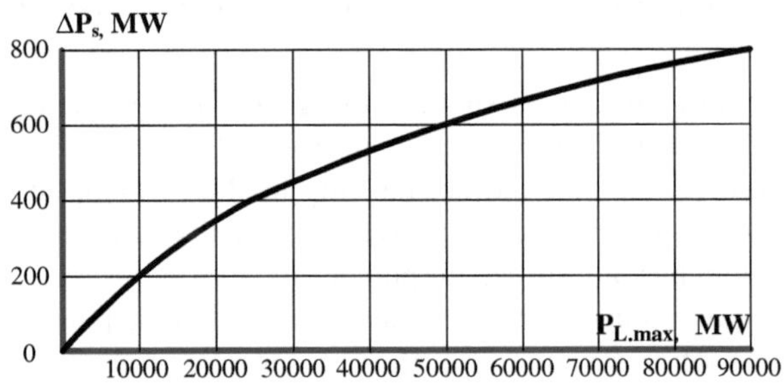

Fig. 14.6 The recommended secondary control reserve as a function of the maximum anticipated load

Considering the reaction of generating sources to the frequency variation it should be taken into account that they can be multi-component. To fast frequency variations corresponding to synchronous oscillations of generators the turbine speed governors do not respond.

Fast-operating turbines of thermal power plants respond to the load-related frequency variations. Hydro-turbines assume load after the transient negative feedback time has elapsed. To cover slower load changes, the secondary control automatics are used.

14.4.3 Algorithms of Automatic Under-Frequency Load Shedding

When a power system is in stable operation at normal frequency, the total mechanical power input from the prime movers to the generators is equal to the sum of all the connected loads, plus all real power losses in the system. Any significant upset of this balance causes a frequency change.

Under-frequency load shedding (UFLS) is a common practice for electric power utilities for preventing frequency drop in power systems after disturbances causing dangerous imbalance between the load and generation [23, 27].

The main goal of UFLS is to gradually shed portions of the load when the system frequency reaches values lower than allowed.

UFLS is the last step in protecting electric power systems from blackouts and severe damages. An UFLS scheme must be reliable, simple, and efficient.

As load shedding is a somewhat drastic control measure, it is usually implemented in steps with each step triggered at a different frequency level to allow the least important loads to be shed first.

When after UFLS operation frequency restores up to rated value the previously disconnected load can be gradually restored. In some power systems the load restoration process occurs automatically, but in some systems manually. Further details about load shedding will be provided in Chap. 17.

14.4.4 Power System Separation

Controlled system separation is widely used for out-of-step elimination [7, 9, 13, 18, 28]; in particular cases it could be applied to prevent voltage collapse or to maintain stability, e.g. in case of several tie-lines' tripping. To ensure satisfactory post-separation system conditions in each island, controlled separation should be accompanied by UFLS and/or other anti-emergency means.

To avoid outages of generating units during under-frequency several generators or power plant may be isolated to ensure power supply not only to its auxiliaries but also to the sufficient load of the corresponding area (island operation). Switching power plant to island operation is made using one or more of the following parameters: frequency decrease, rate of frequency change, change of power direction.

Considering blackouts that have occurred in recent years the special emphasis has to be placed on these automatics. Further details about system separation will be provided in Chap. 16.

14.5 A Monte-Carlo Method as Tool for Risk Indicator Calculation

A flow-chart for a risk indicator calculation algorithm using a Monte-Carlo method is presented in Fig. 14.7. The algorithm consists of the next main blocks:

- Assignment of the initial state of the power system. Two main variants can be used. The first one is based on the use of SCADA data (topology, power consumption, generating power and power flows). In the second variant the power system state is assigned by the person performing the calculation of operational conditions with the help of a specific software. The person verifies the admissibility of the risks for specific operational conditions.

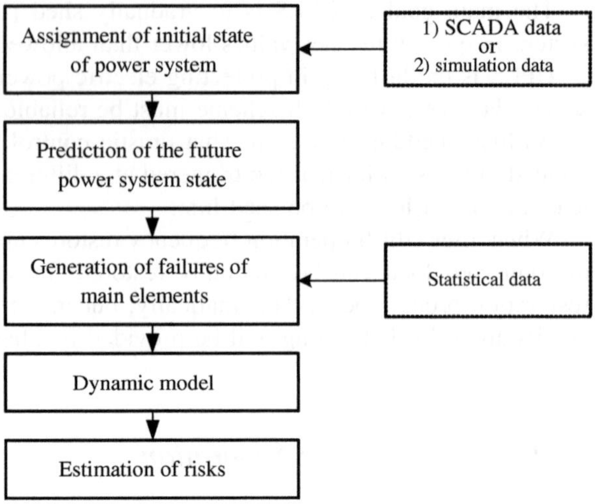

Fig. 14.7 Risk indicator calculation algorithm

- Prediction of the future power system state for a specific time interval. At this stage all elements are assumed as intact and the predicted operational condition is admissible from the point of view of absence the overloads and securing the static stability (otherwise measures are developed for preventing inadmissible conditions).
- Generator, modeling failure of power system elements. Modeling takes place for short circuits and disconnections of power system main elements. Statistical data for the power system are used.
- Dynamic models describing processes in the power system caused by tripping of elements. This model is used for verification of consequences of trippings considering undeliverability of energy.
- Estimation of risks. Probability of development of emergency situations and costs of undelivered energy (the cost is assumed to be known) are considered.

Note that for estimating the risks caused by large blackouts even rare events should be considered. In this case the number of trials of the Monte-Carlo method can be very big (100,000 and more).

Let us assume that during a certain trial, a failure **Fbo** which may incite a blackout may occur with the probability Θ. Let us now find the probability of the failure occurring once during χ mutually independent trials.

The trials are performed until the first failure **Fbo**. In this case, the number of trials X that need to be performed until the first failure, will be characterised by a distribution according to Farri's law (a special case of the Pascal distribution [29]). Then:

$$P(X; 1; \Theta) = \Theta(1 - \Theta)^{x}, \tag{14.6}$$

where *P(X;* 1; Θ*)* is the probability of a failure occurring during the χth trial. Let us find the probability of not more than X trials being necessary for registering a failure:

$$P(X<x) = \sum_{i=0}^{x-1} \Theta(1-\Theta)^i. \tag{14.7}$$

When organizing the trials, it is logical to formulate the problem as follows: Choose the number of trials χ, for which even rarely occurring failures **Fbo** can be registered with a probability that is close to 1. The dependence of χ on P and Θ that corresponds to such a formulation is shown in Fig. 14.8.

The values of plausibility P and probability Θ are to be chosen on the basis of the trial costs and the economic situation that is expected as a result of calculation errors releasing products in case of blackout (possibility of change of structure and parameters of a power system for the purpose of prevention of losses as a result of occurrence of such error is missed).

An approximate value of Θ can be assumed on the basis of the requirement that the probability of a failure caused by calculation errors has to be substantially lower than the probability of a real blackout. Thus, assuming for example that the blackout can arise once in 25 years, the probability of a blackout occurring during one year of operation can be found (supposing that the failures occur according to the exponential law):

$$P = 1 - e^{-\frac{1}{T}} = 0.039, \tag{14.8}$$

which can be taken as the basis when setting the value of Θ in the expression (14.7).

When organizing the above-discussed procedures, we are facing the issue of the number of trials which guarantees the calculation of failures with a preset plausibility.

Taking into consideration that for estimating the probability of events and their sequences, labor-consuming dynamic modeling has to be performed, the application of direct modeling for the determination of failures, their combinations and consequences for a large power system is unrealistic. To avoid the above-mentioned difficulties, it is possible to use an approach that is well-known in the Monte-Carlo method for minimizing the number of trials by excluding those events, the results of which are known a priori. In the considered case, it is necessary to exclude those events, which will not cause system blackouts.

Such events (operational conditions) can be:

- All conditions in a completely operable power system.
- Short circuits followed by an insignificant voltage decline at substation buses (a criterion prescribing a decline smaller than 40 % is assumed for fast acting protective relaying and has been proved in practice).

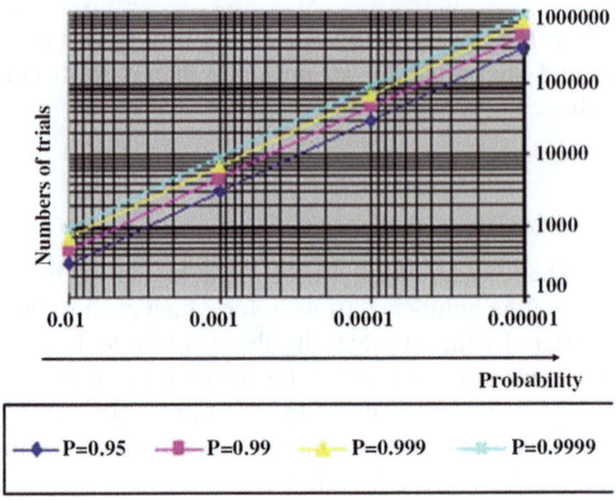

Fig. 14.8 Dependence of the number of trials χ on plausibility P and probability Θ

- Tripping of those generators, lines, and transformers, the capacity of which constitutes an insignificant part of the capacity of the power system under consideration.

Besides this, a significant reduction in the number of trials can be achieved by means of modelling complex events occurring as a result of several simple events that coincide in time. The probability of complex events can be calculated in advance. Examples of complex events can be as follows:

- Short circuit and simultaneous failure of the main and reserve protection devices of a certain power system element;
- Tripping of a plant consisting of several units;
- Simultaneous tripping of two or more power transmission lines.

The probability of the occurrence of complex events can be calculated in advance and stored in the form of a corresponding database. In order to evaluate the probability of the occurrence of complex events, it is possible to use both the Monte-Carlo method and other methods of the reliability theory.

14.6 Example of Risk Assessment

14.6.1 Synthesis of the Loss Function

For determination of the loss function using (14.2) let us assume that losses arising from the frequency actuated emergency situation are proportional to undelivered electrical energy. For estimation of the coefficient b in (14.3) let us use experience of a selection of electrical diagrams of high voltage substations. For selection of the main scheme of substation two criteria are considered:

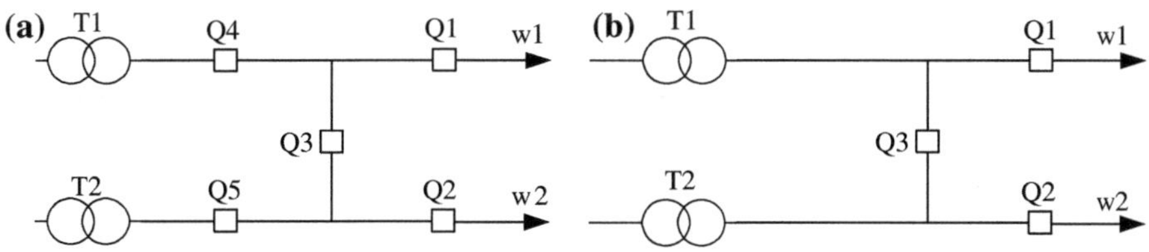

Fig. 14.9 Schemes of substations

Table 14.2 Initial information of event frequencies

Name of event	Frequency, λ (1/year)	Average duration, T (h)
Preventive maintenance of transmission line	$\lambda_1 = 0.15$	$T_1 = 60$
Preventive maintenance of transformer	$\lambda_2 = 0.15$	$T_2 = 80$
Short circuit at transmission line (at 100 km)	$\lambda_3 = 0.25$	$T_3 = 10$
Short circuit at transformer	$\lambda_4 = 0.05$	$T_4 = 50$
Failure of circuit breaker	$\lambda_5 = 0.044$	$T_5 = 16$
Preventive maintenance of circuit breaker	$\lambda_6 = 0.15$	$T_6 = 20$
Preventive maintenance of disconnecter	$\lambda_7 = 0.05$	$T_7 = 8$
Failure of protective relaying	$\lambda_8 = 0.001$	$T_8 = 4$

- Cost of construction;
- Reliability level.

Figure 14.9 illustrates two possible schemes of substations.

Initial information of event frequencies is taken from Table 14.2. Table 14.3 shows events causing interruption of the power supply of substations.

Table 14.4 presents values of undelivered electrical energy and investments necessary for the realization of each scheme. The choice of the expert is the more reliable (and more expensive) substation A. Let us compose an equation for determination of the risks for each substation using (14.2) and (14.3). Taking into consideration the opinions of experts it can be accepted that $a = 1$ and the coefficient b can therefore be determined.

Let us determine the losses functions for two essentially different conditions:

- Normal operational condition. Presume all equipment is in operable condition.
- Emergency situation caused by combination of unfavorable conditions (see Table 14.4) leading to interruption of power supply.

Besides let us assume that the loss function for the considered substations A and B for normal operational conditions can be expressed as:

$$\begin{aligned} L_A &= -E(\mathrm{Pr\,f}_A) \\ L_B &= -E(\mathrm{Pr\,f}_B) \end{aligned} \tag{14.9}$$

where $E(\mathrm{Prf}_A)$, $E(\mathrm{Prf}_B)$ are estimations of the profit.

Table 14.3 Events causing interruption of power supply of substations

Event	Presence of events (+) and their combination							
Interruption for maintenance of w1	+						+	
Interruption for maintenance of w2		+						+
Interruption for maintenance of T1			+					
Interruption for maintenance of T2				+				
Faulted w1			+					+
Faulted w2				+			+	
Faulted T1		+			+			
Faulted T2	+					+		
Failure of circuit breaker Q			+	+	+	+	+	+
Interruption of power supply of substation (Scheme A)	No	No	No	No	No	No	Yes	Yes
Interruption of power supply of substation (Scheme B)	No	No	Yes	Yes	No	No	Yes	Yes

Table 14.4 Comparison of substation parameters

Name of substation	Frequency of substation blackout (1/year)	Undelivered electrical energy, U_{en} (kW h/year)	Costs of substation (M€)	Costs of undelivered electrical energy (€)
Substation A	0.25×10^{-4}	0.02	45	0
Substation B	0.32	2,564	44	41

Both substations fulfil their functions during normal operational conditions and facilitate the sale of energy in full capacity. However substation A is more expensive (see Table 14.4), hence:

$$\begin{aligned} E(\Pr \mathrm{f}_A) &< E(\Pr \mathrm{f}_B),\ or \\ E(\Pr \mathrm{f}_A) &= E(\Pr \mathrm{f}_B) - C \end{aligned} \tag{14.10}$$

where C are the additional annual expenses.

Calculations made for substation A estimate C = 100,000 €. During an emergency situation the economic losses are in both substations proportional to the undelivered energy W_{enA} and W_{enB} (see Table 14.4):

$$\begin{aligned} L_A &= b \cdot W_{en_A} \cdot l \\ L_B &= b \cdot W_{en_B} \cdot l \end{aligned} \tag{14.11}$$

where l is price of electric energy (€/kW h).

Taking into consideration a small probability of emergency situations, let us assume an annual probability p_n of normal operational condition equal to 1. Technical risks for both considered substations are:

$$\begin{aligned} E(L_A) &= -E(\Pr f_B) + C + b \cdot W_{en_A} \cdot p_{\alpha VA} \cdot l \\ E(L_B) &= -E(\Pr f_B) + b \cdot W_{en_B} \cdot p_{\alpha VB} \cdot l \end{aligned} \tag{14.12}$$

where:

$p_{\alpha VA}$ and $p_{\alpha VB}$ are the probabilities of a failure occurring during a year;
$E(L_A)$ and $E(L_B)$ stand for the mathematical expectation of the losses.

Thus, using the data from Table 14.4.

$$\begin{aligned} E(L_A) &= -E(\Pr f_B) + 100{,}000 + b \times \left(0.25 \times 10^{-4} \times 0.02 \times 0.05\right) \\ E(L_B) &= -E(\Pr f_B) + b \times (0.32 \times 2{,}564 \times 0.05) \end{aligned} \tag{14.13}$$

Taking into consideration the opinion of experts, preferring scheme A as a decision, it is possible to state: $E(L_A) < E(L_B)$ and hence, $b > 2{,}438$. The obtained value of coefficient b will be used for estimation of risks caused by the under-frequency situation considered below.

14.6.2 Electric Power Supply of Cities

For energy supply of large cities of Northern Europe centralized heat supply and cogeneration power plants have found wide application. Figure 14.10 represents a diagram of a fragment of such an electric power system with a 100 MW cogeneration power plant (CG-1).

Three main undesirable events can cause disconnection of load in area A:

- Full shutdown of cogeneration plant CG-1 (in case of low transmission capacities of HV-1 and HV-2).
- Simultaneous disconnection of two high voltage transmissions HV-1 and HV-2.
- The loss of a tie with the system (infinite bus power system).

Hence, for estimation of technical risks it is enough to determine the probability for initiation of the mentioned events and the caused losses.

14.6.3 Power Supply Interruption and Losses

In Fig. 14.10 two areas of the distribution network of a large city are shown. Areas A and B are provided with modern UFLS system which can disconnect up to 70 % of the load of each area. Areas A and B are connected by old cables (HV-1 and HV-2), which are subject to relatively frequent failures (1 $\div$ 2 times per year) as consequence of that, after UFLS operation both of areas are staying in operation.

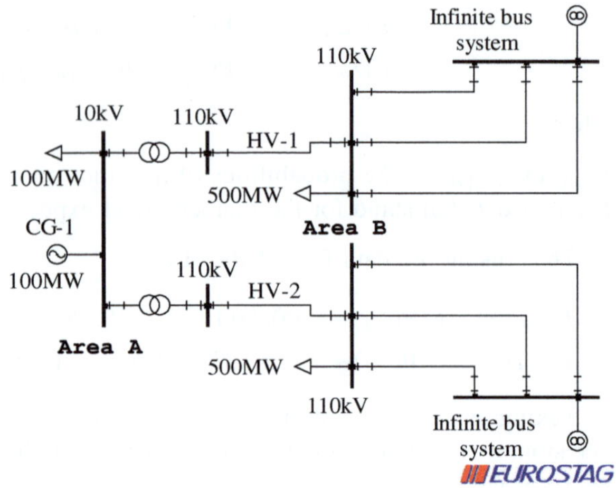

Fig. 14.10 Diagram of electric power system

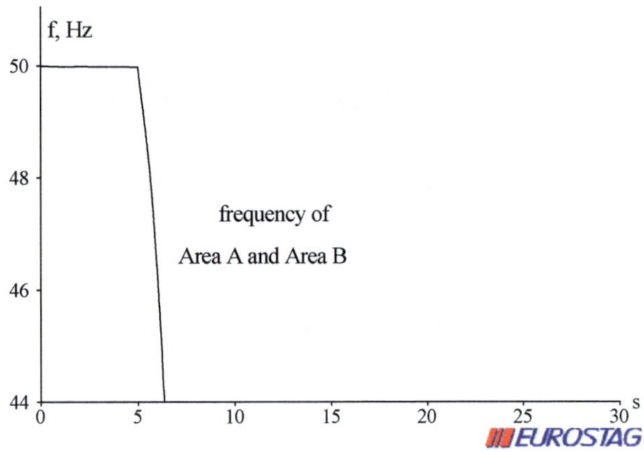

Fig. 14.11 Process of change of frequency during the blackout in the areas A and B at the loss of connection with the infinite bus system

During emergency an under-frequency situation that happened in 2010 all 110 kV lines, which transferred 1,000 MW in the area B from an infinite bus system (Fig. 14.10), were disconnected. After an isolated region was formed in which area A had to supply a power of its own load and also the load that stayed connected to the area B.

In the result of this accident even after UFLS operation one generation source in the area A is not able to supply all the consumers in the areas A and B. A blackout occurs in the considered case (Fig. 14.11).

The losses that arose during the considered emergency under-frequency situation could be decreased by separating area A from the network. The indicator for the necessity to separate area A is fast decreases of frequency in the area A and change of power flows direction in the lines HV-1 and HV-2. Figure 14.12 illustrates behaviour of frequency in areas A and B after separation.

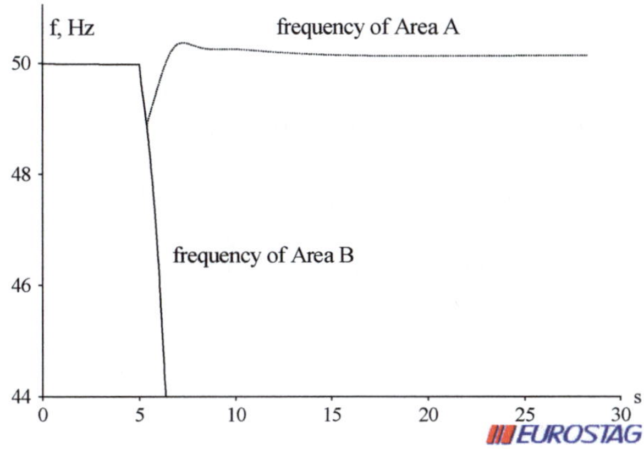

Fig. 14.12 Process of change of frequency in the areas A and B during separation of network

Table 14.5 Comparison of losses

Considering operational situation	Lo (M€)	T (h)	P (MW)	b
Without separation	402.2	3	1,100	2,438
With separation	366.6	3	1,002.5	2,438

Most of cogeneration power plants are equipped with automation for disconnection from the power system during under-frequency situation to save plant's auxiliaries. In this case the cogeneration power plant CG-1 will remain in operation and the 1,000 MW load of area B will be lost.

Losses due to undelivered power to consumer caused by undesirable events are calculated as follows:

$$Lo = T \cdot W \cdot b \cdot l \tag{14.14}$$

where T is the time of power supply interruption (hour); W is the undelivered power (MW); l is a price of undelivered energy (€/MW h) and b is the coefficient. Comparison of the results is shown in Table 14.5.

Calculations show that the value of probability of the loss of tie lines with the system is 0.001/year. In spite of this rather small probability application of separation automation gives significant decrease of technical risk. Using data from Table 14.5 and b = 2,438 it is possible to determine that minimization of technical risk is 35,600 €/year.

14.7 Conclusions

Combination of seldom unfavorable events and conditions in the power system can cause considerable economic and social losses.

Estimation of risks is conjugated with significant difficulties of mathematical (large number of variables, necessity to calculate multiple integrals) and informational (presence of indefinite parameters, complexity of representation the distribution and losses functions) characters.

For determination of the losses function experience of power unit's design and selection of substation's parameters can be used.

Under-frequency load shedding in combination with power system separation automation are powerful means for minimization of the losses rouse during frequency actuated power system emergency.

References

1. Final Report on the August 14, 2003 Blackout in the United States and Canada. U.S.-Canada Power System Outage Task Force, April 2004. Available at https://reports.energy.gov/
2. Final report of System disturbance on 4 November 2006. UCTE, 2006, 84p. Available at https://www.entsoe.eu/index.php?id=59
3. Final Report of the Investigation Committee on the 28 September 2003 Blackout in Italy. UCTE, 2004, 128p. Available at https://www.entsoe.eu/index.php?id=59
4. Final report on the course of events of Power failure in Eastern Denmark and Southern Sweden on 23 September 2003. UCTE, 2003, 24p. Available at http://www.energinet.dk/NR/rdonlyres/BC99F243-304D-4ADD-B769-46351A959C85/0/Powerfailurereportsept2003.pdf
5. Y.V. Makarov et al., Blackout Prevention in the United States, Europe, and Russia, in *Proceedings of the IEEE*, vol. 93(11), Nov 2005. pp. 1942–1955
6. M. DeGroot, Optimal Statistical Decisions (McGraw-Hill Series in Probability and Statistics, New York, 1970)
7. W. Li, Risk assessment of power systems: Models, methods, and applications. IEEE press series, 2005, 325p
8. B. Stott, O. Alsac, A.J. Monticelli, Security analysis and Optimization, in *Proceedings of IEEE*. Dec 1987. vol. 75(12), pp. 1633–1644
9. С.А. Совалов, В.А. Семенов, Противоаварийное управление в энергосистемах. Москва: Энергоатомиздат, 1988, 416 с
10. Task force on probabilistic aspects of reliability criteria, Probabilistic security assessment for power system operations, in *Proceedings of IEEE PES General Meeting*, 2004, vol. 1, pp. 212–220
11. Nordel System Operation Agreement. Nordel, 2006. Available at https://www.entsoe.eu/index.php?id=63
12. M.G. dos Santos, M.T. Schilling, R. Billinton, Bibliography on power systems probabilistic security analysis 1968–2008. Int. J. Emerg. Electr. Power Syst. **10**(3), Article 1 (2009), Berkeley Electronic Press
13. J. Barkans, D. Zalostiba, *Protection against Blackouts and Self-Restoration of Power Systems* (RTU Publishing House, Riga, 2009), p. 142
14. CIGRE TF 38.03.12 (R.J. Marceau and J. Endrenyi, Chairmen), Power system security assessment: a position paper, Electra, Dec 1997, no. 175, pp. 48–78
15. A. Schreiner, G. Balzer, Value at risk method for asset management of power transmission systems, in *Proceedings of IEEE PES Power Tech Conference*, 2007
16. M.J. Beshir, Probabilistic based transmission planning and operation criteria development for the Western Systems Coordinating Council, in *Proceedings of the 1999 IEEE PES Summer Meeting*, pp. 134–139

17. A.M. Leite da Silva, J. Jardim, A.M. Rei, J.C.O. Mello, Dynamic security risk assessment, in *Proceedings of the 1999 IEEE PES Summer Meeting*, pp. 198–205
18. J. Momoh, M. Elfayoumy, W. Mittelstadt, Y. Makarov, Probabilistic angle stability index, in *Proceedings of the 1999 IEEE PES Summer Meeting*, pp. 212–218
19. S. Lee, S. Hoffman, Power delivery reliability initiative bears fruit. IEEE Comput. Appl. Power **14**(3), 56–63 (2001)
20. R. Billinton, R.N. Allan, *Reliability Evaluation of Power Systems* (Pitman Publishing Inc., New York, 1984)
21. A. Bondarenko, V. Gerikh, V. Kreslins, N − 1 reliability criterion interpretation, in *Proceedings of CIGRE 2006*, C2-106
22. A1-Appendix 1: load-frequency control and performance, UCTE. Available at http://www.ucte.org
23. V. Chuvychin et al., An adaptive approach to load shedding and spinning reserve control during underfrequency conditions. IEEE Trans. Power Syst. **11**(4) (1996)
24. Y. Ohura et al., A predictive out-of-step protection system based on observation of the phase difference between substations. IEEE Trans. Power Delivery **5**(4), 1695–1704 (1990)
25. D. Diegel et al, Fulfillment of grid code requirements in the area served by UCTE by combined cycle power plants, Siemens AG, Power Generation, Germany, 2004. www.powergeneration.siemens.com
26. S. Aboreshaid, R. Billinton, A framework for incorporating voltage and transient stability considerations in well-being evaluation of composite power systems, in *Proceedings of the 1999 IEEE PES Summer Meeting*, pp. 219–224
27. A. Sleva, Protective Relay Principles (CRC Press, Taylor & Francis Group, Boca Raton, 2009)
28. A. Sauhats et al, Development of algorithms for prevention of asynchronous operation in high-voltage networks. in *Proceedings of IEEE Power Tech Conference*, BPT03-191, 2003, pp. 1–6
29. G.A. Korn, Th.M. Korn, Mathematical handbook for Scientists and Engineers: Definitions, Theorems and Formulas for Reference and Review. Second, enlarged and revised edition (McGraw-Hill Book Company, New York, 1968)

Chapter 15
Containment of Disturbances

Xiao-Ping Zhang, Xuefeng Bai and Jingchao Deng

15.1 Introduction

A risk estimation concept for black outs was presented in the previous chapter. In order to reduce this risk of black outs, precautional measures can be taken to keep disturbances in the power system more under control. Containment of disturbances in interconnected power systems replies on the Grid enhancement strategies. They may be defined as follows:

1. Strategy for interconnection of bulk power system

Planning for the interconnection of power grids is the most important strategy for power infrastructure to enhance the system stability. For two synchronous areas, which have been developed and operated differently for some considerable length of time, the appropriate interconnection would be helpful to keep major differences in system structure and operation mechanism while improving the efficiency and security of the interconnected system.

2. Strategy for preventive control of grid enhancement

Enhancement measures before the occurrence of disturbance are named as preventive controls, which are focused on security of the current operating point of the grid while improving system stability.

X.-P. Zhang (✉) · X. Bai · J. Deng
University of Birmingham, Birmingham, UK
e-mail: x.p.zhang@bham.ac.uk

X. Bai
e-mail: x.bai.1@bham.ac.uk

J. Deng
e-mail: JXD704@bham.ac.uk

U. Häger et al. (eds.), *Monitoring, Control and Protection of Interconnected Power Systems*, Power Systems, DOI: 10.1007/978-3-642-53848-3_15,

3. Strategy for emergency control of grid enhancement

For interconnected power grids with bulk energy exchange, severe faults may cause stability problems. Emergency control will be triggered immediately on the concurrency of the fault in order to guarantee the system security.

4. Strategy for remedial control of grid enhancement

If the interconnected system loses synchronization or experiences low frequency/voltage for a certain time period after disturbance, remedial control should be taken to decouple the interconnected grid, or Under Frequency Load Shedding/Under Voltage Load Shedding (UFLS/UVLS) will be put into force. Remedial control will be carried out after the fault to keep the synchronization of the system.

5. Strategy for control implementation of grid enhancement.

Innovative technology brings revolutionary change for the implementation of preventive, emergency and corrective control. Application of promising technologies for transmission, monitoring, control and protection will enhance grid stability greatly.

In the next sections these different strategies will be explained in detail at the example of an interconnection between ENTSO-E RG CE and IPS/UPS followed by a study case.

15.2 Strategy for Interconnection of Bulk Power Systems

In order to enhance the interconnected power grid, many factors should be considered for the coupling of grids, such as the power flows of the interties, the actual conditions of power grids, and the hierarchy of system operations, etc. For a synchronous coupling, the grid should be compact and the connection should be strong enough to overcome the potential low frequency oscillation, transient and voltage stability problems. For HVDC interconnection, it is beneficial for power flow control and the interconnected grids could be in different hierarchy and operating philosophy, and it has been well recognized that the natural of HVDC is helpful to deter the propagation of disturbance. There are already some transmission lines between ENTSO-E RG CE and IPS/UPS, but they are not enough to undertake the task as "highway" for electricity energy exchange between these two bulk power systems. With current weak AC links, there may be stability issues if the power flow increases. Since the synchronous coupling is a long term goal, the HVDC/HVAC interconnection is a prior choice in medium term horizon. In this scheme, the HVDC can be used to control the power flow and damp the oscillation caused by the AC line through HVDC modulation [1].

As the power systems grow larger and become more complex, more attention must be paid to the assurance of power system stability and voltage stability, etc. It is important to extensively utilize the power system control or employ HVDC system to enhance AC power systems performance to realize the stable operation of bulk power systems and assure the efficient utilization of power facilities in the future.

15.3 Strategy for Preventive Control of Grid Enhancement

15.3.1 Strengthening of the Interconnected Grids

The ENTSO-E RG CE interconnected system is being operated more and more at its limits, and important variations of generation patterns within the ENTSO-E RG CE systems displace substantial amounts of electricity from one area to another within an hour of time or even shorter because of the optimization of the power market. The limited inter-system transfer capacities within ENTSO-E RG CE are caused by high utilization of the ENTSO-E RG CE transmission systems. Furthermore, the available inter-system transfer capacities in ENTSO-E RG CE are additionally reduced by the priority for renewable generation, in particular wind generation.

With the limited capacity near the cross border and insufficient allocation of the exchanged energy, the interconnection can lead to congestion within the ENTSO-E RG CE network, which could hinder electricity exchange with neighboring regions. So, the interface between ENTSO-E RG CE and IPS/UPS is not expected to be a limiting factor of the power transmission. In most cases power transfers will be limited by congested internal sections of each synchronous zone. This can only be solved with an adequate transmission infrastructure and sophisticated system management methods of the interconnected networks.

After the coupling of the huge ENTSO-E RG CE and IPS/UPS power systems, the power flow over the interface will still require permanent reservation of free transmission capacity to ensure system security in case of decoupling. So, sufficient internal generation capacity may be still needed to avoid over-dependency of imported energy.

15.3.2 Preventive Control for Security Enhancement

The preventive actions will consist of manipulating a coordinated set of "controllable parameters" in the pre-contingency state consisting of the following:

1. Adjustments of operating reserves on certain units
2. Network element switching
3. Adjustment of interface flows across specifically designated interfaces

4. Adjustment of HVDC and FACTS device control set points
5. Voltage/VAR rescheduling
6. Generation rescheduling
7. Shed certain loads.

Mechanisms for getting the final preventive action decisions may consist of one or a combination of the following:

(a) User-suggested preventive actions
(b) Rule-based preventive actions
(c) Preventive actions obtained through optimal power flow or simulation.

The information available from the base-case analysis may provide sensitivity data and limitation of power transfer that are helpful in consideration of preventive actions. The sensitivity data could be in the form of a "sensitivity matrix" that relates incremental changes in the "controllable parameters" to the incremental changes of "output variables". Data on limitation of power transfer can be obtained from specific critical variables, such as interface flows across designated transmission corridors. The limits could be in the form of operating regions, such as upper and lower limits for a given variable.

Once the decision for preventive action is made, a simulation should be carried out to verify that the resulting conditions would be secure. Otherwise, corrections to that decision should be made and the results tested until an acceptable condition is reached.

15.3.3 Application of HVDC and FACTS for Preventive Control

For system stability and transmission capacity improvement of interconnected systems, power electronics technologies are very promising and expected to be utilized. To realize effectiveness of power electronic devices, it is necessary to classify the applicability of AC power system enhancement measures employing power electronics technology for the future AC system.

In order to introduce power electronics technology to ENTSO-E RG CE and IPS/UPS power systems, it would be necessary to identify their effects and limitations, and utilize them in such a manner that they are most appropriate in the interconnected power systems. Therefore, we should evaluate the applicability of such technology and identify the technology development needs by simulation analysis on model power systems and control protection schemes and testing them on simulators. In particular, HVDC system and SVC adopting power electronics technology have been placed into practical use to improve the power system stability. The following is a typical analysis procedure (We take STATCOM, TCSC, UPFC as examples; compare Chap. 9).

First, the FACTS device, such as a self-commutated converter (STATCOM), a thyristor controlled series compensator (TCSC) or a unified power flow controller (UPFC), is selected through evaluation by digital simulations on model power systems. Second, miniature models are developed. The control and protection systems of the equipment are real and could be used practically and immediately. Finally, effectiveness and the realization-possibility of power system enhance measures will be verified through simulation studies with actual controllers parameters.

For the STATCOM model, it is designed and manufactured taking the following aspects into account:

1. Continuous operation capability during system faults
2. Application of damping enhancement control
3. Low losses and harmonics equivalent to actual STATCOM losses.

For TCSC model, it requires continuous and quick impedance controls under severe system stability conditions. Furthermore the protection systems of the TCSC must make it possible to start control smoothly after fault clearances. Operation of the TCSC should not cause system disturbances even when the TCSC is inserted into the power system. The following objectives should be considered:

1. Soft insertion of TCSC into the power system
2. Basic impedance control
3. Stabilization of power system oscillation
4. Suppression of SSR (sub-synchronous resonance)
5. Protection system during AC system faults.

For the UPFC model, it should have fault protection and control functions. Here, typical control functions are transient, dynamic and steady state stability and power flow control and conditioning functions. For internal control of the UPFC, the model should include published methods. In developing the model, the specifications should be decided by considering particular problems of a miniature model such as modeling accuracy, as well as problems for the implementation on the actual power system. And following functions should be implemented:

1. For the basic functions, obtaining required output and smooth starting and stopping without disturbance are needed.
2. When exceeding the current rating because of a power system faults occurrence, a standby mode of operation must be implemented, followed by quick restarting up after clearing the fault.
3. When an internal fault occurs, disconnecting the UPFC from the power system and avoiding the power system separation are needed.
4. A power system stabilizing control function is needed.
5. A power flow control function is needed.

To sum up, the main objective of preventive control is to enhance the power system by setting the appropriate operating point. Preventive actions are applied in the pre-contingency system so that after any credible contingency the system remains secure. Examples include restrictions on interface flows, angle differences across a particular interface, and total generation out of a plant. Besides the traditional measures, HVDC transmission and FACTS devices could play a key role for power flow control and oscillation damping. FACTS devices could enhance system controllability by allowing a more flexible system operation and improving system performances in terms of power transfer capacity, power flow control, stability, and so on.

From the long term point of view, transmission planning of the infrastructure of grids can be regarded as a kind of preventive control in seasons, while the setting of system operation mode is a minutes or hours-oriented preventive countermeasure. Based on the operating scenarios adjustments, the preventive control can ensure system security under most disturbances, such as N − 1 fault, but there may not be a universal operation mode for all kinds of faults because of the complexity of a large grid. Sometimes, the controls for two faults may be totally contradictory. Furthermore, the strict preventive control may decrease the transfer capability of interconnected power systems and prejudice the economic operation of the system. In these situations, emergency control is a more efficient solution.

15.4 Strategy for Emergency Control of Grid Enhancement

Emergency control automatics, which have been applied extensively in IPS/UPS system to guarantee system stability, are those control actions taken following a contingency. Examples include generator or load tripping, and capacitor bank or reactor switching. Such a technical solution is a feasible measure to ensure the system integrity in case of stability problems arising. If preventive control can be defined as the 1st defence line pre-contingency control, emergency control would be the 2nd defence line post-contingency control. The proper action for individual contingencies can be determined separately. The corresponding emergency action may involve shedding different combinations of load groups or dropping generators at one or several buses depending on the contingency, and the actual operating conditions.

Traditionally, preventive and emergency security measures have been developed from numerous off-line simulations. Conventional time domain simulation is the most reliable security assessment approach. The advantages of this approach are that the modeling of components is flexible and the resulting trajectory is highly accurate. Even now, its wide usage by utilities makes it the most standard tool for transient stability analysis in power system engineering. By this way, the operating scenarios and contingency definition are basic raw data for the formation of control strategies, and the pair of them is to be a filter for the detail control measures after a contingency. As the power systems have become more complex due to the continuous upgrading of equipment and the deregulated and unbundled electricity supply industry operated in a non-discriminatory and equitable manner, the combination of scenarios and contingencies is explosive. It is almost impossible to select the exact or closest scenario and contingency pair of off-line simulation to match the real-time system after a disturbance. The emergency control mode of "off-line simulation and real-time match", which is the most prevail control in most utilities, will not work in the future interconnected power grid.

On-line dynamic security assessment tools have been developed for many years. These tools differ in the methodology but they share the same concepts and fundamental blocks. It is based on the strategy of "on-line simulation and real-time match", in which the control measures will be calculated on the real operation point and contingency will be the only clue for countermeasure filtering.

For large complex power systems, it is impractical and unnecessary to perform full detail analysis on the impact of every conceivable contingency. Instead, accurate but fast contingency screening indices can be used to reduce the computation to a manageable level. For successful screening, the indices should be a good measure of system severity in the transient condition.

The contingency screening and ranking method could make use of transient energy functions, expert systems, neural networks, extended equal area criteria, or indices derived from energy properties or fast time domain simulations using simplified models. The common requirements of candidate contingency screening and ranking methods are high speed and accuracy of the final results. While all these methods can be used for contingency screening and ranking, the final limit computations should be done using more accurate methods.

There are three time-domain simulation based indices for ranking of contingencies. They are energy margin index based on hybrid extended equal area criteria (EEAC), transient energy index (TEF), and critical clearing time (CCT). By combining direct stability assessment methods (TEF & EEAC) with an accelerated time time-domain simulator, the advantages of quantitative assessment can be fully obtained without the restrictions on the complexity of power system models.

The output of on-line security analysis can be used for stability control, such as alarming of the correct number of generating units for tripping. This can be thought of outer-loop adaptive supervisory control. High quality digital measurements can improve state estimation and they can be used for monitoring and stability control. Synchronized positive sequence phasor measurements are one type of digital measurements. Phasor measurements may be sufficiently related to dynamic states such as rotor angles and speeds and be useful for stability control.

Formation and implementation of on-line emergency control measures can do great contribution to grid enhancement. While most of the systems follow the N − 1 means is used to overcome the consequences of disturbances (i.e. power imbalances, grid elements tripping or overloads, violations of voltage limits, etc.), protection, re-dispatching and automation actions. Cascading contingencies can be regarded as the combination of N − 1 fault if the time interval is enough. This is complied with the practical evolution of blackout. Take USA power blackout on 13 August 2003 as an example, it took 62 min for the 5 transmission lines to be tripped in succession with 22, 9, 5, 29 min time interval respectively. Obviously, there were at least 4 N − 1 contingency and if some on-line emergency controls were calculated and implemented, the results would have been totally different. By this way, the occasionally fault will be handled timely and would not lead to blackout.

On-line security assessment is the platform of "on-line simulation and real-time match", it consists of the following elements:

- Preprocessing
- Security assessment
- Post-processing.

15.4.1 Preprocessing

The task of preprocessing includes static state estimation, contingency screening and ranking.

A major impediment to online security assessment for inter-area stability problems in large interconnections is the difficulty of state estimation to obtain the on-line power flow base case. For a particular control center, the main difficulty is with the external network model. Considerable inter-utility data exchange is required. A solution for this problem is provided in Chap. 6. There are other difficulties associated with measurement accuracy, unbalance operation, network parameter uncertainty, etc.

External network models can be obtained by selecting from a number of previously stored dynamically reduced system models. Alternatively dynamic reduction techniques can be used in real-time to develop the external model. This will facilitate base case initialization and help maintain the base case within certain size limit so that the computation speed requirements can be met.

In addition to power flow data, other data required may also need to be updated for a new system snapshot. For instance, the settings of a PSS for a pumped storage generation unit may need adjustment for the different modes of operation of the unit. The contingencies may also need update when the network topology or system operation condition changes.

It's impossible to assess all the credible contingencies within the confines of available computational resources and required response time. Therefore, the list of credible contingencies has to be reduced to make it manageable by the security assessment module. An acceptable contingency screening and ranking method should have zero false dismissals and a very low number of false alarms.

15.4.2 Security Assessment

Normally, DSA assesses transient stability of a power system, or the ability of the system to maintain synchronism after a credible contingency. As power systems operate in more and more stressed conditions, another form of angle stability, small-signal stability in the form of sustained or growing oscillations in part or all of the system, may become critically restricting to the system operating limits. This has already happened in some parts of the North American interconnected power systems. The requirement to address this type of stability problem calls for an efficient and reliable method to compute the critical mode of the system. This is still an area for research. One approach based on the time domain simulation technique is to obtain an estimate of the critical mode by post-processing simulation results.

15.4.3 Post-processing

Online implementation of time simulation requires a built-in intelligence for the following:

- Assessing the system dynamic performance (stable, unstable)
- Determining the degree of stability or instability (margin)
- Determining the sensitivity of the margin to key variables (transfer limit and generation tripping)
- Determining the transfer limit or preplanned stability control actions (arming of generator tripping).

Through the evaluation of the ability of the system to withstand contingencies by surviving the transient conditions to acceptable steady-state operation, indications about the emergency actions can be given when necessary.

Two of the main features of the "on-line simulation and real-time match" function are:

1. Contingency screening: to rank a large number of contingencies and select those are likely to cause dynamic security violations.
2. Contingency evaluation: to carry out time domain simulation based transient and dynamic stability assessment, and, if necessary, to propose preventive/emergency actions to improve system security according to the contingency severity.

15.5 Strategy for Remedial Control of Grid Enhancement

The main objective of remedial action determination is to determine appropriate remedial action schemes in case the preventive/emergency action is not adequate to ensure system security. If the system is hazarded by losing synchronization, low frequency or low voltage, remedial control will be taken to avoid the system collapse. It could be the 3rd defense line against blackouts in large power grids.

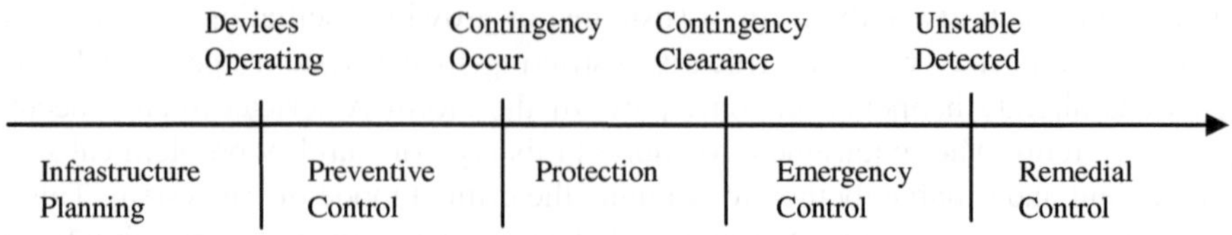

Fig. 15.1 Relationship of defense-lines for interconnected power grid enhancement measures

The main measures for the 3rd defense line include decoupling of the interconnected grids, under frequency load shedding (UFLS) and under voltage load shedding (UVLS). Compared with preventive and emergency control, no definition of contingency is needed here, and the settings of the control devices are independent of security simulation. The results of the control measures can be reflected by system variables, which could be indicators for further action, so remedial control is feedback control. While preventive and emergency control can only be evaluated by simulations and they are no-feedback controls.

Remedial control could be centralized or de-centralized; the difference directly comes from the organization philosophy and structure of the power systems. Basically, remedial control is based on the local measurements and act while the setting valve is satisfied. For interconnected grid, certain remote signals could be introduced to improve the control results, but attention should be paid to the reliability and effectiveness of the remote signals because of factors such as time delay. Furthermore, all remedial control equipments near to the interface of interconnected grid should be inspected and tested in operation. Remedial control is an effective measure to deal with the serious contingency with little possibility.

The relationship of preventive, emergency and remedial controls is shown as in Fig. 15.1.

Infrastructure planning, preventive control and protection constitute the 1st defense line for grid enhancement, whose target is to avoid the loss of load. The infrastructure planning is a kind of preventive control in long terms run. After the operating of new devices, the modulation of parameters will act as the enhancement measure. Once a contingency occurs, protection systems will act and clear the fault. If the relay protection cannot ensure system security, the 2nd defense line will take effect. The measures of emergency control may include the tripping of loads and generators. Finally, system decoupling, UFLS and UVLS take the place as 3rd defense line if unstable characters of the system are detected.

Preventive control guarantee the security of system by decrease the transfer capacity, while emergency can be used to increase the transfer limit. Preventive control, which should be enforced before disturbance, must cover all the conceivable potential contingencies, and may meet contradict control requirements. Emergency control, which will be enforced after disturbance, may execute the exact countermeasures according to the particular contingency. So it is easier to avoid contradict control requirements for emergency control. A practical method is to move the operating point to a scenario that is relatively optimal to overall

Table 15.1 Applicable measures for grid enhancement

Classification		Control technology items
Bulk power transmission		UHV AC transmission HVDC transmission 500 kV Cable transmission Gas insulated transmission
Stabilization	Generator control	Quick response exciter with PSS Fuzzy excitation control Turbine high-speed valve Multi-variable control Variable speed pumping storage
	System control	Series capacitor(TCSC) Phase-shifter(UPFC) Self-commutation SVC (STATCOM) Damping resister high speed fault clear System stabilization relay Power modulation by HDVC system
	System configuration	Multiple circuit looping Multiple transformer banks Intermediate switching station Upgrade to higher voltage BTB DC link
Reliability enhancement		Total automation control Gas insulated substation Application of AI

situation by preventive control. Emergency control is used to secure the system after the contingency. By this way, the selection of operating point is a question for the coordination of preventive and emergency control.

Both preventive control and emergency control are prediction dependent, they guarantee the security of post-fault system by analysis before the fault's occurrence. For off-line control assessment, the operating mode is the index entry for preventive control, and contingency is added as another index entry for emergency control. For on-line security assessment, operating mode is real time and there is no index entry for preventive control, while contingency is the only index entry for emergency control.

Preventive control is taken effect during the process of system operation, and the cost of the control action has nothing with the disturbance details. The cost is in proportion to the time that the control action lasts, so preventive control is not suitable to small probability contingency. While there is no control cost under system daily operation for emergency control, and the control results would be worse if the action is taken later after the contingency. The economic characters are different with these two control measures; the key of the coordination is to minimize the cost of emergency control risk and the cost of preventive control in daily operation.

Emergency control, which is incurred by contingency signal, is rather conservative. Remedial control is activated by the characteristic quantity of power angel, frequency and voltage curves; it is a feedback control and is more accurate than emergency control. But the remedial defense line is executed later than the emergency defense line, so more control variables are needed to improve the result. The decoupling devices in emergency control defense will act if specific contingency is detected. Once the real operating mode is safer than anticipated, the decoupling would be unnecessary. Similarly, the load may be over tripped in emergency control. So over-control is not economical and may even hazard the system security. The qualitative conclusion is that emergency should not be too conservative, remedial control could be a supplement for the control of serious error with small probability.

Some applicable enhancement measures are listed in Table 15.1, they may be used as preventive control, emergency control or remedial control.

15.6 Strategy for Control Implementation of Grid Enhancement

Implementation of the measures for grid enhancement relays on the availability of system information and improvement of technology. In the past, most operators could mainly focus on the stability of their own gird and the use of conventional systems such as SCADA, this has proven not to be sufficient in large interconnected grids any more. It is becoming increasingly necessary for them to see beyond the borders of their grid and to have additional sources of information available providing them with a higher quality of online data. For many incidents like the large blackouts in the US and Europe, the origin lies in a neighboring region or country. Unpredictable changes in physical power flows due to increased energy trading and undeterminable generation from wind farms, could have a severe impact on grid operation within a utility's regional borders.

In reality, many TSOs of interconnected grids face increasing difficulties to build new network infrastructures (lines, substations, etc.). This puts more pressure than ever before on all TSOs to be able to rely on each other via closer co-ordination mechanisms. So it substantially increases transparency of the fundamentals of the TSO rules and therefore the necessary mutual confidence of TSOs among themselves as well as their credibility towards stakeholders.

Operators forced to push transmission capacity towards dynamic limits can hardly rely on static information with update frequencies of one or more minutes. WAMS has therefore become an effective device to supervise changes of load or impacts of disturbances in the own grid as well as across traditional borderlines online. WAMS has proven the value of additional online stability information for use in the control center and to better understand dynamic power system behavior as well as fine-tune system parameter, e.g. in power system stabilizers.

Therefore, it is complementary for existing SCADA system to be integrated with WAMS to effectively prepare and decide on control center interaction before the impacts of a neighboring event can propagate into the own operational area.

Upgrading traditional SCADA to DSCADA by WAMS is beneficial to monitor static behavior and dynamic behavior of the system. It can also be used in system stability margin assessment, fault evaluation, low frequency oscillation analysis, model and parameter verification, etc. Furthermore the accuracy of State Estimation would be improved. For more details on these topics please refer to Chaps. 4–8.

Traditionally, simulations are for the control strategy and measurements are for the action of devices. Based on the WAMS, such relation may be broken up. Security analysis may be done by measurements only, which is independent of system model. The integration of simulation and measurements can be used in determination of stability margin, damping of low frequency oscillation and corresponding sensitivity analysis.

As a consequence, wide area protection systems pose demanding requirements in comparison to wide area monitoring systems in terms of reliability, security, redundancy and processing speed. Therefore, it is necessary to build up a wide area protection system on a well proven and universal high-level automation platform with high performance.

The research on evolution of blackout proves that the loop of information monitoring, on-line security analysis and control strategy are essential to guarantee the system stability. The information should include static SCADA and dynamic WAMS, not only in local area, but also from remote data centers. On-line security analysis can ensure the validity of the controls, which guarantee the system stability after contingence. The defense lines are strategies for enhancement of grid, HVDC, FACTS and WAMS are essential technology to implement the strategies.

15.7 Case Study

The case study is based on the network model [2] presented in Chap. 2. In this section, the critical contingencies and the corresponding emergency control measure will be introduced and the impact of UPFC, TCSC and VSC-HVDC is investigated [3, 4].

In this study, 3 tie lines are defined between the two giant power systems; they are North corridor, Central corridor and South corridor. The interconnection options include pure AC, pure DC and AC/DC hybrid interconnections.

N − 1 contingencies screening have been carried out with the base loadflow and interconnection scenarios. Six cases are found that will endanger the stability of the interconnected power system. For the interconnection scenarios, 4 cases are found with one HVDC interconnection on South corridor and one case is found with pure HVAC interconnection and one HVDC interconnection on North corridor respectively. The locations of the critical contingencies can be classified as:

- Transmission corridor within IPS system
- Bus near the receiving end of north tie-line
- Transmission corridors near the load center Madrid.

In the following the simulation results will firstly be presented in Sect. 15.7.1 without FACTS devices only by use of conventional load tripping as emergency control method. Then, in Sect. 15.7.2 the same situations are presented with use of FACTS devices and in 15.7.3 with use of VSC-HVDC. The main principle of the simulations for each case is to increase the power transfer on the corridor between IPS/UPS and ENTSO-E RG CE until the stability limit is reached.

15.7.1 Emergency Control Study

With the reduced network model, N − 1 contingency simulations show six critical contingencies for an assumed interconnection between ENTSO-E RG CE and the IPS/UPS system with different connection scenarios. Some emergency controls are needed for these contingencies to stabilize the system.

Four of the six severe contingencies are with the AC/DC hybrid interconnection scenario, which means that the 1 HVDC interconnections on South corridor is relatively weak compared with the other interconnection scenarios.

The severe disturbances are located in 3 areas including the high voltage transmission corridors in IPS, the receiving end of the North interconnection corridor and the load center near Madrid. In the following all six emergency control scenarios are briefly described:

- For the 750 kV transmission line from Severoukrain (4SEV5 0) to Chernoblskay (4CHE6 0) with AC interconnection scenario, the transferred power is 1,738 MW and the system will lose synchronism between IPS/UPS and ENTSO-E RG CE if no emergency control measure is applied. To keep stability of the system, emergency control of tripping one generation unit at Smolenskener is essential, which provides the power transferred on the faulted transmission line.
- For the 750 kV transmission line from Brjanskenerg (4BRJ2 0) to Kurskenergo (4KUR4 0) with 1 HVDC on north corridor and 2 AC inter-ties on Central and South Corridors, the transferred power is 3,238 MW and the system will lose synchronization without any control measure. To keep stability of the system, emergency control of tripping one generation unit at Smolenskener is essential, which provides the power transferred on the faulted transmission line.
- For the 500 kV transmission line from Jarenergo (4JAR2 9) to Kostromaener (4KOS3 9) with 1 HVDC on South corridor and 2 AC inter-ties on Central and North Corridors, the transferred active power is 1,265 MW and the system will lose synchronization without any control measure. To keep stability of the system, emergency control of tripping one generation unit at Jarenergo is essential, which provides the power transferred on the faulted transmission line.
- For the 380 kV transmission line from Polaniec (4POL5 1) to Koszienic (4KOS4 1) with 1 HVDC on South corridor and 2 AC inter-ties on Central and North Corridors, the transferred active power is 1,384 MW and the system will

Table 15.2 Emergency controls for severe cases

Transmission corridor		Transferred power (MW)	Voltage level (kV)	Inter-tie scenario	Emergency control measures
From	To				
Severoukrain (4SEV5 0)	Chernoblskay (4CHE6 0)	1,738	750	Pure AC	Generation trip at Smolenskener
Brjanskenerg (4BRJ2 0)	Kurskenergo (4KUR4 0)	3,238	750	AC/DC hybrid	Generation trip at Smolenskener
Jarenergo (4JAR2 9)	Kostromaener (4KOS3 9)	1,265	500	AC/DC hybrid	Generation trip at Jarenergo
Polaniec (4POL5 1)	Koszienic (4KOS4 1)	1,384	380	AC/DC hybrid	600 MW load shedding at Koszienic
Alcarama (4ALC1 1)	Madrid (4MAD1 1)	960	380	AC/DC hybrid	300 MW load shedding at Madrid
Almaraz (4ALM1 1)	Madrid (4MAD1 1)	666	380	AC/DC hybrid	600 MW load shedding at Madrid

lose synchronization without any control measure. For this case, the emergency control of tripping 600 MW load at Koszienic is essential to keep stability of the system.

- For the 380 kV transmission line from Alcarama (4ALC1 1) to Madrid (4MAD1 1) with 1 HVDC on South corridor and 2 AC inter-ties on Central and North Corridors, the transferred active power is 960 MW and the system will lose synchronization without any control measure. For this case, the emergency control of tripping 300 MW load at Madrid is essential to keep stability of the system.
- For the 380 kV transmission line from Almaraz (4ALM1 1) to Madrid (4MAD1 1) with 1 HVDC on South corridor and 2 AC inter-ties on Central and North Corridors, the transferred active power is 666 MW and the system will lose synchronization without any control measure. For this case, the emergency control of tripping 600 MW load at Madrid is essential to keep stability of the system.

To sum up, the emergency control measures are listed in Table 15.2.

15.7.2 Impacts of FACTS on Disturbance Containment

To investigate the impact of FACTS on the containment of disturbances, the impacts of UPFC and TCSC on the critical contingencies with paralleled transmission lines were studied. The UPFC or TCSC is applied on the control path to control the power flow or parameters, the set points will be reset to accommodate more power once the contingencies are detected. Three of the six critical

Table 15.3 Critical contingencies with paralleled transmission

Bus1		Bus2		Voltage level (kV)	Interconnection scenario
Bus ID	Bus name	Bus ID	Bus name		
4ALC1 1	Alcarama	4MAD1 1	Madrid	380	AC/DC hybrid
4ALM1 1	Almaraz	4MAD1 1	Madrid	380	AC/DC hybrid
4KOS4 1	Koszienic	4POL5 1	Polaniec	380	AC/DC hybrid

Table 15.4 Impact of UPFC on emergency controls of critical contingencies

Bus1		Bus2		Load tripping with UPFC (MW)	Load tripping without UPFC (MW)
Bus ID	Bus name	Bus ID	Bus name		
4ALC1 1	Alcarama	4MAD1 1	Madrid	200	300
4ALM1 1	Almaraz	4MAD1 1	Madrid	200	600
4KOS4 1	Koszienic	4POL5 1	Polaniec	600	600

contingencies with paralleled transmission are listed in Table 15.3. For all of the three cases, the interconnection scenario is AC/DC hybrid.

15.7.2.1 Impact of UPFC on Disturbance Containment

Impacts of UPFC on the emergency control are studied in this section. For the 380 kV transmission line from Alcarama (4ALC1 1) to Madrid (4MAD1 1), the transferred active power is 916 MW and the system will lose synchronization without any control measure. With UPFC installed, the emergency control of tripping 200 MW load at Madrid is essential to keep stability of the system.

For the 380 kV transmission line from Almaraz (4ALM1 1) to Madrid (4MAD1 1), the transferred active power is 648 MW and the system will lose synchronization without any control measure. With UPFC installed, the emergency control of tripping 200 MW load at Madrid is essential to keep stability of the system.

For the 380 kV transmission line from Koszienic (4KOS4 1) to Polaniec (4POL5 1), the critical contingency near the receiving end of North corridor, the emergency control of tripping 600 MW load at Koszienic is essential to keep stability of the system. The simulation results are almost the same as the emergency control without UPFC.

The impacts of the UPFC on the critical contingencies are listed in Table 15.4.

From Table 15.5, we can find that the application of UPFC can reduce the amount of load shedded for the contingencies near load center of Madrid. The swing of power angel can be damped greatly, which is beneficial for the system stability. For the contingency near the receiving end of North corridor, the UPFC

Table 15.5 Impact of TCSC on emergency controls of critical contingencies

Bus1		Bus2		Load tripping with TCSC (MW)	Load tripping without TCSC (MW)
Bus ID	Bus name	Bus ID	Bus name		
4ALC1 1	Alcarama	4MAD1 1	Madrid	0	300
4ALM1 1	Almaraz	4MAD1 1	Madrid	300	600
4KOS4 1	Koszienic	4POL5 1	Polaniec	600	600

on the control path doesn't contribute to the system stability. In fact, the parameters of the paralleled lines are not exactly the same; the impedance of the control path is much larger than the faulted line. The high impedance decreases the effectiveness of the UPFC.

15.7.2.2 Impacts of TCSC on Disturbance Containment

Impacts of TCSC on the emergency control are studied in this section. For the 380 kV transmission line from Alcarama (4ALC1 1) to Madrid (4MAD1 1), the transferred active power is 916 MW and the system will lose synchronization without any control measure. With TCSC installed, no emergency control is needed to keep stability of the system. The TCSC can change the parameters of the transmission corridor to accommodate more power.

For the 380 kV transmission line from Almaraz (4ALM1 1) to Madrid (4MAD1 1), the transferred active power is 648 MW and the system will lose synchronization if no control measure is taken. With TCSC installed, the emergency control of tripping 300 MW load at Madrid is essential to keep stability of the system.

For the 380 kV transmission line from Koszienic (4KOS4 1) to Polaniec (4POL5 1), the critical contingency near the receiving end of North corridor, the emergency control of tripping 600 MW load at Koszienic is essential to keep stability of the system. The simulation results are the same as the emergency control without TCSC.

The impacts of the TCSC on the critical contingencies are listed in Table 15.5.

From Table 15.5, we can find that the application of TCSC can reduce the amount of load shedded for the contingencies near load center of Madrid. The swing of power angel can be damped greatly, which is beneficial for the system stability. The impact is especially effective on the case from ALC to MAD, no emergency control is needed.

For the contingency near the receiving end of North corridor, the TCSC on the control path doesn't improve the system stability. The situations are similar to that with UPFC, the high impedance of the paralleled line decreases the effectiveness of the TCSC.

Table 15.6 N − 2 cases study simulation scenarios

Faulted line				Outage line			
Bus ID	Bus name	Bus ID	Bus name	Bus ID	Bus name	Bus ID	Bus name
4ALC1 1	Alcarama	4MAD1 1	Madrid	4ALM1 1	Almaraz	4MAD1 1	Madrid
4ALC1 1	Alcarama	4MAD1 1	Madrid	4LAM1 1	Lamudarr	4MAD1 1	Madrid
4ALM1 1	Almaraz	4MAD1 1	Madrid	4ALC1 1	Alcarama	4MAD1 1	Madrid
4ALM1 1	Almaraz	4MAD1 1	Madrid	4LAM1 1	Lamudarr	4MAD1 1	Madrid
4KOS4 1	Koszienic	4POL5 1	Polaniec	4LEM1 1	Lemesany	4POL5 1	Polaniec

Table 15.7 Emergency control measures for N − 2 cases study

Faulted line		Outage line		Control measures (load shedding)		
Bus1 ID	Bus2 ID	Bus1 ID	Bus2 ID	No FACTS (MW)	UPFC (MW)	TCSC (MW)
4ALC1 1	4MAD1 1	4ALM1 1	4MAD1 1	300	300	0
4ALC1 1	4MAD1 1	4LAM1 1	4MAD1 1	400	300	0
4ALM1 1	4MAD1 1	4ALC1 1	4MAD1 1	600	300	400
4ALM1 1	4MAD1 1	4LAM1 1	4MAD1 1	600	300	400
4KOS4 1	4POL5 1	4LEM1 1	4POL5 1	600	600	600

15.7.2.3 N − 2 Case Study

The objective of this study is to investigate the impact of the FACTS devices on the severe contingencies when the adjacent transmission line is out of service. Several N − 2 cases are studied in this section. The cases are based on the contingencies in previous two sections with one adjacent transmission line outage.

Five scenarios listed in Table 15.6 are simulated and the emergency control countermeasures are listed in Table 15.7. Compared with the simulation results in previous sections, the following conclusion can be drawn:

For the three-phase short-circuit fault on the line of corridor ALC-MAD, if the outage transmission line is in corridor ALM-MAD, the control measures would be the same as that when the line in ALM-MAD is in service. Since the network near Madrid is highly meshed and the power flow of one line in ALM-MAD is not very heavy (648 MW), the power flow in the outage transmission line can be transferred by its paralleled transmission lines. But 100 MW more loads should be shedded if the outage line are in corridor LAM-MAD, because the transmission are heavy loaded (1,015 MW). For the application of FACTS devices, the TCSC is much more efficient than UPFC.

For the three-phase short-circuit fault on the line of corridor ALM-MAD, no matter that the outage transmission line is in corridor ALC-MAD or LAM-MAD, the control measures would be the same as that when the line is in service, because there is not much difference for power flow in the lines of ALC-MAD and LAM-MAD (916 and 1,015 MW respectively). For the application of FACTS devices, the UPFC is a little more efficient than TCSC.

For the three-phase short-circuit fault on the line of corridor KOS-POL, the outage transmission line is in corridor LEM-POL, which transfers most active power in the adjacent lines. The control measure is the same as that when the line is in service, and the situation cannot be improved by the application of FACTS devices. For the contingency near the receiving end of North corridor, the TCSC on the control path doesn't improve the system stability. The reason is the same as in previous sections, the high impedance of the paralleled line decrease the effectiveness of the FACTS devices.

15.7.2.4 Brief Conclusion

The application of FACTS devices can enhance the stability of the interconnected system effectively. The critical contingencies could be resolved or the amount of load shedding could be decreased. The effectiveness depends on the contingency scenarios. The control strategy of FACTS devices is practical with the paralleled transmission lines according to the simulation results.

For the contingency near the receiving end of North corridor, the FACTS devices on the control path doesn't improve the system stability. The high impedance of the paralleled line is the main reason and the construction of new transmission lines may be necessary.

The critical $N - 2$ contingencies simulation shows that the FACTS devices can improve the stability and the situation is much similar with the $N - 1$ analysis. Because the ENTSO-E RG CE network is highly meshed, the power transferred by the outage line can be taken over by its neighboring lines. The application of FACTS devices should be focused on certain severe contingencies on specified critical transmission lines.

15.7.3 Impacts of VSC-HVDC on Disturbance Containment

The application of the latest HVDC technology VSC-HVDC (compare Chap. 9) was also studied. The cases are based on the critical contingencies in $N - 1$ contingency screening; they are mainly relative with the interconnection scenarios of one VSC HVDC interconnection on North and South corridor.

15.7.3.1 Simulation of the VSC HVDC on North Corridor

There is one severe contingency for the hybrid AC/DC interconnection scenario. It is with the 750 kV transmission line from Brjanskenerg (4BRJ2 0) to Kurskenergo (4KUR4 0), on which the transferred power is 3,238 MW. In this case, the CSC HVDC is replaced by VSC HVDC. From the simulation result, we know that the system will lose synchronization without any control measure. The control measure could be the same as the traditional one by tripping one generation unit at Smolenskener. The application of VSC-HVDC does not contribute for this critical contingency because of the heavy power flow transferred through this key transmission line.

15.7.3.2 Simulation of the VSC HVDC on South Corridor

There are four severe contingencies for the hybrid AC/DC interconnection scenario. Two of them are located on the load center of Madrid. One is at the receiving end of north corridor and the last one is with a 500 kV transmission line in IPS.

For the 380 kV transmission line from Alcarama (4ALC1 1) to Madrid (4MAD1 1), the transferred active power is 916 MW and the system will keep synchronism without any control measure with VSC HVDC on south corridor.

For the 380 kV transmission line from Almaraz (4ALM1 1) to Madrid (4MAD1 1), the transferred active power is 648 MW and the system will keep stable with VSC HVDC on south corridor instead of CSC HVDC with emergency control.

For the 500 kV transmission line from Jarenergo (4JAR2 9) to Kostromaener (4KOS3 9) in the IPS/UPS system, the transferred active power is 1,265 MW and the system will keep synchronization without any control measure.

For critical contingency at the receiving end of the North corridor the system will keep synchronization without any control measure.

15.7.3.3 Brief Conclusion

The simulation results show that the VSC HVDC can improve the stability of the interconnected system greatly. The critical contingencies with interconnection scenario HVDC13 can all be resolved with VSC HVDC. Power will be transferred through long distance heavily loaded transmission lines and the independent control of active and reactive power of VSC HVDC can play an important role for grid enhancement. But the emergency control is still necessary for the critical contingencies on 750 kV transmission line with HVDC interconnection on north corridor because of the heavy loads.

15.8 Conclusions

The disturbance containment strategies have been presented in detail, including the coupling options, preventive control, emergency control, remedial control and coordinated control of FACTS devices. The control measures have been applied in the N − 1 critical contingencies analysis, which were proved to be efficient to keep the stability of the interconnected systems.

The application of FACTS devices can enhance the stability of the interconnected system effectively. The critical contingencies could be resolved or the amount of load shedding could be decreased. The control strategy of FACTS devices is practical with the paralleled transmission lines according to the simulation results. But for a special case, the FACTS devices on the control path do not improve the system stability because of the high impedance of the paralleled line and the construction of new transmission lines may be necessary.

The simulation results have shown that the VSC HVDC can improve the stability of the interconnected system greatly. The critical contingencies in the scenario with 1 HVDC on North corridor can all be resolved with VSC HVDC. But the emergency control is still necessary for the critical contingencies on 750 kV transmission line with HVDC interconnection on north corridor because of the heavy loads.

References

1. P. Kundur, *Power System Stability and Control* (McGraw-Hill, New York, 1993)
2. Software Models of EU and Russia Interconnection Networks. Deliverable D1.2 of ICOEUR project
3. Simulation Models for FACTS and Appropriate Equipment, HVDC, HVAC interconnections including Their Control Systems to be Used in Simulation of Large Scale Electric Networks. Deliverable D2.1 of ICOEUR project
4. Tractebel Engineering, Eurostag Theory, Eurostag Manual

Chapter 16
Wide Area Protection

Christian Rehtanz, Nikolai Voropai, Ulf Häger, Dmitry Efimov, Daniil Panasetsky, Alexander Domyshev and Alexey Osak

16.1 Introduction

Emergency control, including emergency operation dispatching and automatic emergency control that provides reliability and survivability of the electric power grids plays an important part in controlling the operating conditions of large Electric Power Systems (EPS)s, as introduced in the previous chapter. Emergency control is performed by the technological (dispatching and automatic) control systems that include the automatic systems of voltage, frequency and capacity regulation, basic automatic systems of EPS elements, relay protection and automatic line control, system emergency control [1].

Many thousands of emergencies and more occur usually in large power systems during a year for different reasons—short circuits, failure of equipment, errors of personnel, etc. Most of these emergencies are eliminated by relay protection devices and automatic emergency control systems. The failures of relay protection

C. Rehtanz (✉) · U. Häger
TU Dortmund University, Dortmund, Germany
e-mail: Christian.Rehtanz@tu-dortmund.de

U. Häger
e-mail: ulf.haeger@tu-dortmund.de

N. Voropai · D. Efimov · D. Panasetsky · A. Domyshev · A. Osak
Energy Systems Institute, Irkutsk, Russia
e-mail: voropai@isem.sei.irk.ru

D. Efimov
e-mail: d.n.efimov@gmail.com

D. Panasetsky
e-mail: panasetsky@gmail.com

A. Domyshev
e-mail: domyshev@gmail.com

A. Osak
e-mail: osakalexey@mail.ru

U. Häger et al. (eds.), *Monitoring, Control and Protection of Interconnected Power Systems*, Power Systems, DOI: 10.1007/978-3-642-53848-3_16,

and emergency control devices and personnel errors lead to cascading development of emergencies. The automatic emergency control system of higher level localizes and eliminates such cascading emergencies. If this high level control system operates ineffectively or unreliably and additional failures and errors occur we will have unique and catastrophic blackouts like in North America and Europe in 2003, Moscow blackout in 2005, European blackout in 2006 and others.

An important trend in electric power industry is towards integration of EPSs and formation of regional, state and interstate interconnected electric power grids. Interconnection of EPSs results in system effects owing to maneuvering the energy resources, generation capacities and power flows.

The integration of EPSs is primarily intended to provide consumers with power and power services of high quality on the whole territory of electric power grid. Thus the electric power industry becomes to a larger extent an infrastructural sector of the economy.

Liberalization of the electric power industry leads to a principal change in the EPS organizational structure that does coincide with its technological structure and the structure of the control system. The wholesale electricity market participants have principally new character of interrelations that are on the market principles. All this requires that the organization of operating and emergency control of EPS must be revised in the context of a market environment without damage to the efficiency of the EPS operation.

As for future EPS we have to consider possible new generation and network technologies, new principles and ideas in the development and construction of electric power industry, and so on.

All these factors essentially complicate power system operation, increase its changeability and unpredictability, raise danger of severe emergencies with undesirable development and massive consequences for system and consumers and, therefore, call for prompter and more adequate response of control systems. This generates the need to improve and develop principles and systems of power system operation control which can be based on [2]:

- New systems for measurement of operating parameters (PMU) and their control (FACTS, energy storage devices, etc.) that significantly improve EPS observability and controllability (compare Chaps. 4 and 9).
- Modern communication systems, new information technologies and artificial intelligence methods, highly efficient computers, which totally change the processes of acquisition, transmission, presentation (visualization) and use of information on power systems.

This chapter deals with new approaches to create a wide area protection system based on new out-of-step prevention (Sect. 16.3) and a multi-agent distributed protection system against voltage collapse (Sect. 16.4). For preparation on these sections protection system modeling is introduced in Sect. 16.2 and finally conclusions will be made in Sect. 16.3.

16.2 Protection System Modeling

For appropriate modeling of EPS behavior for different problems dealing with analysis of emergency and post-emergency conditions, it is necessary to simulate the devices of emergency control and relay protection.

When modeling simple (typical) emergency situations instead of modeling relay protection and emergency control devices it is possible to consider their standard behavior. For example, if consideration is given to a short circuit in the transmission line then it's tripping after the short circuit is taken into account.

When it is necessary to model a complex cascading emergency during which several successive disturbances occur in the EPS with subsequent equipment overload and values of state parameters exceeding feasible limits, then the standard behavior of relay protection and emergency control devices can be taken into account manually but this requires special attention since it becomes necessary to overlook the operation of individual devices of relay protection and emergency control, including their nonselective operation.

In such cases it is suggested to create a mathematical model of relay protection and emergency control devices, which can be used in the calculations of load flow and electromechanical transient processes. Besides this, it is suggested to take into account the main logic and algorithms of relay protection and emergency control devices, and state variables that are controlled by these devices. Moreover, it is suggested not to model the operation of individual relays, instrument current and voltage transformers, analog–digital transducers and microprocessor devices, i.e. there is no need to thoroughly investigate the interior arrangement of each relay protection and emergency control device.

The mathematical model of relay protection and emergency control devices will automatically control state variables and use specified logical rules to determine the time of operation of these devices. This approach will make it possible to reveal the situations:

- of correct and coordinated operation of relay protection and emergency control devices;
- of non-coordinated operation of individual components of relay protection and emergency;
- control as a result of their wrong adjustment;
- in which the emergency is exacerbated due to the absence of relay protection and emergency control devices.

All test cases for the Wide Area Protection Systems presented in the following sections of this chapter are based on modeling of relay protection and emergency control in the ANARES software [3]. The ANARES software includes:

- A module for modeling the relay protection (RP) and emergency control systems (ECS).
- An editor of emergency control systems and relay protection.

The module of modeling the relay protection and emergency control is related to the modules of solving the problems of short-circuit current, electromechanical transient processes, steady state, and estimation of EPS operation security. Relay protection and most of emergency control systems affect the operation of electrical networks by discrete actions (line, load, generation tripping; discrete signals that actuate continuous control of different regulators, etc.). The module analyses the current (calculated) condition and detects the actuated relay or emergency devices which generate control actions to the EPS model.

The control actions include:

- tripping/connection of EPS components (transmission lines, generators, loads, circuit breakers, etc.)
- actions on regulators of power generators that are modeled by change in the generated power.

For convenient input and edit of data on relay protection and emergency control in the editor as well as for the convenience of modeling and analyzing, the information is subdivided (systematized) into the following types:

- primary switching elements (circuit breakers, etc.)
- secondary switching elements (plates, keys, etc.)
- pickups (relay of current, voltage, power, etc.)
- network scheme state, i.e. some state of the network components or operating conditions, under which certain relay protection devices operate; a corresponding table of settings is chosen, etc.
- fault detectors, a condition, whose emergence stipulates triggering of emergency control and relay protection devices.
- control actions generated by emergency control and relay protection devices
- emergency control and relay protection devices that have a set of logical conditions for operation (with indication of the network scheme and the fault detector) with several stages. Each stage is assigned with its own control actions.

The conditions can be specified by setting freely-programmable logic. The freely-programmable logic can be applied for setting the following conditions:

- choice of the network scheme
- actuation of fault detector
- generation of control action
- operation of emergency control and relay protection devices.

The description language of logic is well optimized to provide the maximum description speed of programming, which is very important for modeling transient processes when the operation of fault detectors and other conditions should be checked in each iteration. Therefore, the language has some constraints, but provides good possibilities for description of random logical conditions, expressions, etc.

The editor for emergency control systems has a text editor, a debugger of logic, as well as an editor of stages of relay protection and emergency control system operation.

The data prepared in the emergency control editor are stored in a separate file which is then used in technological programs. In order to involve a prepared set of data on relay protection and emergency control devices in technological programs it is sufficient to choose a file prepared in the emergency control editor. Thus, it is possible to create different sets of emergency control systems, for example depending on the problem to be solved or the extent to which the emergency control is detailed.

For the convenience of data input the emergency control editor has a mechanism for setting patterns. The software has some standard patterns which can be changed or supplemented by the user.

When relay protection or emergency control devices operate during the execution of technological programs a detailed protocol is issued. It indicates the place and time of operation, type of emergency control device, control action magnitude, stage, and an object of control action.

16.3 Out-of-Step Prevention and Elimination

Out-of-step operation in interconnected power systems is one of the most severe emergency conditions. It is related to the loss of stability which may cause damage to equipment, interruption of power supply to consumers and unwanted development of emergency processes with severe consequences for the entire interconnection and its parts [1, 4, 5]. With the future possibility for the ENTSO-E and IPS/UPS power interconnections to operate jointly by the AC tie lines the occurrence of out-of-step operation of the interface between these power interconnections can be dangerous for both of them and result in undesirable consequences for the systems and consumers. Measures are therefore required to detect, prevent and eliminate out-of-step conditions.

Special automatic Out-of-Step Protection Systems (OSPS) have been used in electric power system for reliable, timely and selective detection and elimination of out-of-step conditions [1, 6, 7]. The most effective system is the so called selective OSPS which is based on the angle measurement [2, 7, 8]. Previously the difficulties of measuring angles by the devices for tele-transmission of voltage phase made us use indirect angle calculation. Most popular methods of this calculation are based on determination of current amplitude or complex impedance at the connection point of the automatic system [7, 9, 10]. In this case the system is represented by a two-machine equivalent with regard to the cutset of the ties in which the OSPS is installed. The parameters of the two-machine equivalent are determined on the assumption that the motion of generators in the initial system along both sides of the cutset at issue is coherent [7, 11]. This assumption is based on the fact that kinetic energy of mutual transient oscillations of generators during disturbances passes the kinetic energy of the out-of-step motion between two groups of generators along both sides of the cutset at which the out-of-step conditions occur. Inter-machine oscillations within these two groups of generators decrease essentially.

The easiest way to implement an indirect method for calculating the transmission angle is the use of the angle dependence of the transmission current. The disadvantage of this method is a wide scatter of operating angle values of the OSPS under the assumed current pickup settings due to various possible compositions and structures of ties in the cutset in different schemes and conditions of the power system. Besides, this dependence is nonlinear. The lesser error is obtained by the use of the so called phantom scheme, i.e. by modeling of voltage phasor of a point located at some distance from the site of automatic system placement, for example a receiving end of a transmission line. The phantom scheme is used in the OSPS installed in power systems in Russia. Particularly complicated conditions for selective operation of OSPS occur in the multi-frequency out-of-step conditions along several cutsets [7].

Further development of selective OSPS has resulted in creation of a multi-functional device. The device makes it possible not only to counteract the out-of-step conditions if they have occurred but also to prevent their occurrence. It has two stages of control actions [7]: the control actions of the first stage are intended to prevent the loss of stability and for this purpose generation is disconnected in the surplus part of the system and fast reserve is used (or secondary load is shed) in the deficient part. If these control actions are insufficient and fail to prevent out-of-step operation the control actions of the second stage are triggered and the power interconnection is split.

The use of synchronized voltage phase measurements obtained from PMU offers principally new capabilities of implementing the selective OSPS and Selective Out-of-Step Protection and Prevention Systems (SOSPPS)s [12]. Some SOSPPSs have been lately suggested on the basis of PMU. In order to reveal transient instability the equal-area criterion is used in [13, 14] when representing the system by a two-machine equivalent. Its parameters are determined by the complex values of power system state variables. In [15] the measurements of voltage phases and differences of their first derivatives are used to forecast power system stability losses according to the criterion based on the energy function. In [16] an integrated criterion is suggested to reveal the center of oscillations with the use of estimates of the voltage magnitude projection at some point of the tie line between two parts of the system and current along this tie line when using the two-machine equivalent of power system on the basis of the generators motion coherence in these two parts of the system which is estimated on the basis of currently measured voltage angles.

16.3.1 Scheme of Interrelation Between States and Control Actions

Loss of synchronism in power system operation at a cutset can be caused by two main reasons:

- The maximum admissible transfer capability of the cutset is exceeded and, thus, the aperiodic static stability of the system is lost.
- The transient stability is lost as a result of disturbance on one of the ties or near the considered cutset.

In both cases an indicator for loss of synchronism and the beginning of out-of-step conditions is the difference in voltage phases on the ends of the most critical tie line of the considered cutset. In other words there is some maximum value of the voltage phase difference $\delta_{\lim}^{as}$, whose excess indicates the beginning of the out-of-step conditions.

In order to avoid this critical situation it is necessary to maintain the cutset load at the level not exceeding some admissible level corresponding to $\delta_{\lim} < \delta_{\lim}^{as}$. The difference between $\delta_{\lim}$ and $\delta_{\lim}^{as}$ should take into account irregular variations of flows along the tie lines and the need to ensure transient stability of power systems under standard disturbances. In Russia dispatching centers of power systems use the recommended values of transmission loading margins under normal and post emergency states [17]. In the power systems of ENTSO-E there are no similar explicit recommendations. Nevertheless, setting the value $\delta_{\lim} < \delta_{\lim}^{as}$ can be expedient.

Thus, power systems can have four states (Fig. 16.1): secure, dangerous, emergency (out-of-step conditions) and post-emergency. The secure state of power systems is determined by the condition $|\delta_{ij}| < \delta_{\lim}$. PMU measurements are used to trace the current value of $\delta_{ij}(t)$. The dangerous state of power system occurs at $|\delta_{ij}| > \delta_{\lim}$. For the power system to return to a secure state it is necessary to perform control actions to reduce the loading of the cutset by decreasing the generation of power plants on the transmitting side and by using fast reserve (or disconnecting secondary consumers) on the receiving side. If these control actions are sufficient the power system returns to the normal state. However, if the control actions are insufficient the system passes to the emergency state (out-of-step conditions) which is eliminated by disconnecting the cutset (by dividing the power system). Should the splitting of the power system be unsuccessful, then an emergency situation can develop and the post-emergency state may turn out to be severe and even turn into a blackout. In the event of a successful splitting the generation and load in both subsystems are balanced by generation disconnection in the surplus subsystem and by automatic frequency load shedding in the deficient subsystem. Post-emergency state in this case will be less severe as compared to the previous one.

It should be noted that traditionally in Russia's power system dispatching practice automatic load shedding and OSPS are considered separately due to the

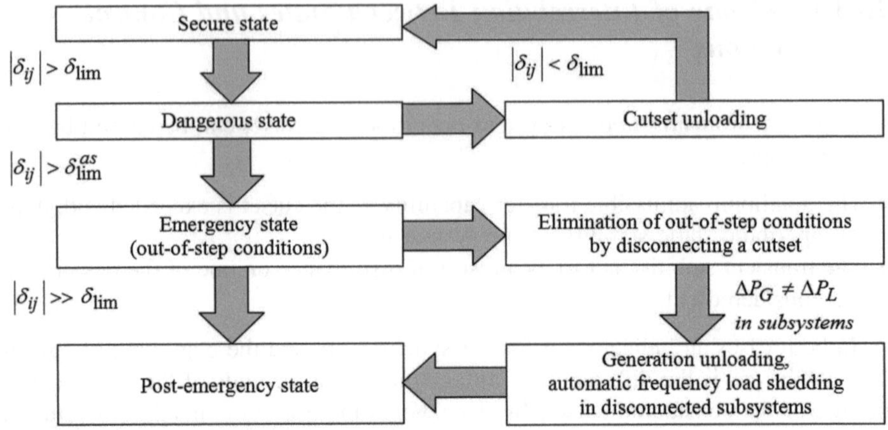

Fig. 16.1 Scheme of interrelation between power system states and control actions in SOSPPS

fact that the former is adjusted and operates using the power flow values as an indicator of overload while the latter is intended for the transmission angle action, with transmission angles being determined by indirect methods [1]. The use of one and the same indicator which is the difference between PMU-measured voltage phases on both ends of power transmission allows one to consider both types of automatic systems as a single integrated emergency control system.

16.3.2 Criteria for Actions of SOSPPS Stages

As noted above the conditions for transition from normal (safe) state to the emergency state is formulated as $|\delta_{ij}| > \delta_{\lim}$. Hence the criterion for action of the cutset unloading stage of SOSPPS will look as:

$$C_{act}^{ul} = \left(\mid \delta_{ij}(t) \mid > \delta_{\lim} \right) \tag{16.1}$$

In the event that the action of the cutset unloading stage is insufficient or inefficient, then the difference in the voltage phases along the critical tie line of the cutset at its overload continues to increase and reaches the value $\delta_{\lim}^{as}$. This indicates the loss of aperiodic static stability of the power system along the considered cutsets and the need to split the system. The criterion for action of the division stage of the SOSPPS will have the following form:

$$C_{act}^{as} = \left(\mid \delta_{ij}(t) \mid > \delta_{\lim}^{as} \right) \tag{16.2}$$

To formulate the criterion for action of the division stage of the SOSPPS according to the conditions of transient instability of power system under large disturbances it is necessary to use the second derivative of the difference between the voltage phases of a critical tie line in the cutset. Decrease in the second derivative of voltage phase difference indicates conservation of transient stability of EPS. System transition to an emergency state (out-of-step condition) is revealed provided that at least for three cycles of measurements by using PMU, the second derivative value of voltage phase difference for the critical tie line at the cutset does not go below a small value $d\delta_{\min}$. In this case each cycle may account for several scores of milliseconds. Theoretically, $d\delta_{\min} = 0$, how-ever practically this value is not equal to zero because of errors and noise in measurements and also inaccurate determination of the second derivative due to discrete measurements. Determination of the acceptable value $d\delta_{\min}$ is an independent problem.

Hence, for the direct power flow through the cutset from node i to node j the criterion for action of the division stage subject to transient stability of the SOSPPS will have the following form:

$$C_{+} = \left(\Delta\delta_{ij}(t) > 0\right) \wedge \left(\frac{d^2\delta_{ij}(t)}{dt^2} \geq d\delta_{\min}\right) \wedge \cdots$$

$$\wedge \left(\frac{d^2\delta_{ij}(t - 2T_S)}{dt^2} \geq d\delta_{\min}\right) \qquad (16.3)$$

where T_S is the cycle length between the PMU measurements.

The corresponding criterion C_{-} for the reverse power flow through the cutset (from node j to node i) is determined in a similar way.

The general criterion of transition to an emergency state (out-of-step condition) and the action of SOSPPS's division stage is written in the following way

$$C_{as} = \left(C_{act}^{as} \wedge (C_{+} \vee C_{-})\right) \qquad (16.4)$$

As a result the EPS is split into two isolated subsystems.

16.3.3 Related Problems

The formulated approach for improving SOSPPS based on PMU measurements includes only its basic principles illustrated by the test example later on. Actually consideration should also be given to the related problems to be solved in the course of practical implementation. Below are the main problems.

As indicated above, it is reasonable to measure voltage phases on the basis of PMUs on the ends of the critical tie line at the cutset. The problem is that the increase of transmitted power along the cutset results in different loading of individual tie lines at the cutset. Such a situation is caused by the parameters of the tie lines and also the structure and parameters of adjacent electric networks. This inhomogeneity of an electric network is revealed in different disturbance-sensitivity of nodes and tie lines at the loaded cutset. In other words those elements affect operation parameters (voltage, power flows etc.) changing to variable extents [18]. Hence, conditions for transition to the out-of-step operation are formed first of all in the most sensitive tie line. The tie line turns out to be critical at the cutset and it is expedient to place PMUs on its ends.

It should be noted that significance of a critical tie line at the cutset requires additional studies. It is explained by the fact that with the start of out-of-step condition change in the voltage angles will be observed in all tie lines of the cutset. It is important to establish the extent to which the change in tie line loading at the beginning of out-of-step condition is significant in terms of the efficiency of SOSPPS operation.

Another problem is the necessity for verification of criterion (16.3). It has two components. The first is associated with available errors and noise in the measurements by using PMU, delays in information transmission, measurement frequency, etc. These technical properties should be studied thoroughly and individually. And the results of these studies will determine specific features of the designs of automatic systems.

The other component of the problem is the accurate determination of the second derivatives of variation of voltage phase differences. It depends on the monotone change of phase differences and the length of intervals between measurements, as well as on measurement errors. The problem is that the second derivatives must be calculated by numerical differentiation of measured parameters. This question also requires thorough additional studies.

The results of studies on both components determine efficiency of using criterion (16.3) and in particular, certainty in setting the value $d\delta_{\min}$.

The issue about selectivity of action of the modified SOSPPS at the multi-frequency out-of-step condition for the case of stability loss at several cutsets remains to some extent open. It seems that selectivity of work of automatic systems should be sufficiently high and acceptable, since the change of voltage phases on the ends of tie lines at the cutset is a quite definite indicator of the beginning out-of-step condition. Here the value $\delta_{\lim}^{as}$ is close to 90° and the value $\delta_{\lim}$ is uniquely determined by the state variables of the critical tie line of the cutset and the required margins of its transfer capability.

16.3.4 Case Studies

Let the test EPS (Fig. 16.2) operate in post-emergency conditions when the tie 8–5 is loaded at 90 % of its maximal transmission capability, and voltage mutual angles equal $\delta_{8\text{–}5} = 36.6°$ and $\delta_{202\text{–}100} = 50.5°$.

Let us consider as a disturbance the unscheduled disconnection of one of the two lines of the tie 8–5. The behaviors of voltage mutual angles and their time-derivatives without any control actions are shown in Fig. 16.3.

Suppose the system is equipped with SOSPPS, and a starting value for action of the cutset unloading stage [see criterion (16.1)] is set as $\delta_{\lim(8\text{–}5)} = 52.2°$. At the

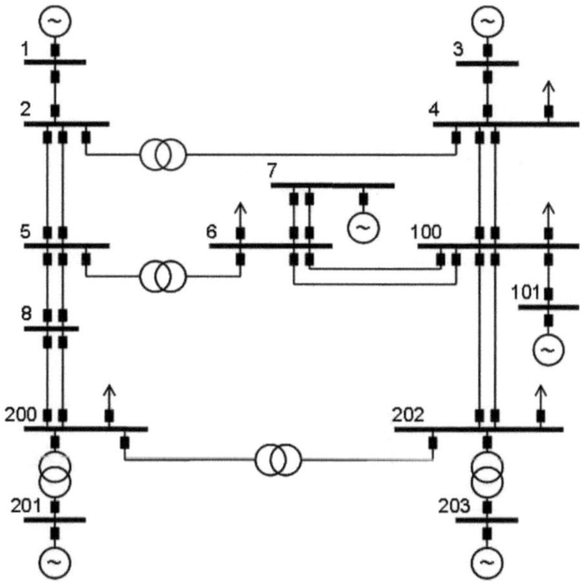

Fig. 16.2 Case study network

time of $t = 0.4$ s the control actions will be realized as partial disconnections (of generation in power surplus part of the system and of load in its deficient part). The values of disconnected power are shown in Fig. 16.4. Suppose larger disconnections are highly undesirable through technical and/or economical limitations.

The behaviors of voltage mutual angles and their time-derivatives with above mentioned control actions are shown in Fig. 16.5.

Figure 16.5 demonstrates the inefficiency of undertaken control actions for providing the system stability. The non-periodic growth of mutual angles in the cutset, which appeared virtually straight after the disturbance, lasts also after these actions (although not so fast as without them). If we suppose (only with a view to exemplify the study) the inadmissibility of further mutual asynchronous motion for the power system, then the starting value for action of the cutset division stage [see criterion (16.2)] in accordance with Fig. 16.5 is to be set as $\delta^{as}_{\lim(8-5)} = 57.3°$. At that case the system is being divided at the time of $t = 0.5$ s (when the first derivative of the angle reaches its maximum $\frac{d\delta_{8-5}}{dt} = 44$ grad/s).

After splitting the system into two separate subsystems (one of them with the surplus and other with the lack of active power) each subsystem faces the challenge of bringing the frequency to admissible level. The solution is further reduction of power (of the generation and load respectively). This reduction may ensue less than what would be required under saving the parallel operation with unified frequency.

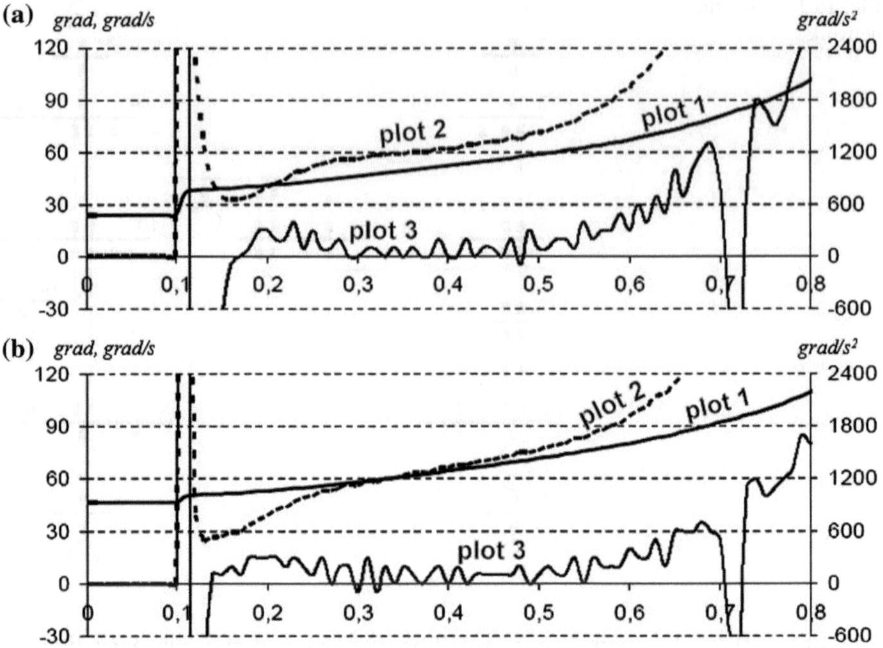

Fig. 16.3 Time behavior of: plot1—δ (*thick line*), plot2—dδ/dt (*dotted line*) and plot3—d2δ/dt2 (*thin line*) without control actions: **a** for the tie 8–5, **b** for the tie 202–100

Fig. 16.4 Generation and load tripping for the cutset unloading

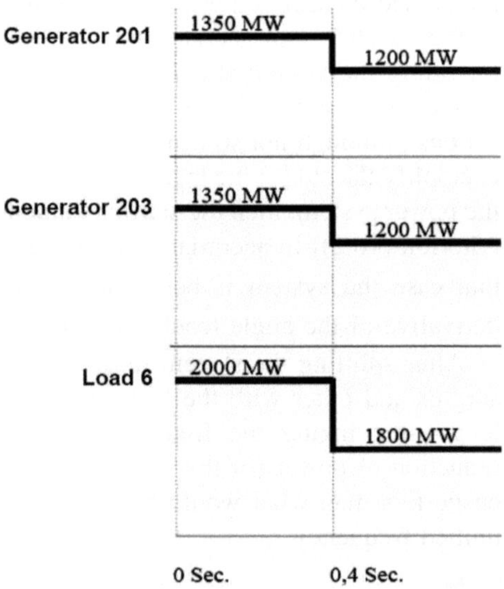

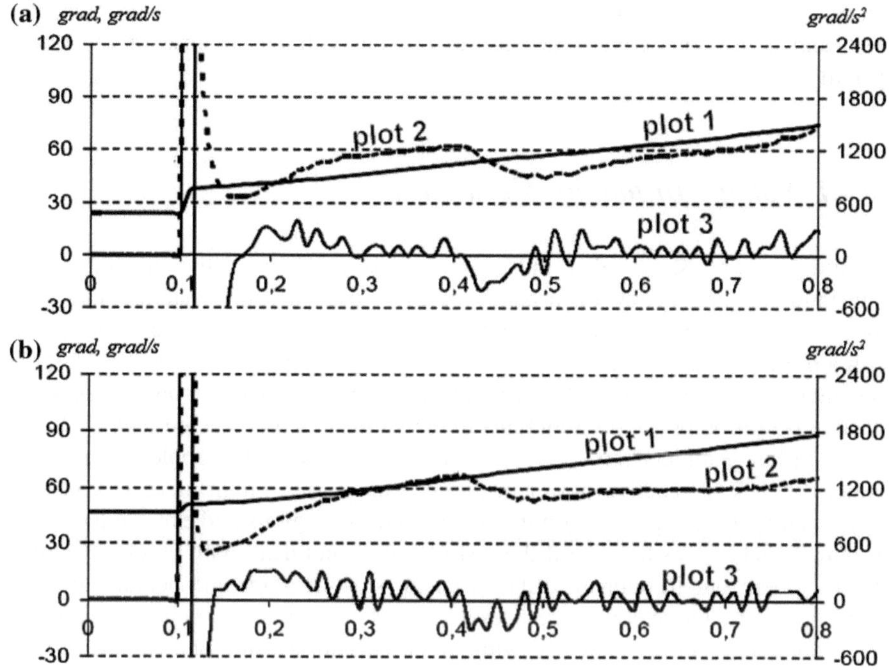

Fig. 16.5 Time behavior of: plot1—δ (*thick line*), plot2—dδ/dt (*dotted line*) and plot3—d2δ/dt2 (*thin line*) at maximum permissible unloading of the cutset: **a** for the tie 8–5, **b** for the tie 202–100

16.4 Distributed Protection System Against Voltage Collapse

Analysis of recent blackouts showed, that the most severe interruptions occurred in highly loaded interconnected power systems due to Extra High Voltage (EHV) line disruption followed by multiple contingencies [19]. These accidents highlighted the deficiency of the existing protection systems that cannot maintain the integrity of the transmission grid during multiple contingencies [20]. Power system behavior in an emergency state is characterized by complex interaction between discrete and continuous control devices. Continuous control devices are Automatic Voltage Regulators (AVR), turbine governors, etc. Discrete control devices are different protection relays, underload tap changers, etc. Currently both continuous and discrete control devices substantially use local signals only and do not coordinate their actions with each other. Absence of coordination between discrete and continuous control devices is the shortcoming of the existing protection system and it may lead to blackouts. The paper presents a control system based on the multi-agent approach. The control system provides coordination of different discrete and continuous control devices to prevent voltage collapse of the power system during the post disturbance period.

16.4.1 Voltage Instability Mechanism

Existing practice shows that most power systems have sufficient stability to withstand the first heavy disturbance in EHV transmission systems. The post-disturbance phase represents a deceptively calm period that lasts several minutes with a normal level of frequency and then voltage collapse that lasts seconds [21]. The first heavy disturbance leads to increase in the reactive power losses and reactive power output of rotating units in the vicinity of the affected region. So, the first disturbance affects only the local region around the initial problem. But after a while, the lack of reactive power in the affected region might increase considerably, leading to voltage collapse in the neighboring regions and even in the whole system. This happens because if the disturbance is not dealt with timely, the after-effects spread out through the EHV transmission network and actuate different control devices. These control devices act at different speed, respond to changes in the immediate vicinity and act without coordination with one another. Their actions in response to the post-disturbance conditions are actually the main cause of power system breakdown; consequently, the timely control of the discrete and continuous control devices under the post-disturbance conditions is the only measure to prevent voltage collapse of the whole system [20]. Undoubtedly, the absence of different control actions coordination during the post-disturbance period can cause different types of instability. But first of all, one should cope with voltage instability because it was the main cause of the recent blackouts. A new protection system philosophy has to be proposed to prevent voltage instability during the post-disturbance period.

16.4.2 New System Protection Philosophy

The new protection system must detect critical situations and coordinate the work of control devices to exclude any possibility of voltage instability. So, how can the new protection system identify critical situations and what kind of control actions should the system use to control the capacity of available reactive power resources?

Reduction of voltages and increase of rotating unit excitation were proposed in different papers to indicate the proximity to voltage collapse [22]. Thus, these two criteria may be used to detect the appearance of critical situations and activate the protection system.

The analysis of recent blackouts showed that rapid load shedding is usually the only way to prevent a collapse of the whole system [19]. On the one hand, load shedding should be as fast as possible, on the other hand, it should be optimal, for instance, with respect to load needed to be shed. Optimal load shedding scheme can be realized by using different optimization procedures, but it is hard to solve an optimization problem for any possible situation in advance, because the number of situations is too big. This means that some optimization computations should be made during the post-disturbance period. In spite of the fact that there is a number of optimization techniques that can be used to calculate emergency control actions quickly, the amount of input data required to solve the problem is usually too big, but load shedding under post-disturbance conditions has to work fast. Hence, the load shedding procedure has to use less complex methods to control the post-disturbance phenomenon. The following simple countermeasures to control the post-disturbance phenomenon were proposed in [21]:

- Countermeasure 1: Fast tap changing on transmission substation transformers.
- Countermeasure 2: Raising terminal voltage on selected synchronous condensers and hydro generators.
- Countermeasure 3: Fast tap changing on selected generator transformers.
- Countermeasure 4: Strategic load shedding at selected transmission substations only if voltage levels and reactive outputs do not meet the requirements, or some transmission lines are overloaded.
- Countermeasure 5: Re-arranging generators active power outputs. Connecting part of the disconnected load.

Countermeasures 1–3 have approximately the same execution time and their main purposes are to impede the sharp increase of series reactive power losses, to increase transmission line charging and to inhibit tap changing on sub-transmission and distribution transformers. Load is shed (Countermeasure 4) only after countermeasures 1–3. This will decrease the amount of load to be shed. Countermeasure 5 considers an optimization procedure. The optimization procedure takes much more time in comparison with countermeasures 1–4 and provides post-emergency operation optimization.

Thereby, countermeasures 1–4 provide fast control of the post-disturbance phenomenon to avoid voltage collapse and countermeasure 5 provides long-time-period post-emergency operation optimization. The proposed control principles can be applied to various parts of the grid that work independently and the proposed control system can be built by using distributed intelligence principles [23], for instance, the Multi-Agent Systems (MAS) approach.

16.4.3 Multi-Agent Control Systems

The proposed Multi-Agent Control System (MACS) provides reactive power control to prevent generator tripping and preserve load bus voltages within the normal range. It consists of two types of agents: Load Agents (LA) and Generator Agents (GA). Any agent has at any time the following set of local data:

- Local state variables (primary and secondary voltages, power flows, etc.).
- Operating characteristics of the local equipment (generator terminal voltage, tap range of the tap changer, excitation current of the generator, etc.).

Any agent has two goals:

- Local goal. It consists in maintaining local state variables and equipment operating characteristics within the normal range.
- Global goal. It consists in voltage collapse prevention.

To make different parts of the proposed MAS work independently, each agent must know only about a limited number of agents, which influence his activity most. For instance, in a test system, as presented later in Fig. 16.11, LA, installed at Bus101–Bus103 in Subsystem A must know much about the agents in Subsystem A, because all these agents can influence them. On the other hand, in spite of the fact that agents in Subsystem B could know much about one another, they must know only about three agents in Subsystem A: LAs, installed at Bus101–Bus103, because these three agents can only influence them. In this case, the subtransmission system produces minimal influence on the transmission system.

16.4.3.1 MAS Ontology

According to Foundation for Intelligent Physical Agents (FIPA) standards, messages exchanged by agents have a number of fields and in particular: sender, receiver, language, ontology, etc. Ontology is the vocabulary of symbols and their meanings. For the effective communication, both the sender and the receiver must ascribe the same meaning to symbols. Ontology can include different elements such as agent actions, terms, concepts, etc. Agent actions indicate actions that can be performed by some agents. Terms are expressions identifying entities (abstract or concrete) that "exist" in the world. For voltage control purposes, the following simple Voltage Control Ontology can be proposed:

Agent actions of the Voltage Control Ontology are:

- Increase Reactive Power.
- Load Shedding.
- Block Tap Changer.

Terms of the Voltage Control Ontology are:

- Owner.
- Voltage Rate.

The Voltage Control Ontology usage principles will be explained in the next sections.

16.4.3.2 Generator Agent

GAs obtain local information about the excitation current of the generator, primary and secondary voltages at the generating substation, active power flows and transformer tap ranges. If the excitation current goes beyond of its normal range, the GA tries to decrease it to exclude the possibility of generator tripping. The GA sends messages to other agents that can decrease the shortage of reactive power in the affected region. The sent messages apply to the Contract Net Interaction Protocol and include *Increase Reactive Power* action of the Voltage Control Ontology. First, the GA tries to eliminate the overload by coordinating reactive power sources, if it can not do this, it goes to the load shedding procedure.

Consider a situation in which the overload of the generator can be eliminated only through coordination of reactive power sources. The sequence diagram for the protocol used in this case by the GA is depicted in Fig. 16.6. First, the initiator locks On-Load Tap Changers (OLTCs) by sending REQUEST messages to LA (step 1 of the protocol). At the same time it sends Call For Proposal (CFP) messages to all agents who can help him—to GA that can increase the AVR set point and to LA which can increase the secondary side voltage on their transformers. Then the initiator receives A PROPOSE and B REFUSE messages (step 2) and sends back A ACCEPT-PROPOSAL messages to those agents which are able to help him (step 3). The requested speeds of reactive power increase (speeds requested in ACCEPT-PROPOSAL messages) are calculated from an analysis of the sensitivity coefficient (influence of the reactive power injection on voltage level at GA node). During the increasing of the reactive power some of the agents may finalize production and notify it by sending C INFORM messages (step 4). Note that $C < A$, this means that after receiving C messages there is still a part of agents which produce reactive power. After a while the joint actions of the agents lead to the removal of the overload and the initiator sends D REJECT-PROPOSAL messages and unblocks OLTCs (send REQUEST messages, see step 5, Fig. 16.6). Note that $C + D = A$.

Now consider a situation when the overload cannot be eliminated only through the redistribution of reactive power. That is, consider a situation in which MAS needs to shed the load, Fig. 16.7. The load shedding procedure starts when the GA "understands" that in the controlled subsystem there are no agents which can increase the production of reactive power (step 4, Fig. 16.7). Thus, the GA always knows when reactive power increase in his subsystem is stopped. Immediately GA sends $E + F$ CFP messages to LA and solicits from them E proposals and F refuses (see Fig. 16.7, step 6). After that, GA accepts J proposals and sends J ACCEPT-PROPOSAL messages (step 7). Note that J is less than or equal to E. Selecting the agents which will shed the load is also based on the sensitivity analysis technique. After receiving ACCEPT-PROPOSAL message, the selected agent starts to shed the load as long as it will not get the REJECT-PROPOSAL message from the initiator, which indicates that the overload has been eliminated. At the same time as in the previous case the initiator unblocks OLTCs (send REQUEST messages, see step 8, Fig. 16.7).

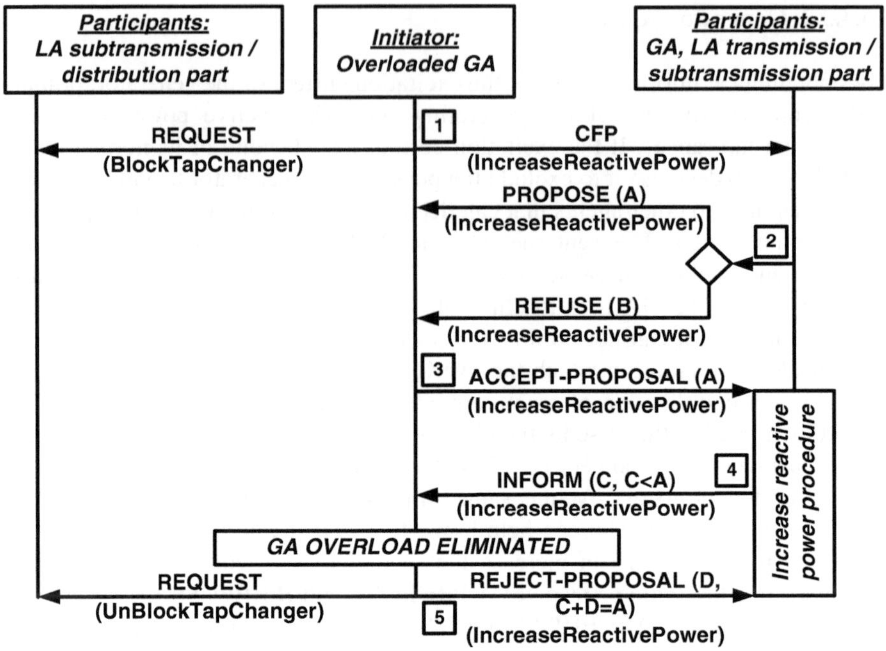

Fig. 16.6 Interaction protocol used by GA. A situation when the overload of the generator can be eliminated only through coordination of reactive power sources

Now consider a situation when a GA receives a CFP message which contains the request for the reactive power production (see step 1, Fig. 16.6). First, it analyzes the operating characteristics of the generator and if they are within normal range it starts to increase reactive power output according to the algorithm, the simplified version of which is presented in Fig. 16.8. Where $U_{\mathrm{GEN\ SV}}$ is the generator secondary voltage, $U_{\mathrm{GEN\ TV}}$ the generator terminal voltage, I_F the excitation current and $I_{F\ \mathrm{MAX}}$ is the highest possible excitation current.

16.4.3.3 Load Agent

LAs obtain local information about primary and secondary voltages at the substation, transformer tap ranges and active power flows. Load Bus agents take part in the Load Shedding procedure (see Fig. 16.7). It also can shed the load independently in case of critical voltage drops. If it is installed at a transmission system substation, a LA can take part in reactive power regulation (see Fig. 16.6). In this case, the LA changes the transmission transformer tap ratio until primary voltage will not decrease or secondary voltage will not increase up to specified values. Changing the transmission transformer tap ratio, a Load Bus agent must coordinate its actions with generators in the transmission system.

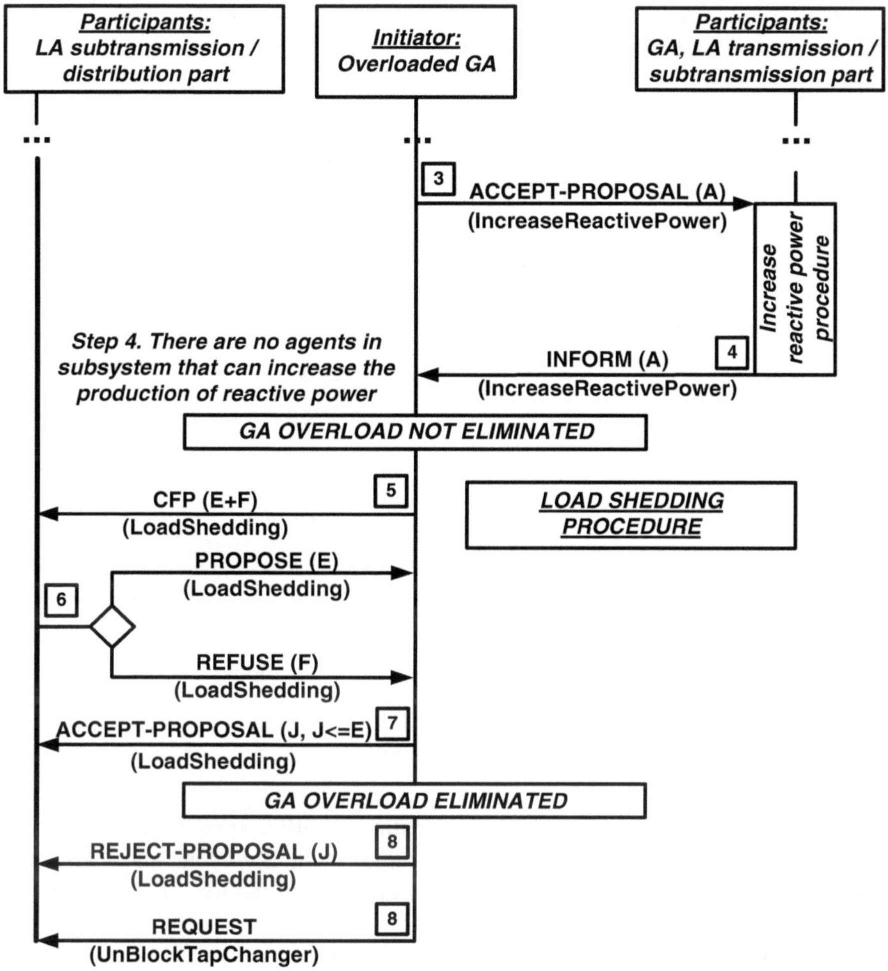

Fig. 16.7 Interaction protocol used by GA. A situation in which the GA needs to shed the load

16.4.4 Multi-Agent Control System Implementation

The computer model of the proposed MAS for power system voltage stability control was implemented in the Java Agent Development Framework (JADE). JADE is based on the object-oriented language Java. Agents, developed for the JADE platform consist of three basic layers: a message handling layer; a behavioral layer; a functional layer. The message handling layer is responsible for sending and receiving of messages from other agents. The behavioral layer provides control when an agent has to carry out a task. The functional layer embodies the actions the agent can perform. JADE provides programmers with the following

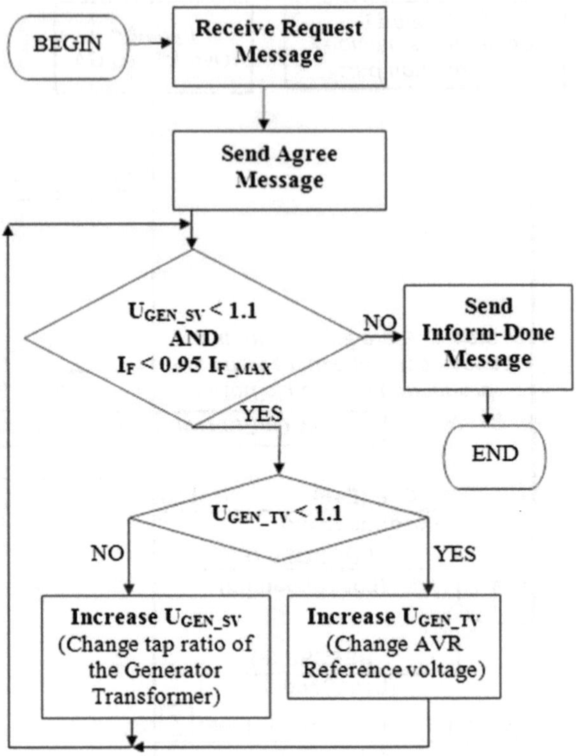

Fig. 16.8 Reactive power output increasing algorithm of GA

ready-to-use functions: full compliance with the FIPA specifications; efficient transport of asynchronous messages; a simple agent life-cycle management; a library of interaction protocols, etc. For further information about the JADE platform, see [23–26].

Necessary power flows and time domain simulations were carried out in a Matlab/PSAT environment [27]. Java capabilities of the JADE environment were used to implement communication between Matlab/PSAT and JADE, Fig. 16.9.

To provide communication between Matlab and JADE, Box Agents are used. Box Agents are Java objects that contain different data structures. During Time Domain Simulation, information about power system operating conditions at each integration step passes from the Matlab environment to JADE by means of Box Agents. After that, agents inside the JADE environment process this information, produce control actions if needed, put information about control actions inside Box Agents and pass Box Agents back to the Matlab environment. Thus, there is no need to use the computer hard disc during the simulation, all computations are performed inside the main memory and the simulation process is faster. The proposed MAS software realization allows one to use complex Matlab/PSAT routines and to model complex behavior of the agents.

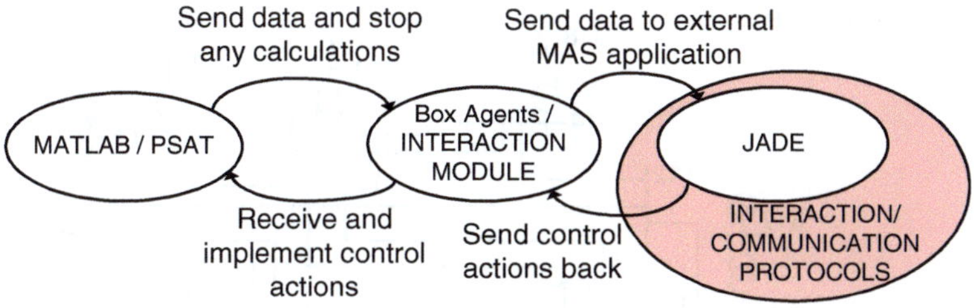

Fig. 16.9 VAS implementation

16.4.5 Test System

A modified IEEE One Area RTS-96 system was used as study case [28, 29]. Initially this test power system contained 24 buses and had no dynamic elements. The modified test system shown in Fig. 16.10 has 53 buses and dynamic elements to represent generators and loads. During modification, the test system structure underwent the following changes:

- All transformers 230/138 kV equipped with OLTCs were installed between the loads of subtransmission and distribution system. The OLTC time delay for the first tap movement is 15 s, and for subsequent tap movements is 5 s. The reactive load levels changed in the 230 kV network with a reactive power surplus and in the 138 kV network with a reactive power shortage. Active and reactive power flows were directed from the 230 kV part to the 138 kV part of the test system. Each load was modeled as an exponential recovery load.
- Each generator was modeled by a six order dynamic model and was equipped with Turbine Governors and AVR. The machines were equipped with Over eXitation Limiters (OXLs).
- For simplicity, it is assumed that the system does not employ Under Voltage Load Shedding (UVLS).

The dynamic model of the modified system was used for time domain simulations.

Thus, the resulting test system can be divided into two subsystems: *Subsystem A* and *Subsystem B*. Active power flow is directed from *Subsystem B* to *Subsystem A*. *Subsystem A* is a low-voltage distribution subsystem being in stressful conditions because of reactive power shortage, which directly points to the possibility of voltage instability. *Subsystem B* is a high-voltage transmission subsystem with surplus of reactive power.

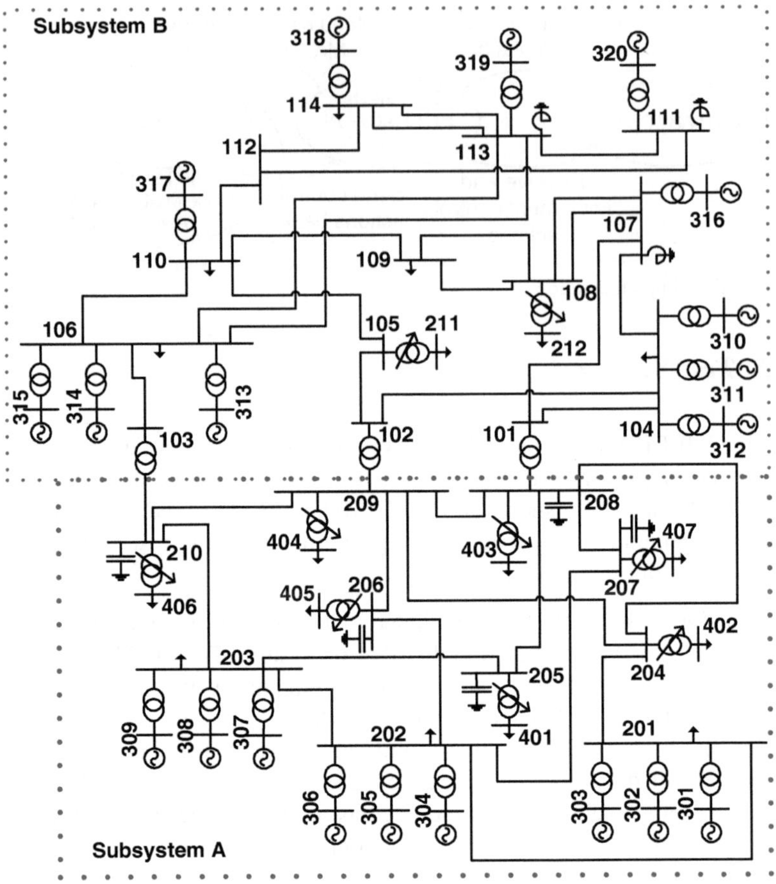

-96 Test System

Fig. 16.10 Modified IEEE One Area RTS-96 Test System

16.4.6 Voltage Stability Control

To demonstrate the proposed approach, the test system is subjected to the following disturbance, Fig. 16.11. At $t = 10$ s a fault free tripping happens of the transformer Bus 101–Bus 208. The fault free tripping can be caused by incorrect adjustment of the relay protection.

It is assumed that two types of automatic control can be used in the power system:

- Conventional automatic control system (CACS): This includes TGs, AVRs and OXLs on each generator, and OLTCs on transformers connected to busses 204–210.

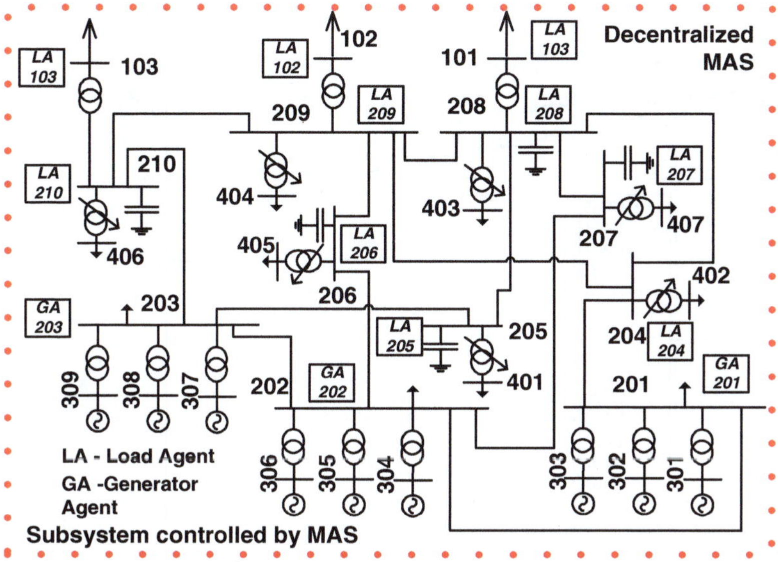

Fig. 16.11 The controlled Subsystem A with the installation locations of the agents

- MACS: in addition to the set of local CACS controllers, it includes OLTCs on transformers connected to busses 101, 102 and 103. Installation locations of the agents are shown on Fig. 16.11.

16.4.6.1 Conventional Automatic Control Modeling

The results of analysis for the CACS are shown in Figs. 16.12, 16.13, 16.14 and 16.15. Almost immediately after tripping of the transformer Bus 101–Bus 208 there is an overload of generators connected to Bus 201 (Generators Bus 301–Bus 303). After about 15 s excitation currents quickly reach the limit (2.5 p.u.), Fig. 16.15. From the moment of the transformer tripping, OLTC start to change the tap-ratio, trying to raise the secondary voltage at the load. The operation of the OLTCs leads to a consistent overload of all the generators in the subsystem. The first overloaded generators are connected to the Bus 202, and to the Bus 203. At time 300 s all the generator rotor currents reach their limits, and the system does not remain any reactive power reserves. Further reduction of voltages makes the situation worse, because of a significant reduction of the reactive power generated by the capacitor banks. Trying to prevent overloading, local OXL continues to reduce the AVR settings (Fig. 16.15), and again this leads to the reduction of the voltages in the subsystem. At time 500 s, through the action of the OLTCs, and due to the lack of reactive power, the primary voltages reach 0.8 Un, while voltages on the secondary side are close to nominal, Fig. 16.12. In this case, UVLS automation installed on the low voltage side, will not work, because it cannot fix

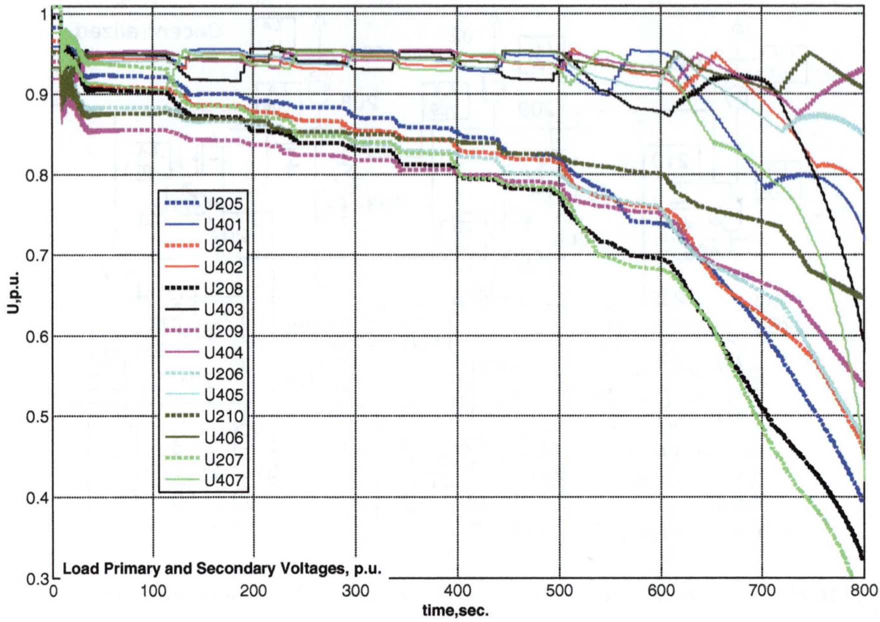

Fig. 16.12 Conventional automatic control modeling. Primary and secondary voltages on load substations

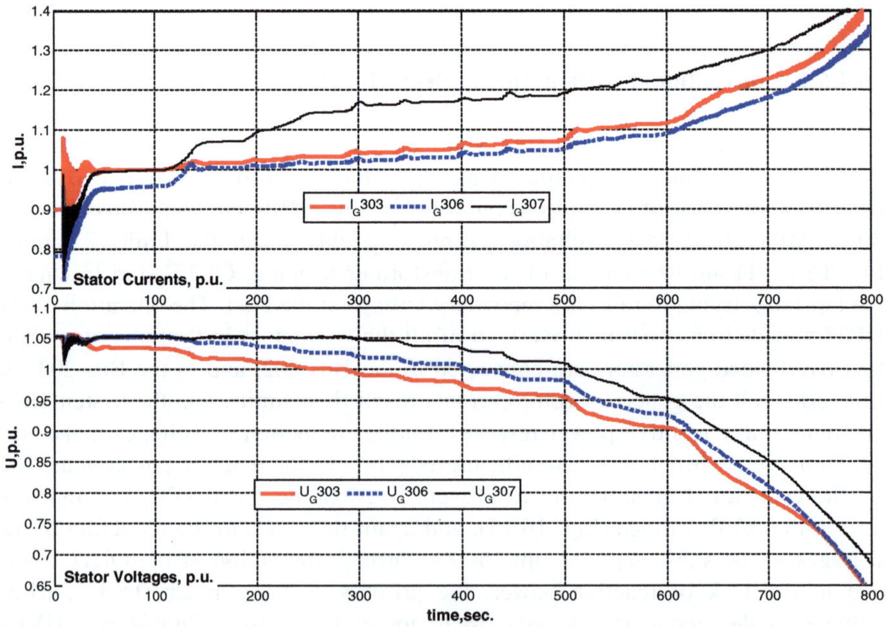

Fig. 16.13 Conventional automatic control modeling. Stator voltages and currents

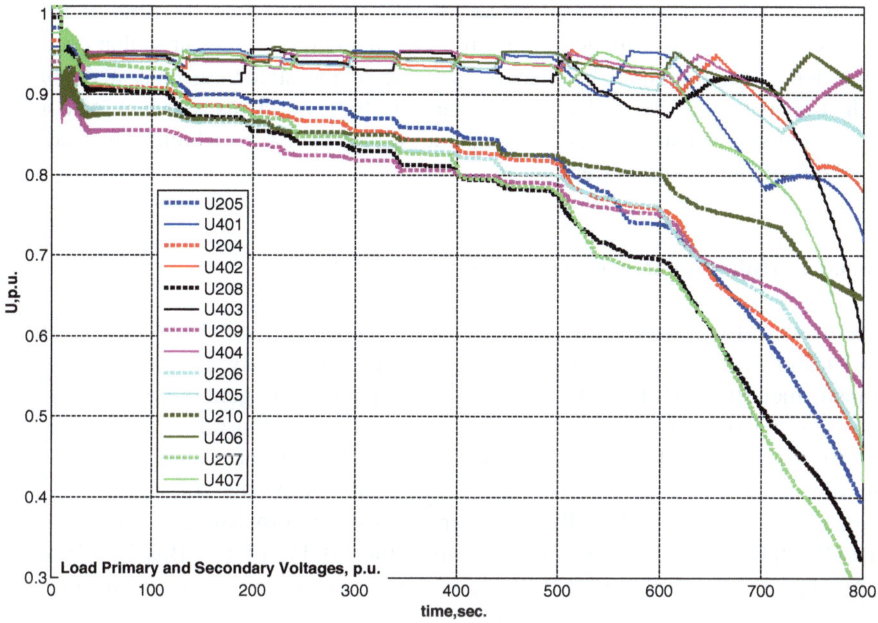

Fig. 16.14 Conventional automatic control modeling. Transmission transformers voltages. Active and reactive loads

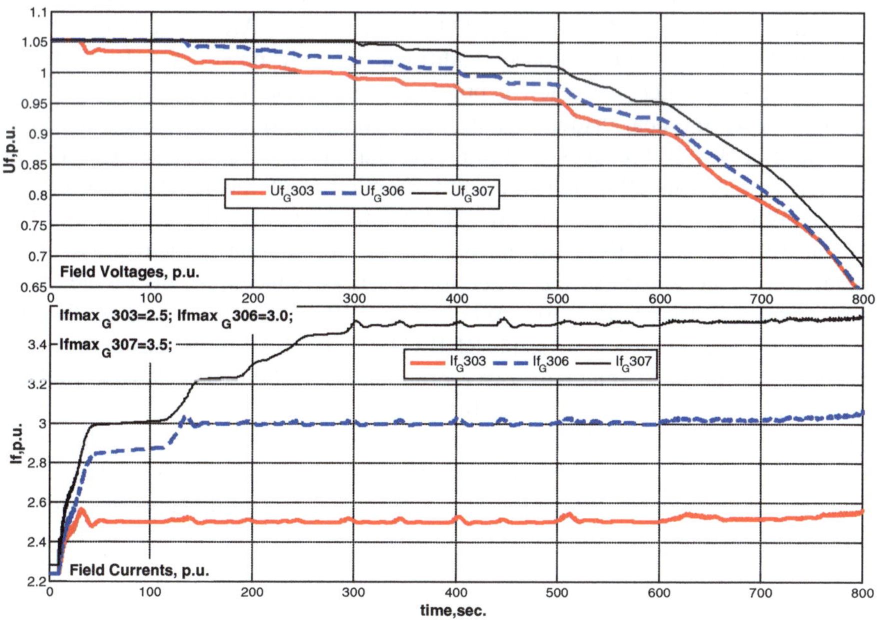

Fig. 16.15 Conventional automatic control modeling. AVR voltages and currents

the fact of the critical voltage drop on the primary side. Since the time of 500 s a sharp decrease of the voltage in the subsystem happens. The collapse that occurs at time 600 s, accompanied by a sharp increase of the stator currents, and voltage reduction (Fig. 16.13), will lead to generators tripping. Thus, the lack of coordination between local devices can cause a voltage collapse, which may lead to a blackout of the whole system.

16.4.6.2 Multi-Agent Automatic Control Modeling

The results of the analysis for the MACS are shown in Figs. 16.16, 16.17, 16.18, 16.19 and 16.20. As in previous case, almost immediately after tripping of the transformer Bus 101–Bus 208 there is an overload of the generators connected to the Bus 201 (Generators Bus 301–Bus 303). Excitation currents quickly reaches the limit (2.5 p.u.), Fig. 16.19. GA Bus 201 sends CFP messages, requesting the possibility of additional reactive power injections to GA Bus 202 and Bus 203, as well as to the LAs Bus 101–Bus 103. Simultaneously, GA Bus 201 sends requests (REQUEST messages, Fig. 16.20) to lock the OLTC of the Bus 204–Bus 210 LAs. In response to the CFP message, every requested agent sends either a PROPOSE (accept proposal) or a REFUSE (reject proposal) message. In particular, in this situation, LA Bus 101 rejected the proposal (REFUSE), because his agent has fixed the transformer tripping Bus 101–Bus 208. After receiving the refuse and accept messages, the overloaded GA (Bus 201) requesting an additional injection of the reactive power from the agents, who confirmed this option. It's doing this by sending ACCEPT-PROPOSAL messages. After receiving the confirmation to start the production of additional reactive power, GA Bus 202 and Bus 203 begin to increase reactive power production as long as their excitation current reaches the near-critical values. In a short period the excitation current of the generator Bus 202 reaches the limit, after that the GA Bus 202 sends INFORM message to GA Bus 201, indicating the completion of the reactive power production. When the voltage on the Bus 102 (primary voltage) decreases to 0.9 Un., LA Bus 102 also sends INFORM message to the completion of the reactive power production. Joint actions of the generator and LAs reduce currents of the generators connected to the Bus 201. When the AVR setting has reached the pre-emergency value, GA Bus 201 sends REJECT-PROPOSAL messages to the agents that have not finished reactive power production. After this, GA Bus 201 also unlocks tap changers of the LA Bus 204–Bus 210 (REQUEST messages). Figures 16.16, 16.17, 16.18 and 16.19 indicate that the implementation of control actions from the MACS allowed to maintain stability only through the implementation of the coordination of reactive power sources, and it is not even required to use load shedding.

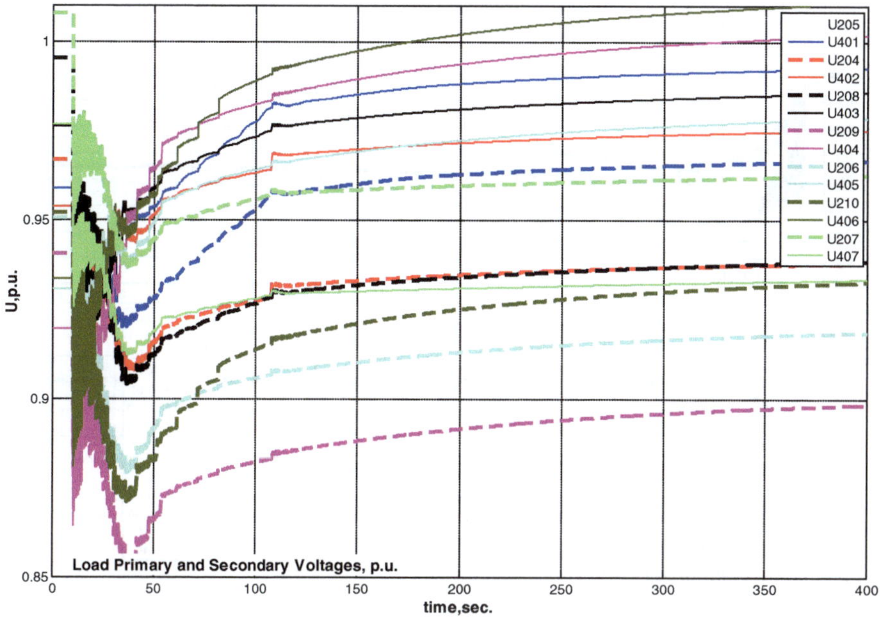

Fig. 16.16 Multi-agent automatic control modeling. Primary and secondary voltages on load substations

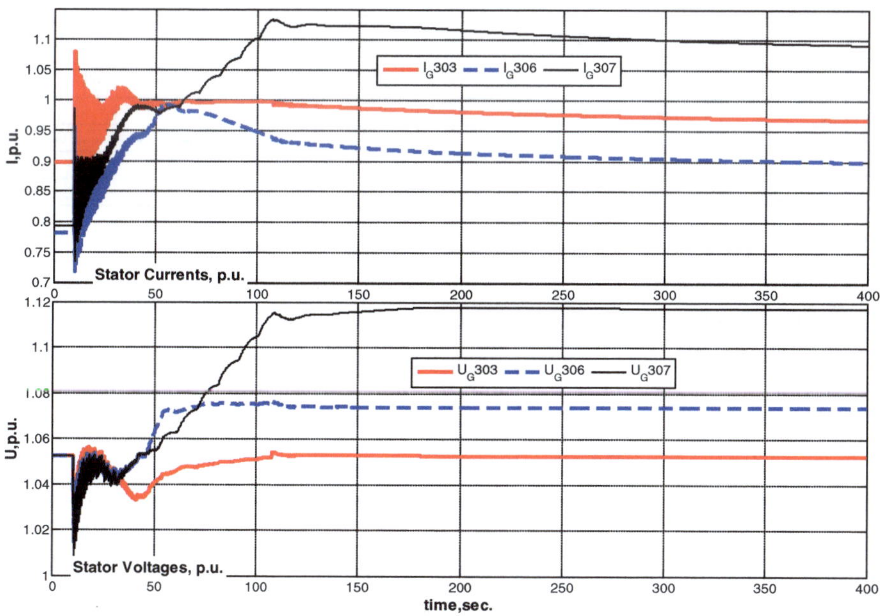

Fig. 16.17 Multi-agent automatic control modeling. Stator voltages and currents

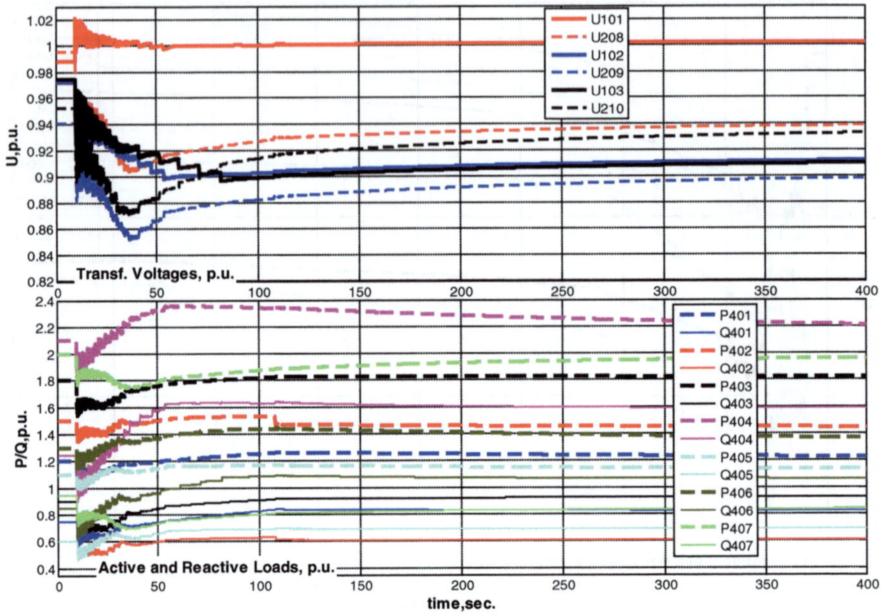

Fig. 16.18 Multi-agent automatic control modeling. Transmission transformers voltages. Active and reactive loads

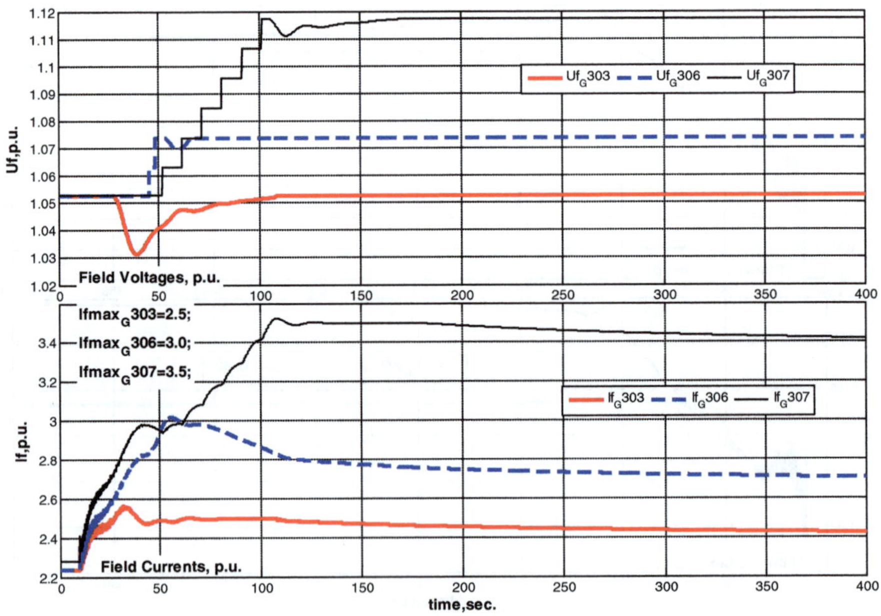

Fig. 16.19 Multi-agent automatic control modeling. AVR voltages and currents

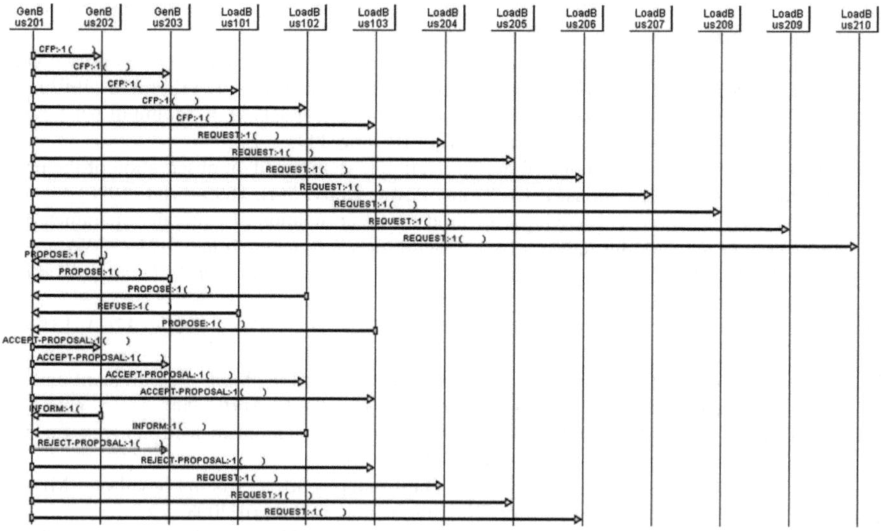

Fig. 16.20 Interaction of agents in MACS during the emergency situation

16.5 Conclusions

Emergency control plays an important role in EPSs and represents an effective means of responding to emergency situations and interrupting a cascading development of emergencies. Structural transformation of electric power grids and their joint operation require improvement in the emergency control principles and means. The improvement should be based on smart concepts and technologies, and advanced devices of information support. Innovative development of emergency control systems in electric power grids should be implemented as a distributed wide area protection system.

References

1. S.A. Sovalov, V.A. Semenov, *Emergency control in power systems* (Energoatomizdat, Moscow, 1988). (in Russian)
2. K.A. Brinkis, V.A. Semenov, Selective partitioning protection at out-of step conditions. *Elektricheskiye stantsii* **2**, 66–69 (1975) (in Russian)
3. A. Osak, A. Domyshev, *ANARES Documentation* (Novosibirsk, Institute of Power System Dispatching, 2010), (in Russian)
4. P. Pourbeik, P.S. Kundur, C.W. Taylor, The anatomy of a power grid blackout. IEEE Power Energy Mag. **4**(5), 22–29 (2006)
5. Yu.V. Makarov, V.I. Reshetov, V.A. Strojev, N.I. Voropai, Blackout prevention in the United States, Europe and Russia. in *Proccedings of the IEEE*, vol. 93, no. 11 (2005)
6. A.B. Barzam, *System Automatic Devices* (Moscow, Energiya, 1964). (In Russian)

7. Ya.E. Gonik, E.S. Iglitsky, *Automatic Elimination of Out-of-step Condition* (Energoatomizdat, Moscow, 1988). (in Russian)
8. V.H. Quintana, H. Muller, Particioning of power network and application to security control in *IEE Proceedings Generation, Transmission and Distribution*, vol. 138, no. 6 (1991)
9. X.M. Wang, V. Vittal, System islanding using nominal cut sets with minimum net flow. in *IEEE PES General Meeting*, Denver, USA, 6–10 June 2004
10. Y. Ohura et al., A predictive out-of-step protection system based on observation of the phase difference between substations. IEEE Trans. Power Deliv. **5**(4), 1695–1704 (1990)
11. V.G. Narovlyansky, A.A. Nalevin, A method for determination of equivalent parameters of power system network during out-of-step condition. *Elektrichestvo* **8**, 15–21 (2005)
12. A.G. Phadke, J.S. Thorp, *Synchronized phasor measurements and their applications* (Springer, Berlin, 2008)
13. V. Centeno et al., An adaptive out-of-step relay. IEEE Trans. Power Deliv. **12**(1), 131–138 (1997)
14. M.C. Bozchalui, M. Sanaye-Pasand, Out-of-step relaying using phasor measurement unit and equal area criterion. in *IEEE Power India Conference*, 10–12 April 2006
15. K.R. Padiyar, S. Krishna, Online detection of loss of synchronism using energy function criterion. IEEE Trans. Power Deliv. **21**(1), 46–55 (2006)
16. Li Li, Liu Yutian, Mu Hong, Yu Zhanxun, Out-of-step splitting scheme based on PMUs. in *DRPT'2008 International Conference*, Nanjing, China, 6–9 April 2008, 6 p
17. *Methodological Guidelines on Power System Stability*. (NTs "ENAS", Moscow, 2004), (in Russian)
18. O.N. Voitov, N.I. Voropai, A.Z. Gamm, I.I. Golub, D.N. Efimov, *Analysis of Inhomogeneities of Electric Power Systems* (Nauka, Novosibirsk, 1999). (in Russian)
19. CIGRE Defense Plan Against Extreme Contingencies, in *CIGRE Task Force C2.02.24*, April 2007
20. W.R. Lachs, Controlling grid integrity after power system emergencies. IEEE Trans. Power Syst. **17**(2), 445–450 (2002)
21. W.R. Lachs, Voltage instability in interconnected power systems: a simulation approach. IEEE Trans. Power Syst. **7**(2), 753–761 (1992)
22. C.W. Taylor, D.C. Erickson, Recording and analyzing the July 2 cascading outage. Comput. Applicat. Power Syst. **10**(1), 26–30 (1997)
23. F. Bellifemine, G. Caire, D. Greenwood, *Developing Multi-Agent Systems with JADE* (Wiley, England, 2007)
24. S.D.J. McArthur, E.M. Davidson, V.M. Catterson, A.L. Dimeas, N.D. Hatziargyriou, F. Ponci, T. Funabashi, Multi-agent systems for power engineering applications: part I. IEEE Trans. Power Syst. **22**(4), 1743–1752 (2007)
25. S.D.J. McArthur, E.M. Davidson, V.M. Catterson, A.L. Dimeas, N.D. Hatziargyriou, F. Ponci, T. Funabashi, Multi-agent systems for power engineering applications: part II. IEEE Trans. Power Syst. **22**(4), 1753–1759 (2007)
26. F. Bellifemine F, G. Caire, T. Trucco, G. Rimassa, JADE Programmer's Guide. CSELT & University of Parma, (2000), Available: http://www.jade.tilab.com/doc
27. F. Milano, An open source power system analysis toolbox. IEEE Trans. Power Syst. **20**(3), 1199–1206 (2005)
28. D.A. Panasetsky, N.I. Voropai, A multi-agent approach to coordination of different emergency control devices against voltage collapse, in *Proceedings of 2009 IEEE Bucharest Power Tech Conference*
29. C. Grigg et al., The IEEE reliability test system-1996. A report prepared by the reliability test system task force of the application of probability methods subcommittee. IEEE Trans. Power Syst. **14**, 1010–1020 (1999)

Chapter 17
Interface Protection

Lazar Bizumic, Rachid Cherkaoui and Ulf Häger

17.1 Introduction

A coupling of the electrical networks of ENTSO-E and IPS/UPS would result in a power system unprecedented in size and complexity, and potentially could have severe effects on operation and stability [1]. With the growth of the interconnection size, the number of events that could initiate disturbances in both systems grows (compare Chap. 14). At the same time, thanks to the larger propagation area, the plausible consequences of those disturbances are even more severe. This means that in order to ensure stable interconnected system operation during contingencies, advanced protection solutions and interconnection strategies are necessary (Chap. 15). The right choice of interconnection concept and protection actions should ensure that severe disturbances are not propagated from one system to another. Ideally, the overall stability and reliability of one large interconnected system should be enhanced in comparison to the two separate systems.

In this section, two different interconnection concepts and their interface protection and control actions are discussed. The first one is a large AC, synchronously interconnected system, while the other one is based on a pure HVDC asynchronous interconnection. The systems will be subjected to the serious disturbances and the results will be analyzed. The out-of-step protection system introduced in Sect. 16.2 is used as one of the options for interface protections discussed in the present chapter.

L. Bizumic · R. Cherkaoui (✉)
EPFL, Lausanne, Switzerland
e-mail: rachid.cherkaoui@epfl.ch

L. Bizumic
e-mail: lazar.bizumic@gmail.com

U. Häger
TU Dortmund University, Dortmund, Germany
e-mail: ulf.haeger@tu-dortmund.de

U. Häger et al. (eds.), *Monitoring, Control and Protection of Interconnected Power Systems*, Power Systems, DOI: 10.1007/978-3-642-53848-3_17,

17.2 Network Model

For the purpose of simulations, the enhanced ENTSO-E and IPS/UPS network model as described in Chap. 2 has been used. This model consists of 545 nodes and its level of detail, especially in the interface area, is sufficient to fulfill realistic dynamic results during transient processes.

Although, in reality, at the moment, these two power systems are not interconnected, for the purpose of the analysis, they have been connected by three 3,000 MVA, 750 kV transmission lines. Later, for the purpose of simulation of a pure HVDC asynchronous interconnection, three Back to Back (B2B) convertor stations have been installed on these three lines.

These interconnection lines have been modeled based on the data of the existing but not operational lines between the systems. These transmission lines were operated as an integrated part of IPS/UPS and the power system "Mir" until 1995 when Poland, Hungary, Slovakia and Czech Republic were synchronously interconnected to IPS/UPS. For a synchronous coupling of ENTSO-E CE and IPS/UPS some of these lines would need to be refurbished and partly reconstructed.

17.3 Simulation and Control Tools

The simulations presented in this chapter were performed using the dynamic power system analysis tool, EUROSTAG. Also, the EUROSTAG API (Application Programming Interface) feature has been extensively used. This feature has been developed and it enables implementation of non-classical control loops, modeling the behavior of the power system with complex special protection schemes, WAMS, optimization actions, Multi-Agent Systems, fuzzy logic control loops and discrete event triggered behavior.

Basically, an external system monitoring and control application has been created using a scripting programming language (Python) and then it has been connected with the EUROSTAG computational core by use of API. This control application drives the EUROSTAG simulation, interrupting it periodically to extract monitored variables values, makes calculations and if necessary changes the set points or states of desired elements in the power system model. The general architecture of the API integrated in the EUROSTAG computational module is shown in Fig. 17.1.

It is worth noting, that all variable values (measurements) and control actions collected, calculated and sent using the EUROSTAG API are obtained in without delay. In reality, there is a small delay associated with the time necessary to receive and processes measurements (communication delay), determine control actions and send control signals. For this reason, all centralized control actions are delayed by a random delay time of 50–250 ms has been added.

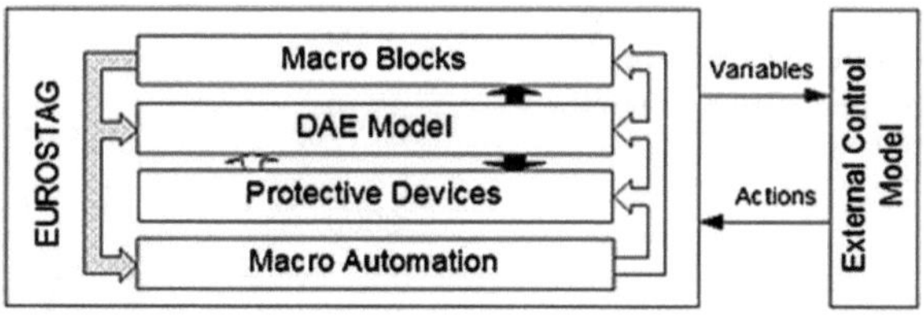

Fig. 17.1 Architecture of the EUROSTAG API

17.4 Simulation, Control Actions and Results

In this section analysis of disturbances for systems interconnected by pure AC and pure HVDC technologies will be discussed. Two simple scenarios will be discussed for both cases. In both scenarios, one typical system operating condition was simulated. In this operating condition, around 6,500 MW of active power is exported from the UPS/IPS system to Europe. For the interconnection interface described in the previous section, this power transfer represents a relatively high value. On the other hand, taking into account the sizes of two interconnected systems and energy transferred by other means (e.g. pipeline), this value represents relatively modest power exchange and in real operation even higher values might be needed.

In scenario 1, at $t = 10$ s, a fault has been simulated on one of the interface lines. 200 ms later, the line gets tripped and then 600 ms later it has been reclosed by automatic reclosing. 200 ms later, the line gets tripped permanently. In scenario 2 at $t = 5$ s, a generation loss of 4,000 MW in central Europe has been simulated.

17.4.1 AC Interconnection

In this section the pure AC interface is analyzed. Except from the usual overcurrent protection, the advanced protection scheme for controlled separation of two power systems as introduced in Sect. 16.2 has been applied to the system. Results with and without this advanced protection are presented. Basically the control center monitors interconnected areas for their phase angles and their second derivatives. Once the system detects that loss of synchronism is imminent by observing voltage phasors, it initiates controlled system separation.

17.4.1.1 Scenario 1

The results of the simulations for the scenario 1 for the interface protected only by conventional relay protection equipment are presented on Figs. 17.2, 17.3 and 17.4.

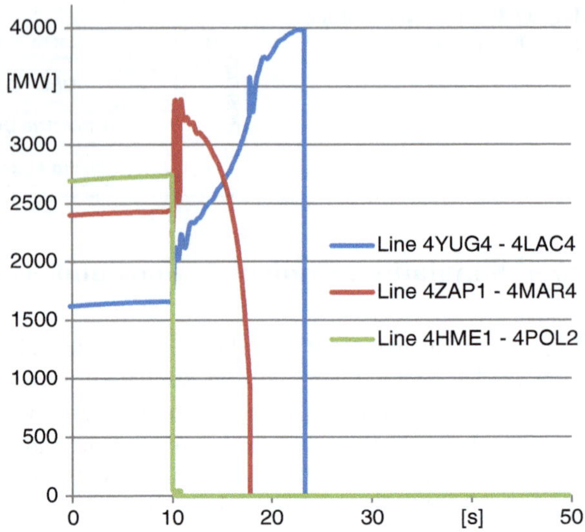

Fig. 17.2 Active power transfer on interface lines without advanced interface protection

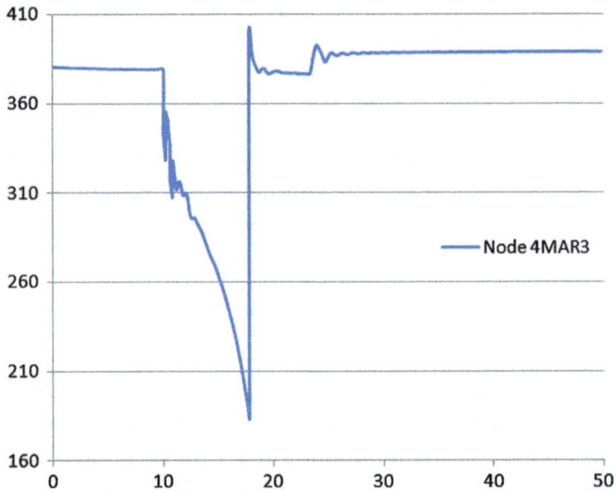

Fig. 17.3 Voltage magnitude at observed node without advanced interface protection [kV]

As can be seen from the charts above, the other two lines become overloaded after the fault on the first line, leading to increased currents and finally in over current protection action. In this case, the fault on one line resulted in cascading effect and uncontrolled separation of interconnected systems. In addition to the interruption of transfer of 6,500 MW between the systems, the nodes in the interface area suffered severe voltage drop during several seconds with voltage falling below 0.5 of its nominal value. This could have very serious consequences resulting in power plants at those nodes to be disconnected from the system due to under voltage protection. Eventually this could result in a further propagation of the disturbances through the network with cascading effects.

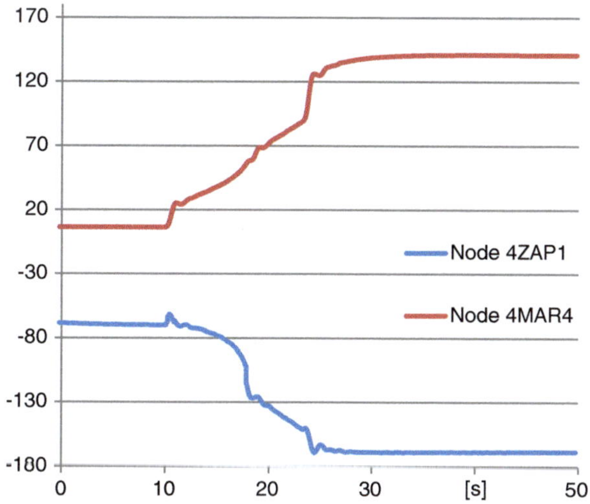

Fig. 17.4 Voltage angles at observed nodes without advanced interface protection [deg]

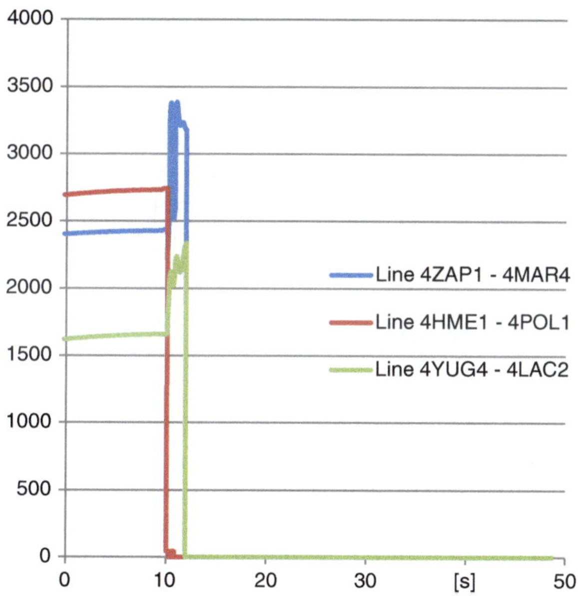

Fig. 17.5 Active power transfer on interface lines with advanced interface protection [MW]

The same scenario has been simulated one more time, but this time the interface is controlled by advanced protection scheme. The results of the simulation are shown in Figs. 17.5 and 17.6.

As can be seen from the graphs, about two seconds after the fault, the control system recognized that the two systems will not return to stable operation and that loss of synchronism is imminent. Immediately the full system separation has been executed. By this timely system separation, the large voltage drop in the interface

Fig. 17.6 Voltage magnitude at observed node with advanced interface protection [kV]

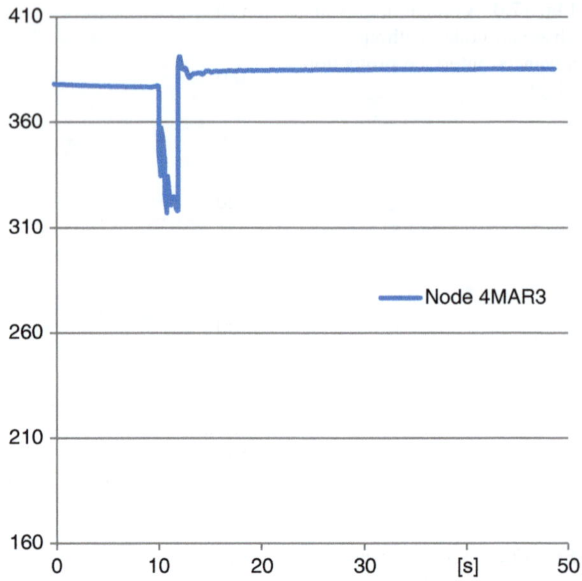

area has been prevented so the disturbance propagation should be stopped. It is important to notice that even in this case the loss of 6,500 MW of transferred power from one system to another has happened. Since this amount of power excides the primary frequency control power in ENTOSO-E (3,000 MW), additional measures like load and generation shedding might be necessary.

17.4.1.2 Scenario 2

The second scenario has also been simulated with and without advanced protection scheme. The results for scenario 2 are shown in Figs. 17.7, 17.8 and 17.9.

The 4,000 MW generation loss in the European system caused interface load flows to change dramatically. Overall power transfer between the systems was increased by around 2,000 MW. This value is expected and it is completely in accordance with the results of one of the previous ENTSO-E—IPS/UPS interconnection studies [2], where it was concluded that in case of a 3,000 MW generation loss, almost 1,500 MW of regulating power flow will cross the interface area. These interface lines become overloaded and voltage phasors at the end of the interconnection lines started separating. Finally, this results in complete uncontrolled separation of the systems and large voltage drops in the interface area.

Again, the same scenario has been repeated, but this time a protection scheme has been used. Results are presented in Figs. 17.10, 17.11 and 17.12.

As can be seen from the results, the defense plan acted one more time, couple of seconds before the uncontrolled system separation would happen. Once again, the

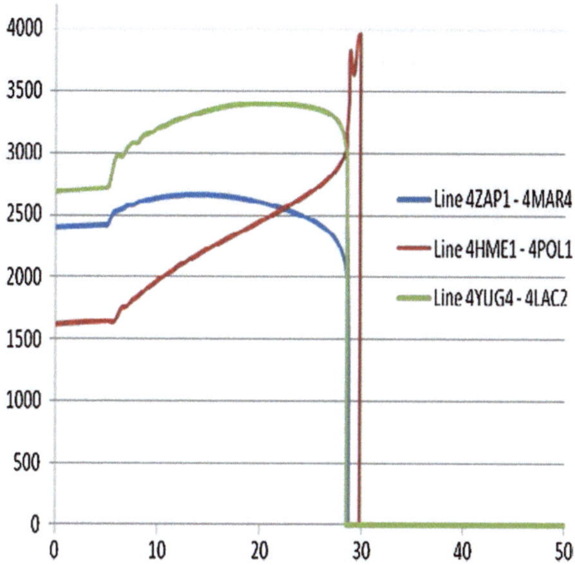

Fig. 17.7 Active power transfer on interface lines without advanced interface protection [MW]

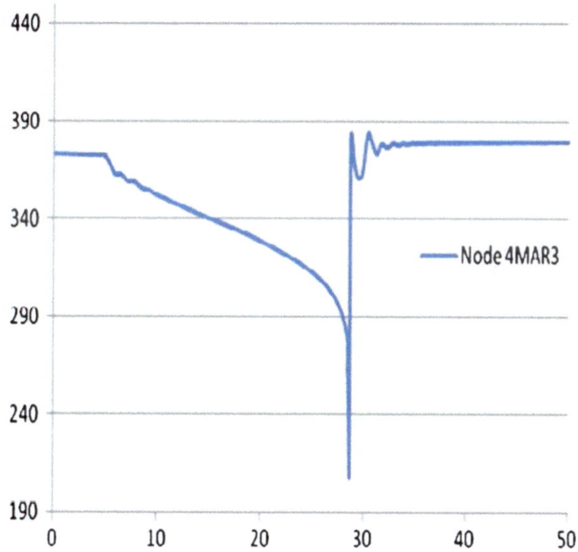

Fig. 17.8 Voltage magnitude at observed node without advanced interface protection [kV]

voltage drop in the interface is much less severe and the disturbance impact has been reduced. Still, in this scenario, European system has been left with a deficit of around 10,000 MW which requires additional measures in order to return to normal operation.

Fig. 17.9 Voltage angles at observed node without advanced interface protection [deg]

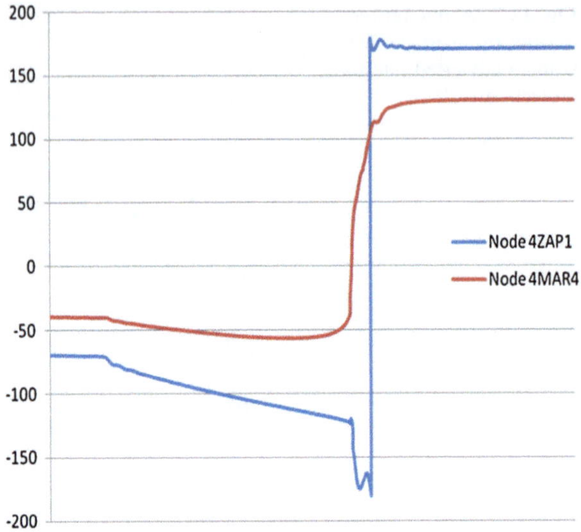

Fig. 17.10 Active power transfer on interface lines with advanced interface protection [MW]

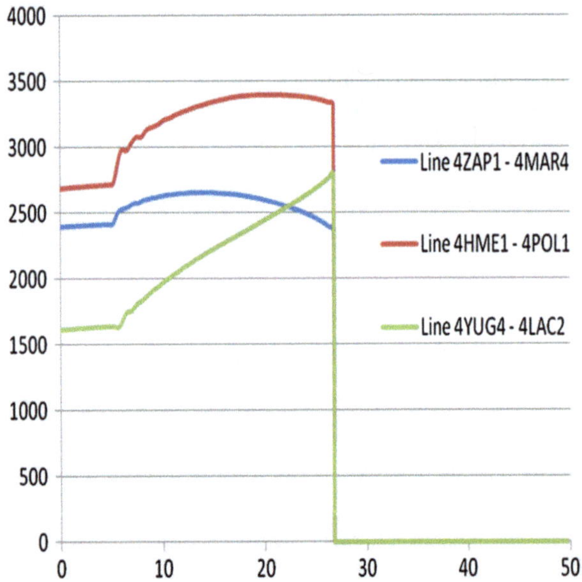

17.4.2 HVDC Interconnection

In this Sub-section, the same two scenarios will be simulated, but this time the systems are asynchronously interconnected by means of B2B HVDC convertor stations. Also, during the disturbances advanced HVDC control actions will be applied. This kind of interconnection has been chosen as a preferred.

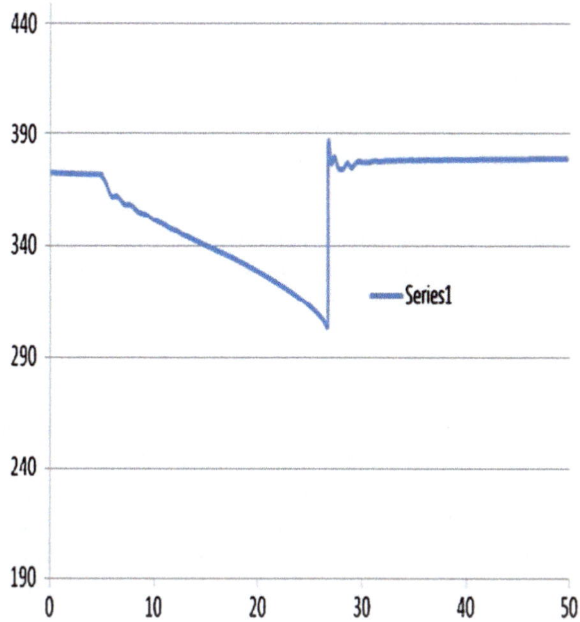

Fig. 17.11 Voltage magnitude at observed node with advanced interface protection [kV]

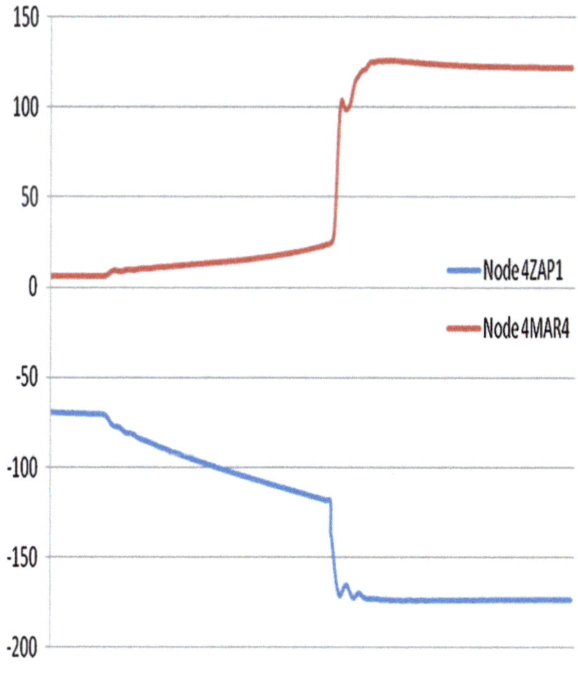

Fig. 17.12 Voltage angles at observed node with advanced interface protection [deg]

For the purpose of the simulations, it is important to choose properly the HVDC technology. In short, the HVDC transmission systems can be generally divided into two categories (compare Chap. 9):

- Classic, line or capacitor commutated converter HVDC (LCC and CCC HVDC)
- Voltage Source Converter HVDC (VSC HVDC).

The classic HVDC transmission system is a fully mature technology that has been in use for more than 50 years with over 80 large installations globally, including overhead lines and submarine cables. In comparison the VSC HVDC is a relatively new technology that was introduced in the 1990s. Also, VSC offers a number of advantages. Some of those important advantages are the black start capability (supply of passive networks), separate control of active and reactive power flows and multi-terminal systems.

For the purpose of this book chapter, the HVDC VSC technology has been chosen. The reasons for this choice are the recent breakthrough in the development of the VSC technology and the expected further development of this technology. Existing HVDC VSC projects are already achieving power ratings of about 1,000 MW and the VSC technology for 1,200 MW per link is already available. Even more, recent reduction in conversion losses for the VSC technology to below 2 % (1 % per converter) made these losses comparable with the HVDC classic losses eliminating one of the major drawbacks of the VSC technology. Also, HVDC classic links might be very vulnerable to faults and short circuits close to its converter stations, which could result in commutation failures and inability of the HVDC line to operate and transfer power during short circuits and voltage drops. This drawback of the classic HVDC technology could result in serious problems for the interconnected systems since they tend to shut down when mostly needed. On the other hand, HVDC VSC would not only be able to continue its operation during the fault duration but also to offer a voltage support for the endangered systems greatly improving the stability of the interconnected systems. Although the power limit for VSC B2B station currently is around 1,800 MW, the B2B stations in this test case have been modeled as 2,500 MW. The reason for this is that strong further development in this field is expected and by the time of plausible construction these power ratings should already be available.

In order to fully exploit advantages of HVDC technology, we can imagine HVDC links monitoring systems frequencies at both ends of the link (Fig. 17.13) and adjusting their active power flows according to the measured frequency difference between the interconnected nodes. Basically this would mean that HVDC terminals would respond to the generator loss in one system similarly to the other generators, increasing or decreasing power flow to or from the affected system.

The regulating power would be drained directly from the adjacent cluster. Of course, changing the transferred power from the adjacent cluster would result in a frequency change in that cluster also. Consequently, generators in that cluster would respond, very similarly like in a case of AC connection. Here, the important difference is that the HVDC lines should not put the other cluster in danger and should limit the effect of the disturbance on the other cluster to an acceptable level.

Fig. 17.13 Frequency difference between both ends of the link

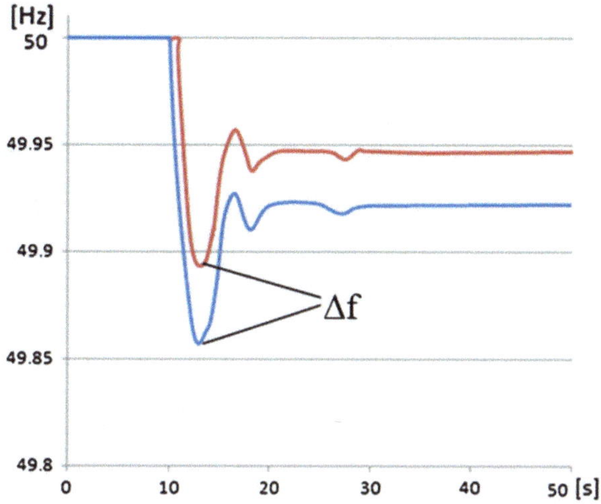

Also, a control curve could be set in such a way, that the HVDC lines do not act at all for small disturbances (dead band characteristic). On the other hand, in a case of severe events their action would be in direct correlation with frequency values. If frequency in the affected system approaches a first under-frequency load shedding value, HVDC lines should offer all possible support, increasing their power flows to the maximum levels that are still safe for the adjacent system. The simplified HVDC control characteristic is shown in Fig. 17.14.

The values of P_{max^+} and P_{max^-} could represent either the maximum power ratings of the HVDC links, or the maximum acceptable values based on the internal limitations of the interconnected systems. In the latter case, since it cannot be determined locally, this value would have to be provided from the remote centralized control center as described in the following section. Additionally, in order to not endanger the unharmed system, if the frequency in the adjacent system falls below a certain value (e.g. 49.2 Hz), the further support actions of HVDC links should be suspended at their current values.

The important advantage of the proposed control is that all control is localized in individual HVDC controllers and since there are no complicated calculations involved it is very fast. Thanks to this, there is no need for any new communication lines, complicated control centers or sensitive network data sharing.

Theoretically, thanks to this control, all the advantages of both AC and HVDC interconnections could be combined and preserved. The systems operate interconnected and during the disturbances they can help each other, but the risk of severe disturbances and their consequences (e.g. large scale blackout) are practically limited to one system only. Also, once the frequencies in both systems are restored to their original values, the power transferred over HVDC lines would automatically return to its original set point values since $\Delta f = 0$.

This means, that real system separation practically should never happen. The only imaginable scenario where the power transfer between two systems is fully

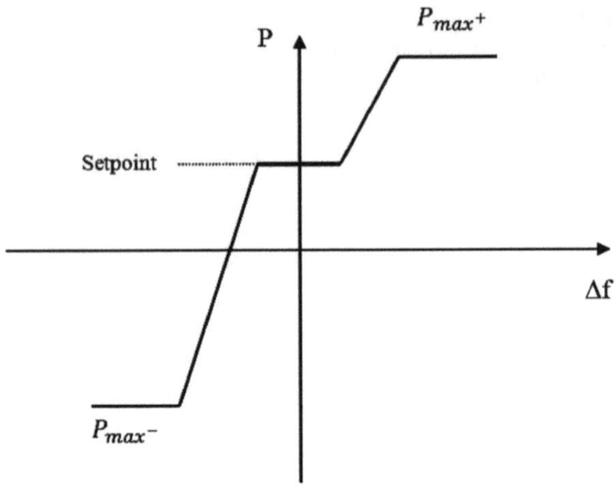

Fig. 17.14 HVDC control droop characteristic

terminated is that both systems are simultaneously in severe contingency state, so that none of the two systems is able to help the other one. This situation is very unlikely, and in this case none of the interconnection solutions or interface control strategies would be able to make any difference. It could also be said, that the strength of the interface is constantly adjusted during the contingency state.

Also, during the operation of the HVDC interface, a centralized interface control center is monitoring power flows on all HVDC lines and if one of the lines is tripped for some reason, the system tries to compensate its loss and automatically increases the power flows on the remaining HVDC lines trying to maintain the agreed power transfer unchanged. The scenario 1 has been simulated and results are presented in Figs. 17.15 and 17.16.

As soon as the first line transferring 2,500 MW is tripped, the control center is trying to compensate for the deficit in power transfer and increases the load flows on the remaining two HVDC lines. Since the full power transfer restoration was not plausible due to the power limits of the HVDC lines, the flows on the remaining two lines have been set to maximum. This way, only 1,000 MW of transfer has been restored, but here one very important advantage of HVDC technology comes to light, HVDC lines cannot be overloaded. This way, both systems continue their operations with just slight disturbance in power exchange (1,500 MW). This way, the impact of the line fault has been minimized and both systems continue their operation without serious disturbances.

The final simulation, once again deals with the scenario 2. This time the interface consists only of HVDC lines and abovementioned control algorithms have been applied. The results are presented in Figs. 17.17 and 17.18.

As can be seen from the results, as soon as the frequency in the European part of the system start falling, as a result of generation loss, all HVDC links automatically increased its power transfers in order to support the endangered system. Since two of these HVDC lines where operating close to their power limits, they increased the power transfer to maximum, but still without the plausibility of

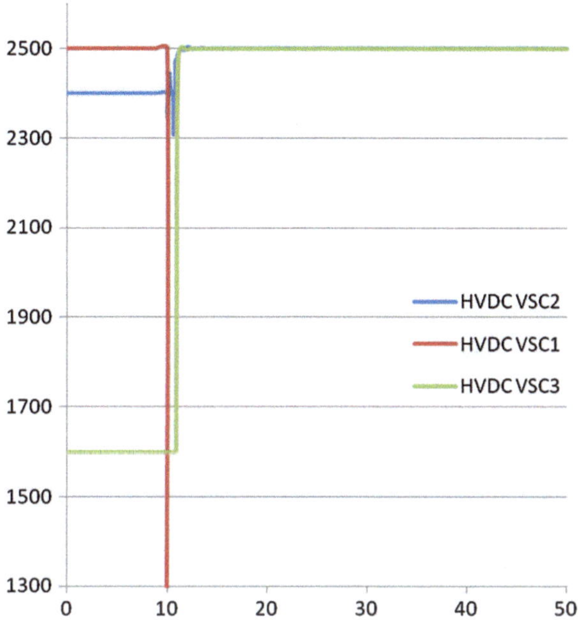

Fig. 17.15 Active power transfer of interface lines at scenario 1 [MW]

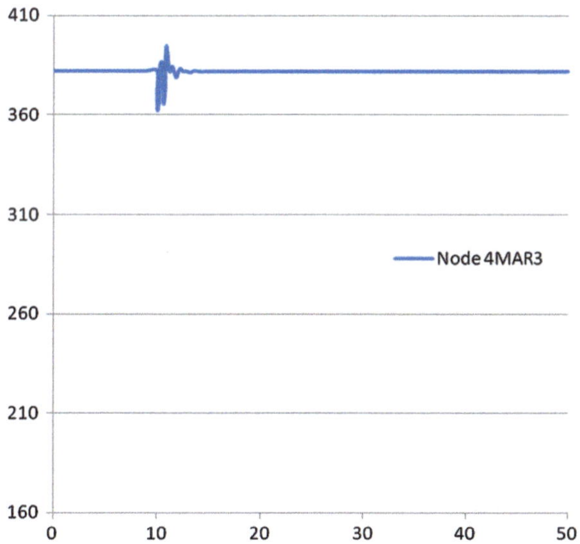

Fig. 17.16 Voltage at observed node at scenario 1 [kV]

overload. As soon as the frequency is stabilized, the amount of power support is reduced, and once the frequency of both systems returns to the dead band, the power transfers will automatically return to their original agreed values. In this particular case, the overall increase of power transfer from one system to another was around 1,100 MW, but this value can be easily adjusted just by tuning the

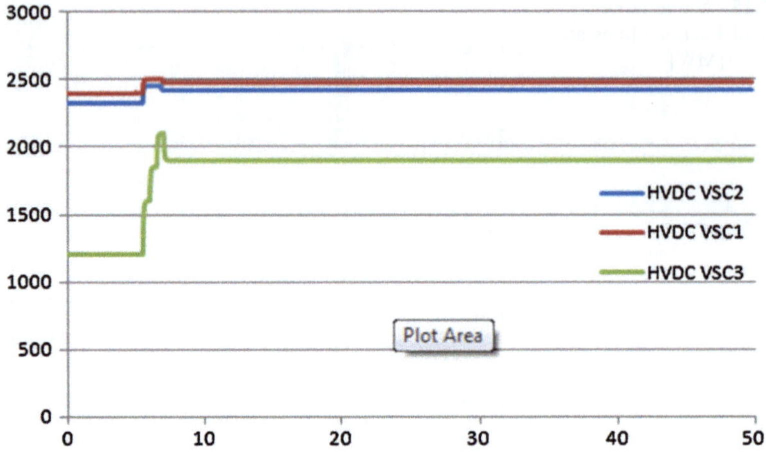

Fig. 17.17 Active power transfer of interface lines at scenario 2 [MW]

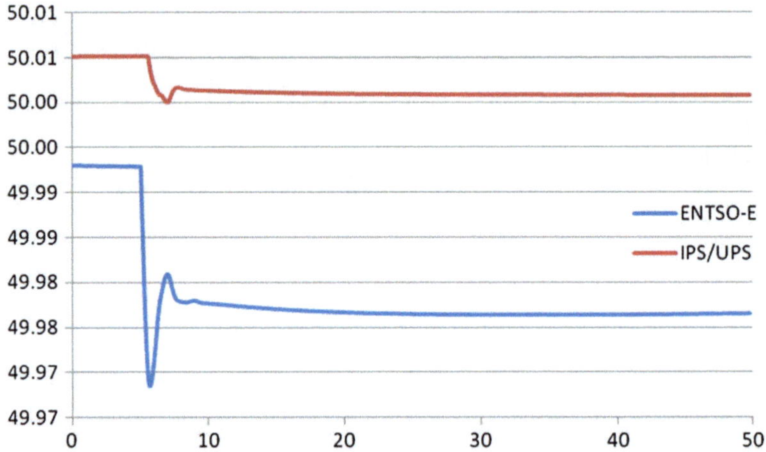

Fig. 17.18 System frequencies at scenario 2 [Hz]

control parameters and frequency deviation sensitivity. Practically, if the state of the system allows, the pure HVDC interconnection can provide even stronger support than synchronous AC interconnection.

17.5 Conclusions

Based on the results presented in this chapter, it is evident that by using the advanced AC protection schemes and emergency system separation, it is plausible to greatly enhance security of large interconnected power systems. Unfortunately, due to the fact that in this case systems are completely separated, consequently power transfer between the systems is stopped instantly. This can be very problematic, and can have devastating effects to the involved systems, especially in case of large power transfers. In order to prevent such events, HVDC technology combined with the presented advanced interface control strategies could be used. This way, all of the benefits of system interconnection might be achieved, while abovementioned problems could be mitigated.

References

1. UCTE, IPS/UPS, in *Synchronous Interconnection of the IPS/UPS with the UCTE: Summary of Investigations and Conclusions*, 2008
2. C. Kim, V.K. Sood, G. Jang, S. Lim, S. Lee, in *HVDC Transmission, Power Conversion Applications in Power Systems*, 2009

Chapter 18
Under-Frequency Load Shedding System

Vladimir Chuvychin, Antans Sauhats, Vadims Strelkovs and Eduards Antonovs

18.1 Introduction

During emergency situations in the power system caused by generating power deficiency frequency decline can happen. Dynamics of underfrequency during deficiency of generation in the power system can have very different character. It depends on the value of disturbance, response of emergency automation and the governor system.

For gradual increase in load, or for sudden but mild overloads, units' governors will sense speed change and increase power input to the generator. Here the so called primary and secondary frequency control is activated. Extra load is handled by using spinning reserve from the unused capacity of all generators operating synchronously to the system.

Severe system disturbances can result in fast frequency drop, which makes fast governor and boiler response impossible. To restore frequency load shedding or spinning reserve involvement may be activated. If the governor action cannot activate spinning reserve fast enough to restore the system to its normal operating frequency, frequency actuated automatic underfrequency load shedding (UFLS) serves as a last-resort tool to prevent the system from collapse.

In most power systems up-to-date automatic load shedding systems practically foresee disconnection of the load at underfrequency without time delay or with small delay part. The numbers of load shedding steps and the value of load to be

V. Chuvychin (✉) · A. Sauhats · V. Strelkovs · E. Antonovs
Riga Technical University, Riga, Latvia
e-mail: chuvychin@eef.rtu.lv

A. Sauhats
e-mail: sauhatas@eef.rtu.lv

V. Strelkovs
e-mail: vadims.strelkovs@rtu.lv

E. Antonovs
e-mail: antonovs@eef.rtu.lv

U. Häger et al. (eds.), *Monitoring, Control and Protection of Interconnected Power Systems*, Power Systems, DOI: 10.1007/978-3-642-53848-3_18,

shed vary for different power systems. Some power systems use rate-of-change of frequency as additional factor to shed a load.

Existing UFLS automation has drawbacks, which limit the adaptability of emergency automation to a change of underfrequency situation in a power system. UFLS tripping frequency settings are selected for some specific emergency situation, which is considered as more probable for a specific power system. It is not possible to foresee all situations that can occur in the power system. UFLS operation will be effective only for the pre-calculated emergency cases. Problems related to the value of a load to be shed is also very typical. Redundantly tripped load can create overfrequency situations, which sometimes is more dangerous than underfrequency.

A very important task for operators of large scale interconnected power systems is the analysis of methods and tools for assessing the stability of large transmission systems.

This chapter shows results of the analysis of frequency behavior during emergency situations in a united power system of ENTSO-E and IPS/UPS. The different parameters of underfrequency load shedding automation in each united power system can cause nonselective load shedding and hence, underfrequency or overfrequency emergency situations.

18.2 Theoretical Background

When a power system is in stable operation at normal frequency, the total mechanical power input from the prime movers to the generators is equal to the sum of the entire connected load, plus all real power losses in the system. Any significant upset of this balance causes a frequency change.

Frequency deviation during emergency situations is a dynamic process. Let us use the swing equation for estimation of situation in the power system during emergency [1, 2]:

$$J\frac{d\omega}{dt} = T_m - T_e \qquad (18.1)$$

where

T_m mechanical torque in N·m,

T_e electromagnetic torque in N·m,

J combined moment of inertia of generator and turbine, kg·m^2,

ω angular velocity of the rotor, mech. rad/s.

Another more convenient form of the swing equation can be written [2]:

$$M\frac{d\omega}{dt} = P_m - P_e \qquad (18.2)$$

where

P_m shaft power output of the machine less rotational losses in W,
P_e electrical output power in W,
M inertia constant of the machine.

For gradual increases in load, or for sudden but mild overloads, the units' governors will sense the speed change and increase power input to the generator. Here the so called primary and secondary frequency control is activated. Extra load is handled by using spinning reserve—the unused capacity of all generators operating and synchronized to the system.

Severe system disturbances can result in fast frequency drop, which makes fast governor and boiler response impossible. If the governor action cannot activate spinning reserve quickly enough to restore the system to its normal operating frequency, underfrequency load shedding (UFLS) serves as a last-resort tool to prevent the system from collapse. Simultaneously spinning reserve is activated as soon as possible. Now both types of emergency control are acting independently which may prolong the frequency restoration process or may cause overfrequency conditions.

18.2.1 Hazards of Under-Frequency Operation

There are two main problems associated with the operation of power systems at a low frequency, both related to thermal generating units.

The first problem is concerned with the vibratory stresses on long low-pressure turbine blades [3–6]. An example of European standards of frequency–time zone determined by parameters of thermal generating units is shown in Fig. 18.1.

The zone for thermal generating units can be described by four points with the coordinates 45.0 Hz and 0 s, 47.0 Hz and 20 s, 48.5 Hz and 60 s and 49.0 Hz continuously. The limitation from the upper side is 53.0 Hz and 0 s, 51.0 Hz and 10 s and 50.5 Hz continuously (red lines in Fig. 18.1) [6]. The frequency–time zone for nuclear power plants is narrower. The blue lines in Fig. 18.1 show an example of the restriction zone of the former Lithuanian Ignalina nuclear power plant [7].

The second problem is concerned with the performance of plant auxiliaries driven by induction motors. At low frequencies the plant capability may be severely reduced because of the reduced output of boiler feed pumps or fans supplying combustion air or water to boilers. This is also dangerous for consumers having rotating load, fans and pumps.

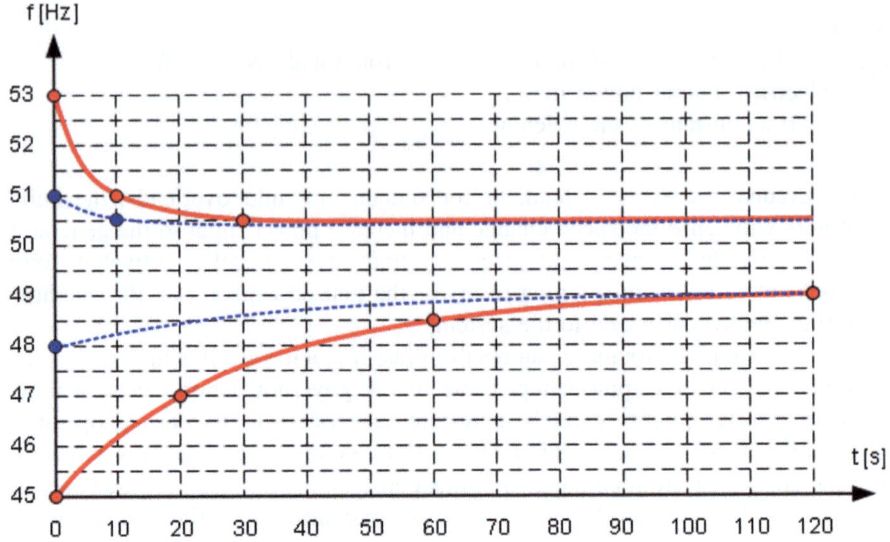

Fig. 18.1 Allowed frequency–time zone—*red* [6], *blue* [7]

18.2.2 Main Parameters Influencing the Character of Frequency Behavior

For analysis of the frequency behavior during emergency situations a few main parameters of the power system should be taken into account. These parameters are:

- Parameters of turbine and generator. They include the time constant of the generator and parameters of the turbine governor;
- Power system inertia constant (H constant) or inertia constant of the machine;
- Load-damping constant D.

Figure 18.2 illustrates the power frequency decline characteristic during underfrequency situations for different power deficiency values ΔP, several combinations of inertia constants (4, 7, 10), assuming a load-damping constant of D = 2.

18.2.3 Frequency Actuated Load Shedding as the Mean for Preventing Deep Frequency Decline

Deep frequency decline can be prevented either by increasing output mechanical torques of the turbines or by decreasing the load power. The first approach is not

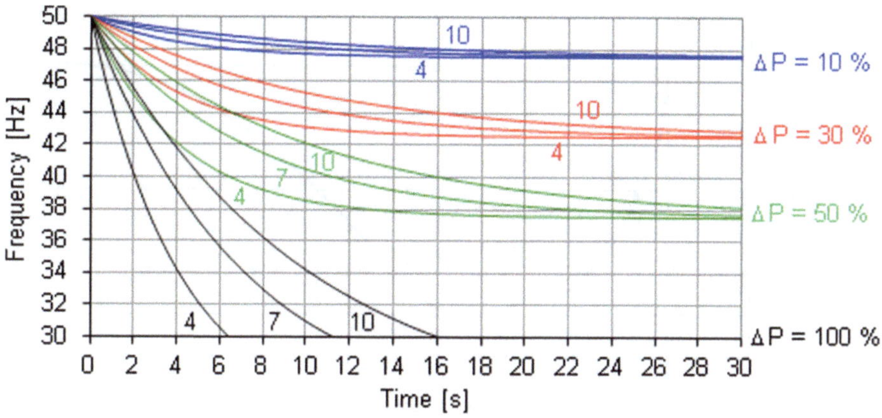

Fig. 18.2 Frequency decline character for different power deficiency values

possible due to the slow reaction of the turbine governor. Logical solution is to shed part of the load on under-frequency situations.

The main requirements to the frequency actuated load shedding (UFLS) are:

1. The frequency actuated load shedding system should prevent power system frequency decline below the level which is dangerous for power system equipment. The frequency should stay at the allowed frequency–time zone. To fulfill this requirement sufficient volume of the load should be connected to an automatic load shedding system to achieve a maximal possible power deficiency value for a specific considered power system.
2. Load, which is connected to the automatic load shedding system, should be equally distributed along the power system. During the under-frequency situation the power system can be separated into few areas. In this case each area should have sufficient load shedding capacity.
3. After preventing a frequency decline, the frequency should be restored to a value close to the rated level. The frequency restoring algorithm should foresee means for preventing an overfrequency situation.
4. After power system frequency restoration it is desirable to re-connect all previously disconnected loads automatically.

Additional requirements:

1. In some cases during an under-frequency situation power system voltage decline can take place. This phenomenon can be used as additional load shedding feature.
2. During the separation of a power system into several parts optimal load shedding algorithms can be different for each situation.
3. About 70 % of under-frequency situations are characterized by multiple frequency decline and restoration behavior (so called cascade type of under-frequency). Load shedding systems should behave adequate in each case.

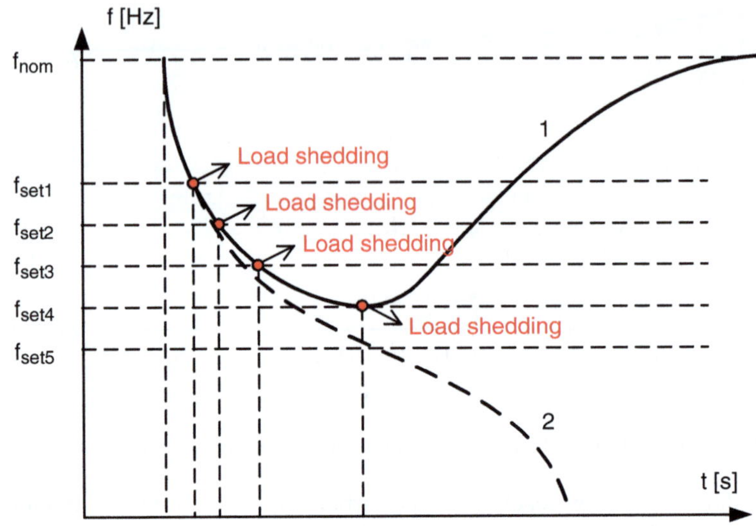

Fig. 18.3 Power system frequency variations during the UFLS operation

18.2.4 Main Parameters of Under-Frequency Load Shedding System

In most power systems up-to-date automatic load shedding systems foresee disconnection of part of the load due to under-frequency without time delay or with small delay [8–11]. The numbers of load shedding steps and value of a load to be shed vary for different power systems [6, 7]. Some power systems use the rate-of-change of frequency as additional factor to shed a load [12].

Figure 18.3 presents an example of a frequency variation during operation of the frequency actuated load shedding system. The point $f < f_{nom}$ corresponds to the moment of active power deficiency appearance in the power system. From this moment the frequency drop starts. When the power system frequency reaches the level of a first load shedding setting f_{set1}, the first part of the load is disconnected. The next part of the load will be disconnected when frequency reaches second load shedding setting f_{set2}. With each next load shedding step rate-of-change of frequency decline is cut down and after a certain moment the increase of frequency takes place. Such logics of the load shedding operation apply to most of the power system utilities.

The joint power system of the former USSR has more advanced frequency actuated load shedding automation. The first system is the fast-acting automatic load shedding system with different frequency settings called UFLS-I. The aim of this system is to prevent deep frequency drop.

The second system is an automatic load shedding system with one frequency tripping setting for all steps and different tripping time delays called UFLS-II.

A time interval between the time delay settings is $\Delta t = 5$ s. The aim of UFLS-II is to restore the frequency to a level close to the rated frequency value. This is a rather advanced automatic load shedding and frequency restoration system, but the system is not adaptive enough to all possible emergency situations [8].

For developing an algorithm of the load shedding system the below mentioned problems should be taken into consideration:

1. Selection of upper frequency setting level.
2. Selection of intervals between nearest frequency settings. Frequency interval depends on the frequency relay's frequency measurement accuracy.
3. Maximal capacity of a load connected to the load shedding system should be determined. This problem is solved specifically for each power system.
4. Selection of the number of necessary steps for the load shedding.
5. Selection of tripping time for a load shedding automatic relay.

During emergency situations caused by deficiency in the interconnected power system under-frequency load shedding automation is activated in all interconnections. Due to the different philosophy of UFLS automation operational conditions in the interconnection can be unpredictable.

As every power system usually has its own UFLS scheme, prior to interconnection it is essential to carry out thorough research work to ascertain the possible problems that may arise due to diversity of the number of steps, the frequency levels and the amount of load to be shed at each step.

The initial period of interconnected operation is characterized by relatively weak intersystem ties. Therefore unnecessary power oscillations due to un-coordinated UFLS schemes can endanger stability of the interconnected system.

18.2.5 Maximal Capacity of a Load in the Load Shedding System

The maximal capacity of a load connected to the UFLS system is determined assuming a minimal allowed frequency level after UFLS-I operation f_{min} [13]

$$P_{UFLS} = \frac{\Delta P_G - P_{L,0} \cdot \left(1 - f_{min}/50\right) \cdot D}{1 - \left(1 - f_{min}/50\right) \cdot D} \qquad (18.3)$$

where

ΔP_G is power system deficiency;
$P_{L,0}$ is initial load capacity;
D is load-damping constant.

Taking into consideration that D is a variable quantity it is assumed as minimal value for calculation of the UFLS capacity.

18.2.6 Frequency Rate-of-Change as Additional Factor for Load Shedding

It is obvious from Eq. (18.3) that the severity of misbalance at the power system at the first moment after a disturbance can be characterized by the value of rate-of-change of frequency df/dt. Curves of Fig. 17.2 also illustrate that phenomenon. In case of a large deficiency value in spite of the load shedding operation a dangerous drop of the power system frequency is possible. Using the rate-of-change of frequency as additional factor for the detection of emergency situations can prevent a large frequency drop in some cases. The initial value of rate-of-change of frequency depends on many factors—the value of power deficiency, the time constant of generators and load.

Hence, the rate-of-change of frequency value cannot be used as main load shedding parameter, but for large system disturbances many power systems use this parameter as additional load shedding factor [9, 12, 14–16].

The authors suggest a more advanced approach of an adaptive load shedding system with the use of a phase plane [17]. For realization of the adaptive UFLS dynamic programming is used at the phase plane with the aim to get the optimal frequency regulation law at the emergency. The new method of UFLS allows for dynamic control of the frequency change at emergency situations by selected dynamic characteristic at the phase plane. The new method using the frequency change dynamic is independent on the power system dynamics parameters, but depends only on the adaptive UFLS automatic influences to the power system.

The device, which operates according to the new principle, controls all connected loads of a substation. The device realizes the following main functions: UFLS-1 function, UFLS-2 function and the Dynamic Breaking (DB) function.

The main purpose of the DB function is to prevent the frequency overregulation possibility by the UFLS automation. This function is developed to reconnect back the disconnected load of the UFLS automation.

With UFLS-1, UFLS-2 and DB functions it is possible to dynamically control the frequency change by the selected dynamic characteristic.

The Fig. 18.4 illustrates all adaptive UFSL (UFLS-1, UFLS-2 and DB) operational zones at the phase plane.

18.3 The Goal of Analysis of Frequency Behavior During Emergency Situation in the Power System

A very important task for operators of large scale interconnected power systems is the analysis of methods and tools for assessing the stability of the transmission system. This sub-chapter shows results of an analysis of frequency behavior during emergency situation in united power systems of ENTSO-E and IPS/UPS (see Fig. 18.5). The different parameters of under-frequency load shedding automation

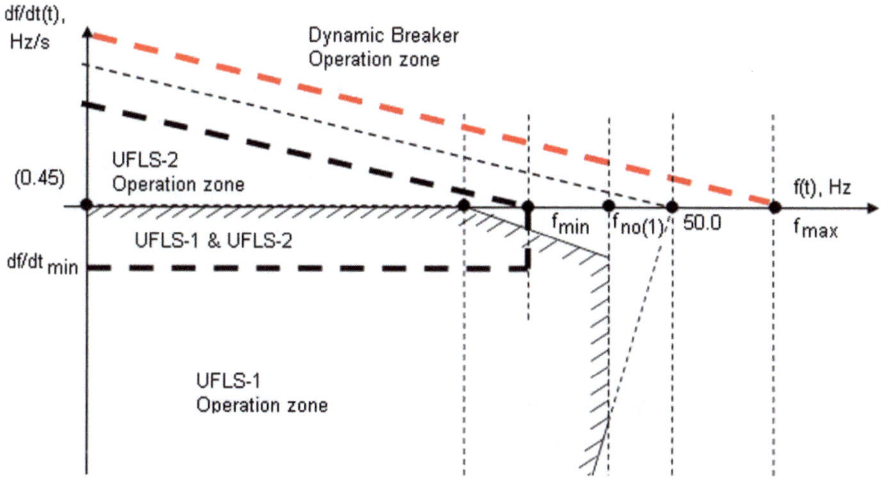

Fig. 18.4 Operational zones of adaptive load shedding system

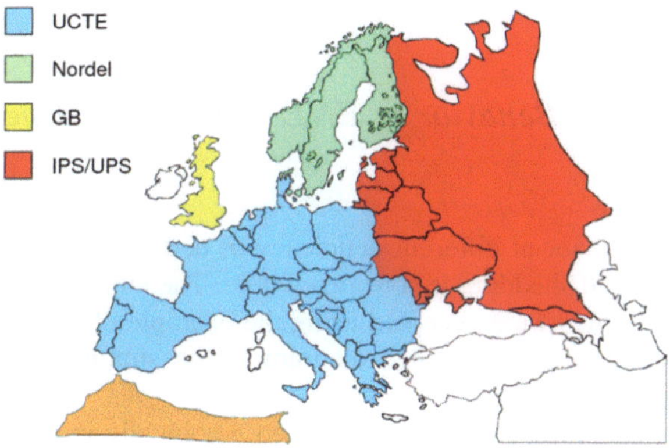

Fig. 18.5 Synchronous areas in Europe

in each united power system can cause nonselective load shedding and hence, under-frequency or overfrequency emergency situations.

The goal of the analysis is to find operational conditions which can cause emergency situations.

The model of first category under-frequency load shedding is presented in Fig. 18.6.

The created model of UFLS-II by its structure and functional description of blocks is similar to the model of UFLS-I described above. The difference between these two models is that the model of UFLS-II has different operational time delays.

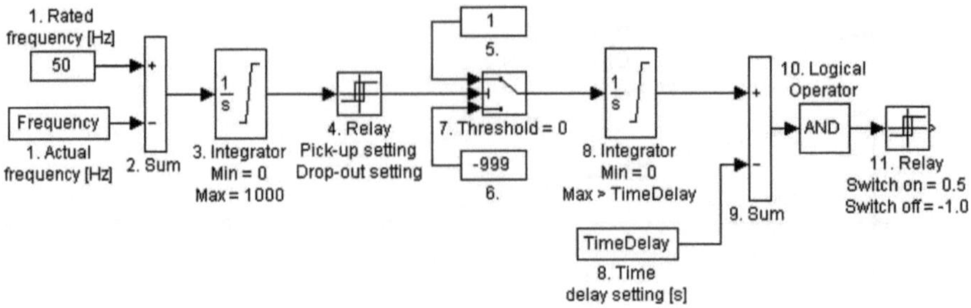

Fig. 18.6 The structure of UFLS-I model for one step

18.4 Analysis of Frequency Behavior for Different Algorithms of UFLS

As it was mentioned above, the main objectives of the UFLS is to prevent decline of the system frequency. This chapter presents results of analysis of under-frequency load shedding operational effectiveness for different emergency situations caused by loss of part of the generation.

18.4.1 Frequency Behavior in ENTSO-E During Operation of UFLS

The frequency behavior at different deficiency of active power with the usage of under-frequency load shedding is considered here. The number of UFLS steps and their threshold settings were accepted using the example of one of the ENTSO-E countries (Spain) [18]. A simplified diagram of the simulated power system is presented in Fig. 18.7.

Part of the calculation results are presented graphically in Fig. 18.8.

In Fig. 18.8 the frequency behavior is shown with a deficiency of active power equal to 5 and 10 % of the total generation. The frequency does not restore to its nominal value after the under-frequency load shedding operation because the total disconnected load by the steps of UFLS is twice smaller as the deficiency of active power in both cases. Thus for the second case two tripping steps (49.5 and 49.3 Hz) were initiated with 5 % of disconnected load. Load-damping was assumed to be constant and equal to 2. This parameter provides smooth restoration of the frequency up to the low level of 49.6 Hz.

Figure 18.9 presents results of calculation of emergency under-frequency situation caused by loss of 10.8 % of generation (P_{def} = 10.8 %). The over-frequency situation after UFLS operation happens because the total disconnected load of the steps of UFLS is two times as large as the deficiency of active power in the network. Three steps of UFLS (49.5, 49.3 and 49.0 Hz) disconnect 20 % of the load. An

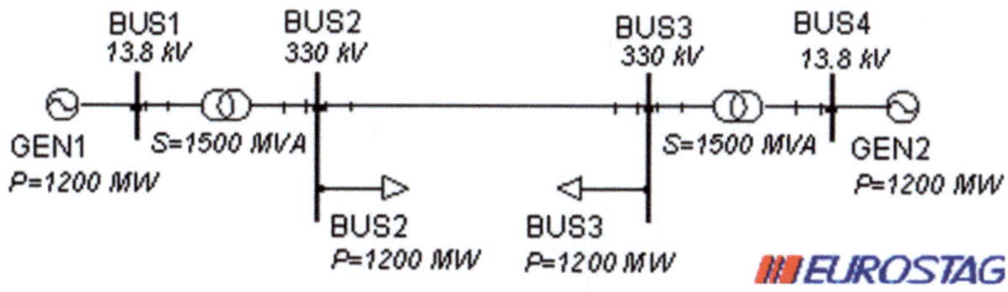

Fig. 18.7 Simplified diagram of simulated power system

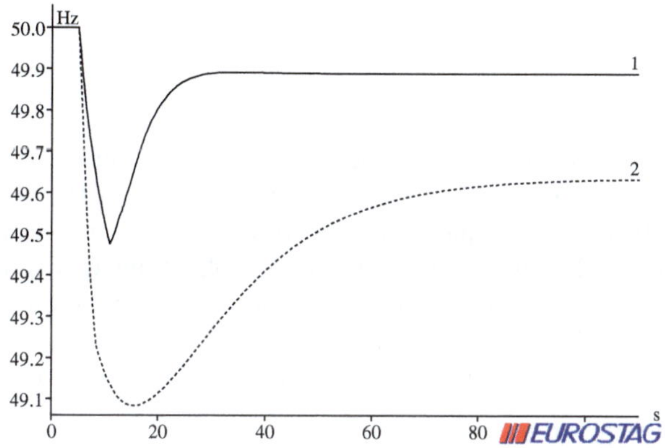

Fig. 18.8 Frequency behavior for 5 % (*1*), 10 % (*2*) of active power deficiency

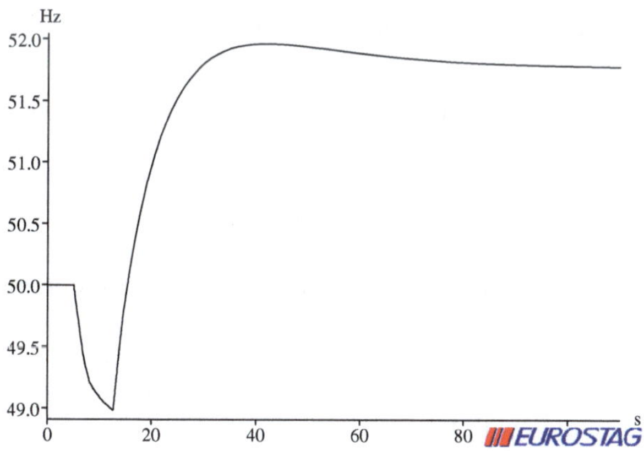

Fig. 18.9 Frequency behavior for 10.8 % of active power deficiency

over-frequency situation was discovered which is very dangerous for the power system.

Very often under-frequency situation in the power system has so called iterative character when after first loss of generation a second generating source can be lost.

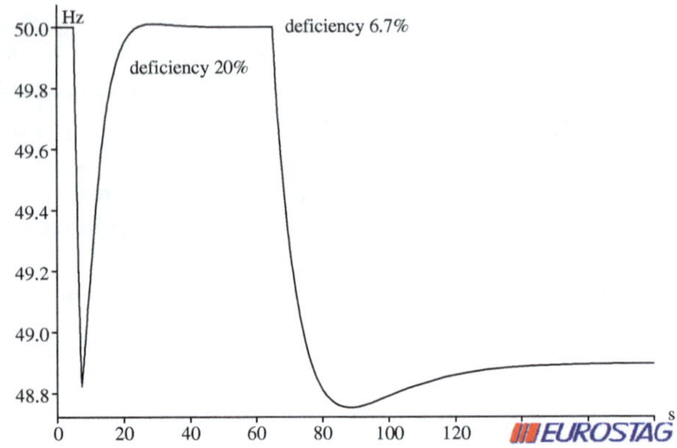

Fig. 18.10 Frequency behavior for 26.7 % of active power deficiency

The frequency behavior at a load deficiency of 20 % and at an iterative deficiency of the active power of 6.7 % is shown in Fig. 18.10. During cascaded events the stabilization of frequency can occur at a dangerously low level without being noticed by UFLS. In some professional publications this situation is called as "frequency hovering". It can happen if the active power disconnected by UFLS is not sufficient to bring the frequency back to the permissible range nor it decreases below the next activation level. In the considered case the frequency stays at 48.9 Hz.

18.4.2 Frequency Behavior in the Power System of the Baltic States During Operation of UFLS

This chapter presents results of the analysis of operational effectiveness of under-frequency load shedding for different emergency situations using the combined model of UFLS.

UFLS-II category, is used for frequency restoration after UFLS-I operation [13]. While restoring the frequency with the help of UFLS-I and UFLS-II at the same time trying to disconnect as few loads as possible the combination of these two algorithms' operation is used.

The number of combined UFLS steps and their threshold settings were accepted as an example of one of the Baltic country (Latvia). This model of UFLS has 74 steps and 11 thresholds with total disconnected load of 66.9 %.

In Fig. 18.11 the frequency behavior at the deficiency of 5 % of active power (1) is shown. For such a small value of power deficiency the frequency decline was stopped by the governor's operation. At the deficiency of 7.5 % (2) one step of UFLS-I has operated but the disconnected power is 2.5 times smaller than the deficiency of power in the network, hence, the frequency restores close to its nominal value, but stays in the rate of the marginal permissible range. In the case

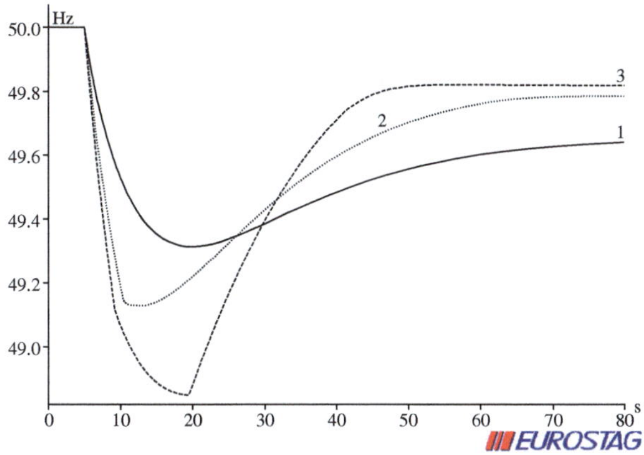

Fig. 18.11 Frequency behavior for 5 % (*1*), 7.5 % (2) and 8.9 % (*3*) of active power deficiency

when the deficiency of the power in the network is equal to 8.9 % (3) one step of the UFLS-I and two steps of UFLS-II were operated.

Investigations show that the application of combined UFLS-I and UFLS-II exclude over-frequency and under-frequency situations in the power system. This is an advantage of the load shedding system in the Baltic States' network.

At the iterative accident the frequency restores close to its nominal value and stays in the area of the normal admissible range. Application of UFLS-II model allows providing such proper operation.

More than 20 different variants of deficiency were considered. For all considered cases the maximal deviation of frequency after its restoration is in the range of −0.22 to +0.023 Hz.

The frequency behavior at a deficiency in the power system of 20 % and at the iterative accidents with deficiency of 6.7 % (1) and 7.5 % (2) is shown in the Fig. 18.12.

At the iterative accident the frequency restores close to its nominal value and stays in the area of the normal admissible range because of the operation of the combined model of UFLS. At the iterative deficiency the steps of UFLS-II combined with UFLS-I and UFLS-II are operating. The over-frequency or under-frequency situations, as it was shown in Sect. 18.4.1 are impossible using the combined model of UFLS.

18.4.3 Frequency Behavior of Joint Power System Operation with UFLS and Power Deficiency at the Baltic State Side

This chapter presents results of the analysis of operational effectiveness of the under-frequency load shedding for different emergency situations using the combined model of UFLS for the united power system of the Baltic States and

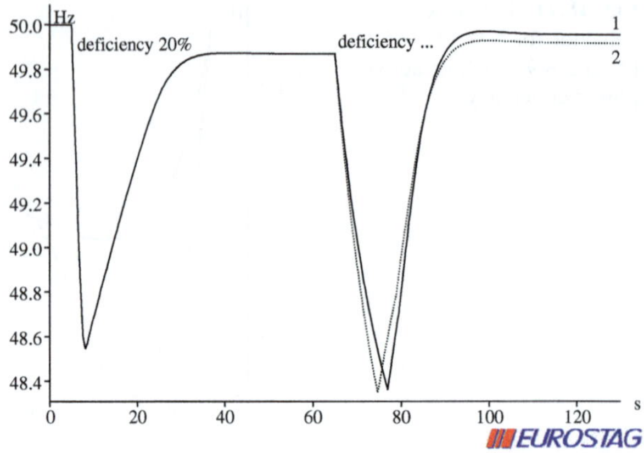

Fig. 18.12 Frequency behavior for 26.7 % (*1*) and 27.5 % (*2*) of active power deficiency

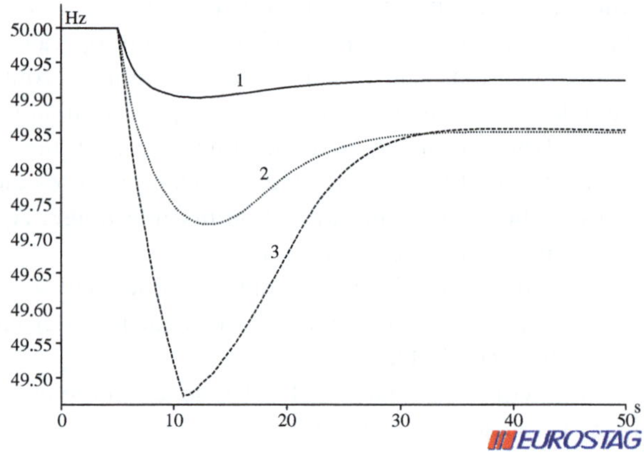

Fig. 18.13 Frequency behavior for 5 % (*1*), 10 % (*2*) and 15 % (*3*) of active power deficiency

ENTSO-E. Let us except the initially availability of a strong tie between the united power systems.

In the Fig. 18.13 the frequency behavior is shown at deficiencies of active power equal to 5, 10 and 15 %. At the deficiency of 5 and 10 % the frequency restores only due to governor operation. The value of the disconnected load by the steps of UFLS are three time smaller as the value of the power deficiency in the network but the frequency restores and stays in the area of the normal permissible range at the deficiency of 15 %.

Figure 18.14 illustrates the frequency behavior for the emergency situation with the iterative accident. When iterative deficiency takes place frequency hovering is possible (curve 1) because the frequency in the second accident does not decrease as lowly as at the first accident and the thresholds of UFLS operation are still not restored. The load reduction factor D is a parameter which stopped the frequency decline. The steps of both UFLS may operate at the deficiency of 65 % (2) and 70 % (3), because the frequency in the second accident decreases more as in at the first accident.

Fig. 18.14 Frequency behavior for 61.7 % (1), 65 % (2) and 70 % (3) of active power deficiency

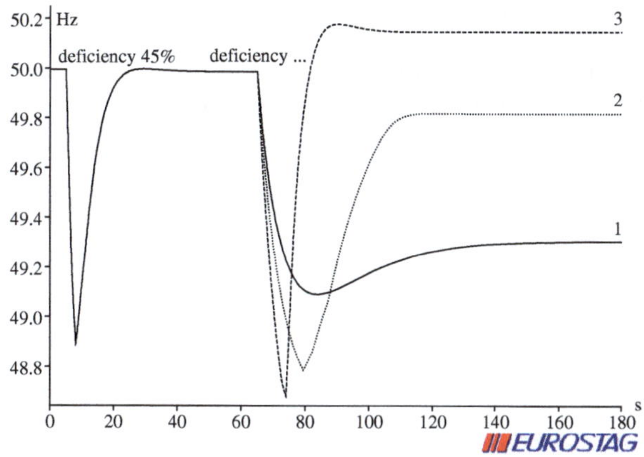

18.4.4 *Frequency Behavior of Joint Power System Operation with UFLS and Power Deficiency at the ENTSO-E Side*

The frequency behavior was analyzed for different emergency under-frequency situation in the united power system of the Baltic States and ENTSO-E when deficiency of power is at the ENTSO-E side. During the united power system operation when the deficiency of active power by its value exceeds the value of the load shed by the steps of UFLS of one of the systems, the frequency restores close to its nominal value and stays in the area of the normal admissible range. Frequency hovering is possible at the iterative accident.

For all considered cases the disconnected load is mostly located in the deficient part of united system.

When both power systems are operating in united manner the possibility of a blackout is minimized (Fig. 18.15).

18.4.5 *Simulation of Joint Power System with Weak Intersystem Ties*

The initial period of interconnected operation of power systems usually is characterized by relatively weak intersystem ties.

The more uniform control action is taken throughout the power system, the better.

The problems that can arise in case of non-uniform UFLS schemes will be shown by the help of the following three simulation cases.

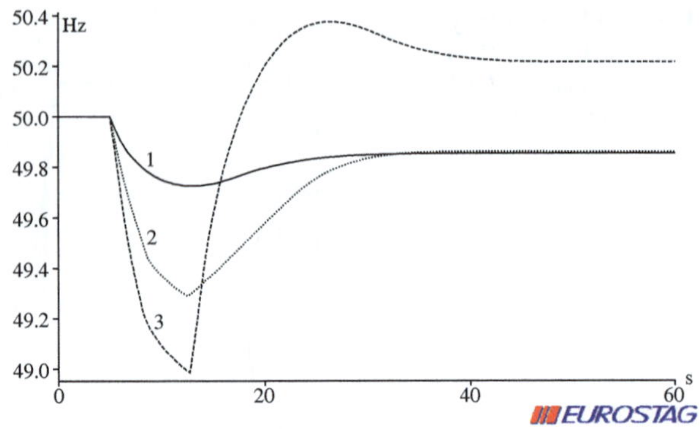

Fig. 18.15 Frequency behavior for 5 % (*1*), 10 % (*2*) and 15 % (*3*) of active power deficiency

For all study cases it will be assumed that the transmission capability of the intersystem tie is 10 % from the installed capacity of the joint power systems and the magnitude of the disturbance is 14 %.

The one-line diagram used in the simulations is presented in Fig. 18.7, where both systems are considered to be equal and there is no initial power flow via the intersystem tie. Such homogenous system will allow us to concentrate on the operation of UFLS.

18.4.5.1 First Case: The Same Type of UFLS Schemes (That Correspond to the One Used in the Spanish Power System)

Here the UFLS is applied in both power systems. The steady-state frequency is 50.24 Hz. It can be seen that at a given deficiency (14 %) an over-frequency situation occurs due to activation of an extra load shedding step.

The frequency that is shown in Fig. 18.16 does not represent any shortcomings of the given UFLS. Depending on power system state and magnitude of disturbance, all kind of frequency deviation scenarios can take place for any UFLS scheme, even a perfect match between the magnitudes of disturbance and control action can happen, bringing frequency back to 50 Hz.

18.4.5.2 Second Case: The Same Type of UFLS Schemes (That Correspond to the One Used in Baltic Power Systems)

UFLS used in the Baltic power systems has a large number of steps and relatively small amount of load to be shed at each step. Frequency at given deficiency (14 %—the same for all three simulation cases) does not over-shoot 50 Hz, though it reaches lower values during the transient process in comparison with the previous case as the smaller amounts of disconnected load result in a slower frequency restoration. Lower frequency values during the transient process can be

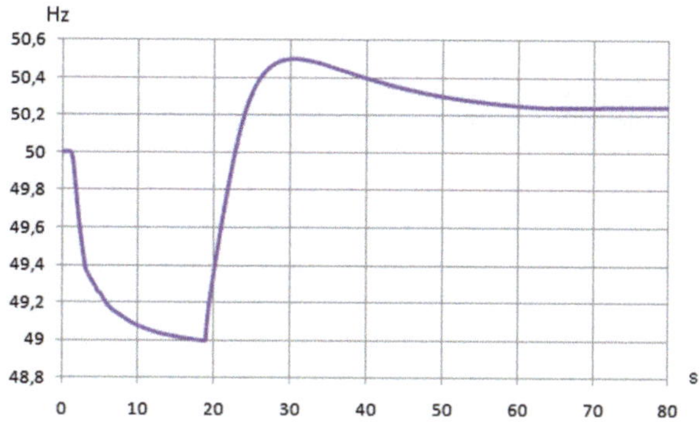

Fig. 18.16 Frequency versus time for the first simulation case (frequency in both systems is the same)

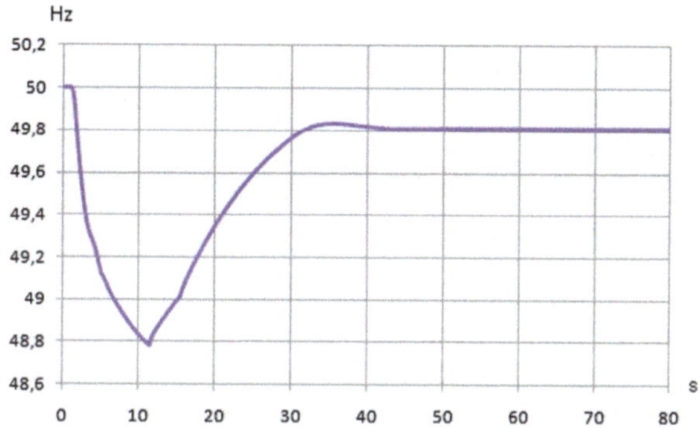

Fig. 18.17 Frequency versus time transient for the second simulation case (frequency in both systems is the same)

attributed also to the fact that the frequency setting of the first load shedding step in the Baltic power systems is lower than in ENTSO-E's UFLS systems.

The steady state value in Fig. 18.17 is 49.81 Hz.

18.4.5.3 Third Case: Different Types of UFLS Schemes

In the previous simulation cases the same UFLS scheme was applied throughout the joint power system. Even though the magnitude of the disturbance (14 %) exceeded the capacity of intersystem tie (10 %), both systems kept working in parallel without loss of stability.

Let us examine the case when there are two different types of UFLS operating together (both types that were used for the previous two simulations). One of them is placed in the first system and the other in the second system. The systems are connected via the same intersystem tie.

Now the 14 % active power deficiency that was successfully dealt with in previous cases leads to asynchronous operation of both systems with different UFLS types due to a weak intersystem tie. One system continues to operate with a frequency close to the rated frequency and the other experiences a frequency

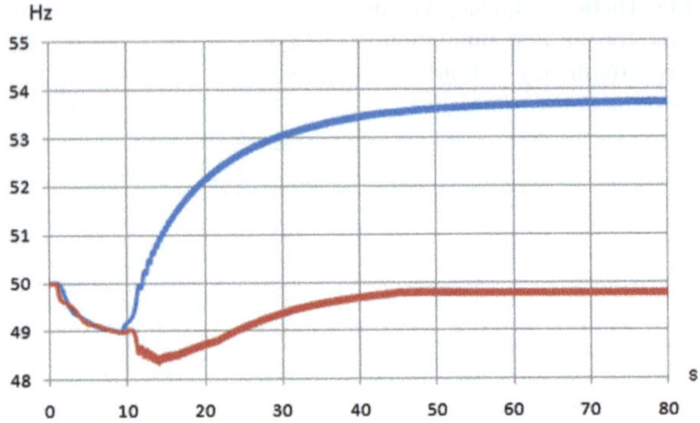

Fig. 18.18 Frequency versus time transient for the third simulation case (both systems operate with different frequencies)

increase towards 54 Hz. The differences between frequency settings and the amount of load to be shed in both systems led to unnecessary active power flows that at one point resulted in the loss of stability (see Fig. 18.18).

18.5 Conclusions

The frequency behavior was analyzed for different values of deficiency in the power system. Two cases were considered—when deficiency is divisible and not divisible by the value of the load disconnected by the UFLS automation.

Over-frequency situations were discovered as well as hovering situation for cases when the deficiency is not divisible by the value of the load disconnected by the UFLS automation. Situations with significant over-frequency or frequency hovering can be caused by iterative emergency situations.

To prevent the power system from dangerous over-frequency situations specific automation algorithms should be used. There is need for additional automation for restoration of frequency close to the nominal value.

For a more reliable frequency restoration process a uniform philosophy of the load shedding system should be preferable.

The frequency behavior was analyzed for different under-frequency emergency situations in the united power system of the Baltic States and ENTSO-E when deficiency of power is at the Baltic side as well as at the ENTSO-E side.

During emergency situations caused by deficiency in the interconnected power system under-frequency load shedding automation is activated in all interconnections. The frequency restores and stays in the area of the normal admissible range in all cases of single emergency under-frequency situations considered. Frequency hovering is also possible at the iterative deficiency.

For some considered cases the disconnected load is mostly located in the normal, but not deficient part of the united system due to the different philosophy of UFLS automation.

References

1. P. Kundur, *Power System Stability and Control* (McGraw-Hill, New York, 1993)
2. J.J. Grainger, W.D. Stevenson, *Power System Analysis* (McGraw-Hill, New York, 1994)
3. ANSI/IEEE Standard C37.106 (1987), IEEE Guide for Abnormal Frequency Protection for Power Generating Plants
4. Applied Protective Relaying (1978), Westinghouse Electric Corporation. Relay-Instrument Division, Newark, N.J
5. M.S. Baldwin, H.S. Schenkel, *Underfrequency Operation of Westinghouse Power Generation Equipment* (Westinghouse Publication, New York, 1974). (GEN 74-009)
6. Standard GOST-13109. (1997). International power quality standard for electric power systems. Intergovernmental standardization council. Moscow-Minsk
7. Technological maintenance rules for Ignalina Power Plant with RBMK-1500 reactor. (1994). Lithuania
8. M.A. Berkovich, V.A. Gladyshev, V.A. Semenov, *Avtomatika energosistem.* M. Energoatomizdat (in Russian) (1991)
9. IEEE Status Report on Methods Used for System Preservation During Underfrequency Conditions (1975), IEEE Transactions on Power Apparatus and System, Vol. PAS-94
10. P.V. Subramanian, M. Viswanathan, V.T. Kairankonda, Frequency trend and discrete underfrequency relaying practices in India for utility and captive power applications. IEEE Trans. Power Deliv. **7**(4), 1878 (1992)
11. J.E. Kochler JE, Control actions to be taken under emergency conditions in the New York power pool in emergency control practices. IEEE Trans Power Apparatus Syst. PAS-104 (1985)
12. C.J. Durkin, J.E. Eberle, P. Zarakas, An underfrequency relay with frequency decay rate compensation. Proc. IEEE **88**(6), 812 (1969)
13. R.S. Rabinovich, *Frequency Actuated Automatic Load Shedding System* (Energia, Moscow, 1980). (in Russian)
14. M. Mandozzi et al., Recent Improvements of Emergency Automatic Control of ENEL Power System in Interconnected and Isolated Operation. CIGRE report 39-302 (1992)
15. Y. Halevi, D. Kottick, Optimization of load shedding system. IEEE Trans. Energy Convers. **8**(2), 207 (1993)
16. M.S. Sachdev, M.M. Giray, A digital frequency and rate of change of frequency relay. *Transactions of the Engineering And Operating Division, Canadian Electrical Association*, vol. 17 part 3, pp. 78-sp-145 (1978)
17. V. Čuvičins, N. Gurovs, S. Rubcovs, *Adaptive load shedding and load restoration systems.* International Scientific conference PowerTech'05, St. Petersburg
18. Final Report, System Disturbance on 4 Nov 2006. Union for the co-ordination of transmission of electricity. http://www.enstoe.eu

Part VI
Integrated System

Chapter 19
Comprehensive Simulation Framework for Power System Operation

Sven Christian Müller, Hanno Georg and Christian Wietfeld

19.1 Introduction

In the previous chapters, newly designed applications from the field of Wide Area Monitoring, Protection and Control (WAMPAC) systems have been introduced and results have been presented for individual applications. However, it is of particular importance to evaluate these applications in interaction in order to analyze their joint performance and their interdependencies. Furthermore, power systems become more and more dependent on Information and Communication Technologies (ICT) for information exchange and data processing at local (intra-substation) and remote (inter-substation/control center) levels. Especially time-critical WAMPAC applications require powerful and reliable ICT systems. Consequently, the real-time performance of these applications needs to be evaluated in detail taking into account the ICT system. Simulations are a common way for analyzing the suitability of new approaches for power and ICT systems, but as both networks are highly interdependent and mutual effects can occur, simulations need to address two aspects comprehensively: electromechanical effects modeled within a power systems simulation and communicative effects modeled within a communication networks simulator.

In this chapter, we present a hybrid simulation architecture that enables the comprehensive analysis of the interplay of monitoring, protection and control applications taking into account both power and ICT systems. The co-simulation framework is based on the IEEE 1516—High-Level Architecture (HLA) and is currently under development in DFG research unit FOR1511.

S. C. Müller (✉) · H. Georg · C. Wietfeld
TU Dortmund University, Dortmund, Germany
e-mail: svenchristian.mueller@tu-dortmund.de

H. Georg
e-mail: hanno.georg@tu-dortmund.de

C. Wietfeld
e-mail: christian.wietfeld@tu-dortmund.de

U. Häger et al. (eds.), *Monitoring, Control and Protection of Interconnected Power Systems*, Power Systems, DOI: 10.1007/978-3-642-53848-3_19,

19.2 Motivation

Smart grid applications are one of the major trends of power system technologies. Driven by political decisions at international levels, the cross-border market integration and the replacements of fossil by renewable energy sources (RES) are increasing the base level and the volatility of power flows within the European electrical transmission system. Furthermore, feed-in by RES and new developments at the consumer level (e.g., demand side management, electric vehicles) cause new dynamics in the system operation. To guarantee a reliable and secure power supply, new approaches are necessary to deal efficiently with these challenges and to operate the transmission system close to its reliable limits. Here, especially Phasor Measurement Units (PMUs) and high-speed communication networks are offering new benefits for both centralized and decentralized management of future power systems by making information available at a wide area level, so advanced applications and intelligent control applications can profit from this data [1] (compare also Chaps. 4 and 5 of this book). For evaluating new approaches based on the technologies introduced previously, simulations are a common method, but as both power systems and communication networks are simulated using dedicated and specialized tools, it is difficult to analyze mutual effects within both systems. For this reason, a new simulation environment is needed taking into account detailed models of both power system and communication network to enable a detailed evaluation of real-time performance of applications and the dynamic system response. A particular challenge constitutes the combined analysis of discrete-time (power system) and event-based (ICT) simulations. Moreover, it is of major importance to analyze the joint performance of the various monitoring, protection and control applications that interact when deployed in system operation. Therefore, a comprehensive simulation environment is needed that enables the integration of a variety of applications and that allows for the simulation of the interplay of different entities and applications including ICT processes.

In the following, we summarize our proposal for a hybrid simulation architecture for power and ICT systems designed for analyzing new WAMPAC applications developed in DFG research unit FOR1511. For this, the chapter is structured as follows: First, an overview on related work is given in Sect. 19.3. In Sect. 19.4 the design of the hybrid simulator architecture is summarized based on the detailed outlines presented in [1, 2]. Within these references, detailed insights in the hybrid simulator architecture as well as simulation results have been presented. Section 19.5 discusses exemplary simulation results in order to explain the analysis capabilities and the interaction of the simulators. Finally, Sect. 19.6 provides an outlook on upcoming developments and a conclusion is drawn.

19.3 Related Work

Since novel WAMPAC solutions relying on modern communication solutions have been introduced, real-time capability and failure tolerance have been an issue for protection and control algorithms. Hence, various approaches have been developed to evaluate the performance impact of communication networks to monitor, protect and manage the power systems. Here, powerful tools are available in the respective fields and some approaches for realizing a combined simulation of communication networks technology and power systems have been presented. In this section, first a brief overview on available tools for power system and communication networks simulation is given in Sects. 19.3.1 and 19.3.2. Afterwards, existing approaches for realizing a combined simulation environment derived from the field of distributed computing will be presented in Sect. 19.3.3. Finally, existing concepts for integrating the simulation of power systems and ICT networks are presented in Sect. 19.3.4.

19.3.1 Simulation of Dynamic Power System Operation

Power systems constitute highly complex systems and the common way for analyzing system dynamics and stability is simulation. Simulation naturally requires modeling and thus abstraction. For this, we distinguish between three levels of the real-world system as it is done similarly in common categorizations of operational power systems management (e.g., [4]):

- Primary equipment (e.g., transformers, lines) is located at the bay level. At this lowest level measurements as well as actions are carried out, and local Intelligent Electric Devices (IEDs), such as local protection devices, are in place.
- At the *substation level*, data from several bays is collected by Local Area Networks (LAN), decentralized data processing is carried out, and a gateway to a Wide Area Network (WAN) is provided.
- Beyond the substation level, a variety of smart grid participants such as control centers or power plants can be located which are connected to a WAN and which we categorize as belonging to the *wide area level.*

Simulation tools for power systems provide detailed analysis functionalities for all processes at the bay level. Common commercial simulators—such as DIgSILENT Power Factory [5] which is used in FOR1511—typically enable steady-state, electromechanical and electromagnetic analyses. For analyzing WAMPAC applications and their interaction with communication and power systems in large transmission systems, steady-state analysis is not applicable because dynamic phenomena such as voltage angle acceleration or oscillations are of interest. Electromagnetic simulations, on the other hand, are too computationally intensive and difficult to equip with reasonable data for representing a realistic large-scale system. For this reason, we choose the electromechanical simulation as the appropriate type for the hybrid simulator. Thereby, the dynamic behavior of the electrical network (such as voltages and currents) can be analyzed taking into account the mechanical dynamics (in particular inertia) of electromechanical machines. The electromechanical simulation is based on the numerical calculation of a system of differential–algebraic equations in discrete time-steps Δt (for our test cases we chose $\Delta t = 10\,\text{ms}$ but it can be chosen even smaller).

Simulation of processes at the substation or wide area level is typically not foreseen in power system simulators (with the exception of global controllers or algorithms). In the hybrid simulator, we explicitly represent these levels as it will be described in Sect. 19.4. This also enables the investigation of ICT impacts on the interactions between the various entities at the bay, substation and wide area level.

19.3.2 Simulation of Communication Networks

Evaluating the performance of communication networks by simulation is probably the most common way to analyze the impact of communication technologies in specified fields of application. For this, Discrete-Event based Simulations (DES) are commonly used to model the technologies and protocols of the communication network. Within the concept of DES, interactions between nodes, protocols and process models are modeled as events. Each event has a timestamp attached, which specifies the sequence of execution. At the beginning of a DES simulation, system state variables are initialized, the simulation clock is set to zero and initial events are scheduled within the centralized event list. Next, the simulator executes each event in the event list, scheduled for the current simulation clock. Each event executed can schedule new events, e.g., the arrival of a message at the destination node. If each event scheduled for the current timestamp is executed, the simulator advances its simulation time and repeats this step until either the end of the simulation is reached or the event list is empty. For simulating communication networks, various simulators exist, each under different software license conditions with different models implemented. An overview of availability and performance analysis of various simulators is provided in [6]. In our approach, we choose the OPNET Modeler [7] as communication network simulation environment.

19.3.3 Distributed Computer Simulation Systems

Driven by military research projects, various approaches for realizing a distributed computer simulation system have been presented and optimized during the last years. These approaches, initiated by the US Department of Defense, have resulted in the IEEE Standard 1278 [8]—*Distributed Interactive Simulation* (*DIS*) in 1993 and its successor which has been called *High-Level Architecture* (*HLA*) (IEEE Standard [9–11]). Since the latest HLA standard has been released in the IEEE 1516-2010 (*HLA evolved*), HLA provides generic interfaces to establish a time-synchronized and object-oriented distributed simulation environment. Although this concept has been widely adopted within military research projects, examples for applications in the context of power system or communication network simulations are rare. In general, a HLA based simulation environment consists of the following components: the *Run Time Infrastructure* (*RTI*) providing the core services for time-synchronization and object management, the subsimulators (so-called *federates*), the HLA rules defining and describing interactions with the simulation network, and interface specifications for communicating with federates and the RTI. Finally, an object model is provided defining rules for each object-oriented simulation data element which is divided into the *Object Model Template* (*OMT*) and the *Federation Object Model* (*FOM*).

19.3.4 Integrated Simulation of Power Systems and ICT Networks

For evaluating the mutual impacts of both networks, an integrated simulation of the power system and the ICT system is necessary to make reliable predictions on the real-time capability of control algorithms for the power system and the communication technologies, respectively. Here, especially the different approaches used in the specialized simulators are posing challenges for engineers: while for the power system discrete time-steps are used to simulate the behavior of the network, the simulation of communication networks is based on discrete event-based time representation. For this reason, interconnecting both simulators constitutes a high demanding requirement for object and time-synchronization. To fulfill these requirements two generic approaches are available:

- **Comprehensive Simulation Environments**: In a comprehensive simulation environment both power system and communication network simulators are merged into one development environment. Here, either power system behavior is modeled inside the communication network simulators environment or vice versa. By this, there is no need for time-synchronization as both models are acting in the same time domain using the same logical time implementation and have access to the same data objects within the simulation. As the major disadvantage of this approach, detailed simulation models for either power system technologies or communication technologies need to be rebuilt from scratch. An example for this concept has been shown in [12] where the authors connected a MATLAB driven distribution system simulation with the OMNeT++ framework built for simulating communication networks.
- **Co-Simulation**: Another approach for analyzing the mutual impact of both networks is to set up a combined simulation environment. Here, both networks

are simulated with their specialized simulators using another framework to keep both simulators synchronized for time- and object-management.

The *Electric Power and Communication Synchronizing Simulator* (*EPOCHS*) [13, 14] has been the first approach to combine simulators for power system and communication networks. Here, the authors connected the two commercial simulators PSLF and PSCAD/EMTDC to the open source *Network Simulator 2* (*NS2*) using a RTI based on IEEE 1516 to realize the interconnection between simulators.

However, in comparison to the proposed architecture in this chapter, EPOCHS does not fulfill the requirements for analyzing the interplay of WAMPAC applications developed in FOR 1511, e.g., the system is focused on simulating communication among software agents and does not take into account IT execution delays and nowadays communication standards in the power systems. Additionally, no information regarding the computational performance has been published or found to our knowledge.

In [15] another approach is presented for realizing a Co-Simulation of power systems and communication networks simulators. In this paper, the authors linked the open source simulators OpenDSS and NS2 for simulating the transmission of dispatching messages in NS2 and generating OpenDSS scripts to run the power system simulation. Although this concept is still work in progress, it will be difficult to extend the environment to use other subsimulators.

19.4 Hybrid Simulation Design

In this section our proposal for a hybrid simulation architecture design is introduced and detailed. First, in Sect. 19.4.1 our modeling approach for the communication architecture is introduced. The model for the power system architecture is presented subsequently, in Sect. 19.4.2. This section closes with the integration concepts for both ICT and power systems networks in Sect. 19.4.3.

19.4.1 Communication Architecture Model

The system architecture, as seen from the communication network, can be divided into three layers; the centralized monitoring and control layer, the wide area communication layer and the local process layer. Each layer can be used to describe the integration aspects as well as interactions occurring between power systems and ICT simulators. The connectivity between these layers is shown in Fig 19.1 and can be described in detail as follows:

- **Centralized Monitoring and Control Layer**: The top layer models the centralized monitoring components realized in control centers or in centralized

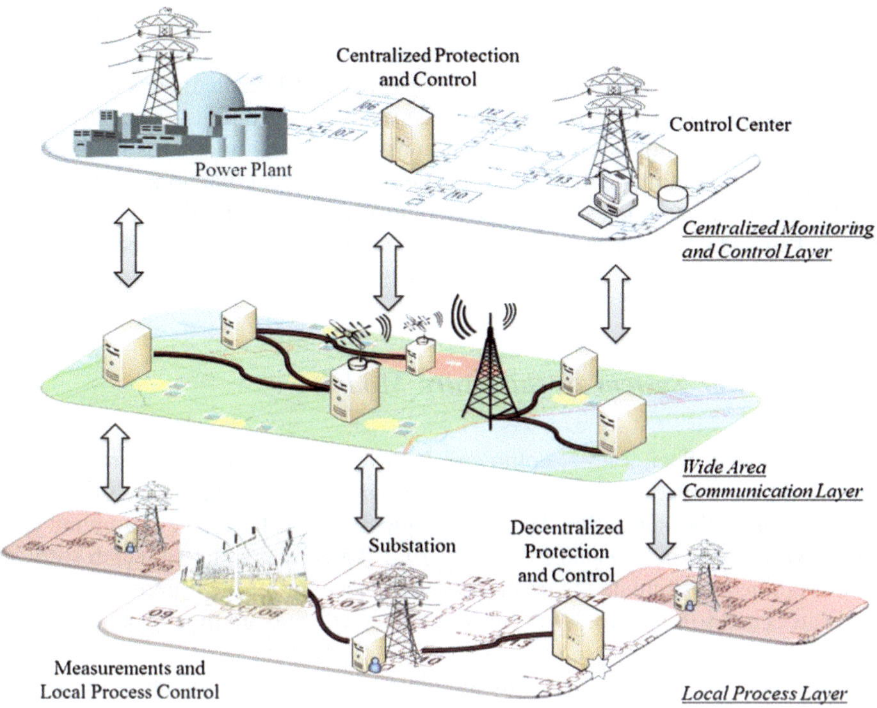

Fig. 19.1 Integration concept from the ICT system perspective, based on [3]

database infrastructures for monitoring the state of the overall power system. Additionally, centralized control components like, e.g., human machine interfaces inside a control center or automated control algorithms for the power system are applied. Interactions and communication between components at this layer are not available directly. Instead, components will be assumed to be separated geographically, so communication needs to be transmitted over the wide area layer beneath.

- **Wide Area Communication Layer**: The wide area communication layer models the communication network using cable, optical and wireless network technologies for linking components on the centralized layer above. Additionally, local process components, e.g., various substation instances, on the layer beneath are linked to centralized components using heterogeneous communication networks.
- **Local Process Layer**: The local process layer handles communication traffic generated within substation and bay level. This traffic consists of local monitoring, measurements and local process control information which is transmitted using dedicated local networks realized by Ethernet links. Communication between components on this layer is available while communication between different substations and to centralized components on the upper layer has to be transmitted over the wide area layer.

For modeling the wide area network links, different technologies and topologies (e.g., fiber along the transmission system lines) can be investigated and compared. According to the communication architecture presented before, nodes are modeled at the substation level representing decentralized operations, e.g., distributed control and protection systems. At the substation level, nodes for mapping the local process communication are modeled using local network technologies. Additionally, as candidates for fallback solutions wireless broadband technologies (e.g., Tetra, LTE) can be taken into account both as dedicated or shared infrastructures.

19.4.2 Power System Architecture Model

As discussed in Sect. 19.3.1, commercial power system simulators are primarily designed for simulations at the bay level. In our hybrid simulation architecture we represent also the substation level, the wide area level and the communication links explicitly as it is visualized in Fig. 19.2. The colors indicate which software tool is applied to simulate the processes and to account for the different delays. All simulators and software tools are run simultaneously and are time-synchronized by the HLA.

All processes at the bay level are simulated with the commercial power system simulator DIgSILENT PowerFactory as it provides a large library of validated models and proven functionalities. Via the HLA, the other levels are connected with the power system simulator. Entities that process data, e.g., execution of control algorithms, at either the substation level or beyond (e.g., at control centers) can be represented explicitly as HLA federates. Currently, these entities are modeled in JAVA but also other secondary software tools such as MATLAB or GNU R can be linked into the simulation. The communication between all federates can be simulated in detail by the communications network simulator OPNET. Within the network simulator, communication delays at the wide area level are the most relevant for the analysis of WAMPAC applications. However, simulating the local intra-substation traffic between bays is available as well at any of the federates, it can be accounted for their timely impact on real-time behavior. Communication latencies $t_{\mathrm{CT,WAN}}, t_{\mathrm{CT,LAN}}$ as well as execution times of algorithms $t_{\mathrm{IT,calc}}$ and waiting for additional input data before the execution of an algorithm $t_{\mathrm{IT,wait}}$ can be accounted for. Furthermore, the response of power system equipment to a change of control values (e.g., mechanical change of a tap position of a transformer) is subject to characteristic time delays t_{PE}. Last, the power system responds to events (e.g., control actions) following a certain trajectory that needs to be simulated and the time until a certain state of the power system is reached (t_{PS}, e.g., until oscillation have decayed below a certain limit) results from the power system simulation. The interplay of this representation of all entities of interest enables a comprehensive analysis of time-critical applications that rely on communication and interaction with other entities.

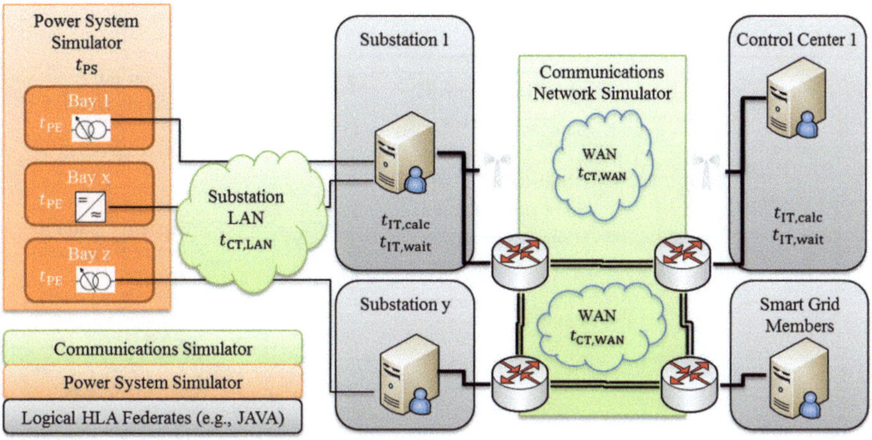

Fig. 19.2 Overview of representation of power system levels and deployed software Tools, based on [2]

19.4.3 Integration Concept for ICT and Power Systems

The section presents the integration concept by introducing the JAVA-based *Substation Data Processing Unit* (*SSDPU*) model to connect both communication and power system network. For this, the SSDPU represents a node within both power system and communication network at the substation level, equipped with additional information needed for redirecting the requests to the corresponding network. The concept of the SSDPU is visualized in detail within Fig. 19.3.

For describing the various parts and functions of the SSDPU, the components are described individually in the following:

- **Interface to the Power System Simulation**: Interactions between the substation unit and the power systems simulator are realized using an industrial standard for substations (OLE for Process Control—OPC). Via an OPC server, data can be exchanged between the power system simulator and the SSDPU. In particular, current state variables of the power system simulation (e.g., measurements) can be accessed by the SSDPU and control actions (e.g., disconnection of a line) can be executed by the SSDPU by triggering events in the power system simulation.
- **Interface to the Communication Network Simulator**: Communication requests and notifications are modeled using HLA interactions. Hence, the communication network simulator is connected to the SSDPU using the HLA interface. E.g., for transmitting a message from one substation to another, the sending SSDPU triggers a HLA interaction, then the communication is simulated in the OPNET communications network simulator which in turn delivers the message after the simulated delay to the receiving SSDPU by another HLA interaction.

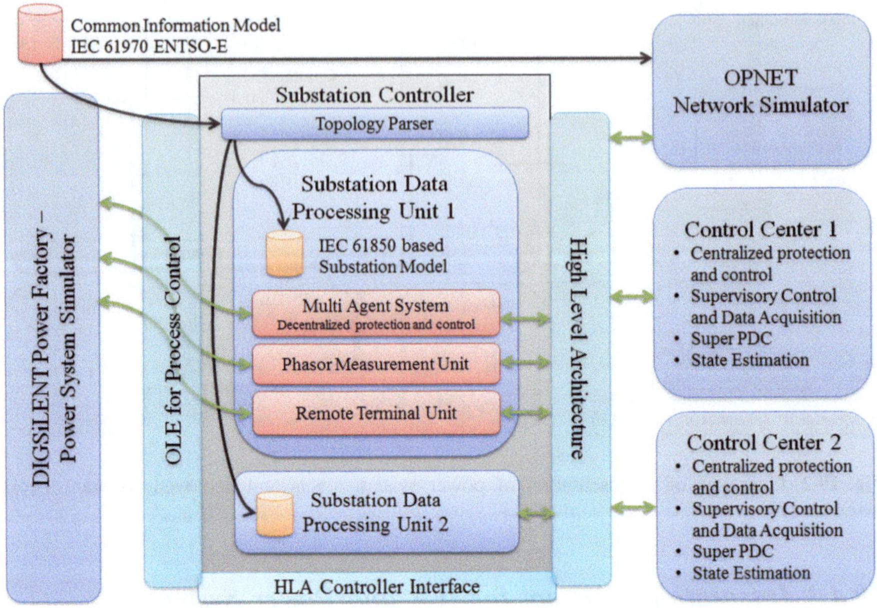

Fig. 19.3 Substation Data Processing Unit (SSDPU) for coupling of Simulators, based on [3]

- **Model partition:** In order to ensure compatibility with a variety of secondary software tools and modern standards, the modeling is based on a Common Information Model (CIM—IEC 61970) of the power system. The substation model within the SSDPU is created by the so called Topology Parser by reading the given CIM model and creating an IEC 61850 based model description within the substation controller. For this, the power system model is split into multiple substation instances containing detailed model information for local process control. Additionally, the power system topology given by the corresponding CIM model is translated into a communication network topology for scenarios close to the real-world.
- **Instance generation:** After the model partitioning has finished, the previously generated sub models are used for instance generation with each substation described within the CIM model being instantiated in a single SSDPU instance, carrying the IEC 61850 based model for local process control. Once the SSDPU instance is started, the interfaces to each subsimulator attributes inside the local substation model are mapped to the corresponding elements: OPC items for accessing the values of the power system simulation and HLA objects within the HLA federation, generated and published according to the Federation Object Model.
- **HLA control interface:** After the interfaces to the subsimulators are mapped and initialized, the HLA control interface is set up, managing time-synchronization with the HLA federation and implementing the update functions on part

of the OPC interface to the power systems simulation. For the attributes updates, different approaches are available to be used by the given protection algorithm: a continuous update at each time-step, a static update on initialization or an event-driven update.

19.5 Simulation Results

As exemplary results, a proof of concept study simulating a wide area control system for Phase Shifting Transformers (PSTs) as presented in [2] will be discussed in the following. The simulation is performed in the IEEE 39-bus 10-machine system ("New England Test System") with four PSTs and one HVDC transmission line added to enable the analysis of power flow control concepts. The transmission system topology is shown in Fig. 19.4. For modeling the wide area network links, we assume a full fiber coverage existing along the power system transmissions lines, which is available exclusively for usage of power system operation.

The aim of the study is to explain the analysis capabilities of the hybrid simulation and to verify the interaction of the simulators by a test case. For these reasons, only a simple wide area control scheme is investigated using the following setup: A decentralized control system being located at the substation at node 4 controls the tap position of PST 3. The objective of the control algorithm is to keep the loading of the transmission line between node 4 and 14 (*TL0414*) below 80 %. Whenever the loading is detected to exceed this threshold, a control message is sent to adjust the tap position of PST 3 by one tap as a countermeasure. As the PST tap position has to be adjusted mechanically, it takes an additional delay of $t_{PE} = 6\,\mathrm{s}$ until the new control value is set. For detecting the overload, PMU measurements being collected at substation 14 are used as input, which needs to be delivered by wide area communication. The reporting frequency between the substations is assumed to be set at a maximum of 0.1 Hz, thus the SSDPU at substation 14 waits for $t_{IT,wait} = 10\,\mathrm{s}$ before it sends another measurement to substation 4. For the communication network infrastructure we assume to have DS1 (*digital signal 1*) links available for wide area communication exclusively with a data rate of 1.55 Mbit/s. Each substation is assumed to have a local substation controller, a wide area network access router and a local network switch installed being connected by 10 Mbit/s Ethernet. Furthermore, a proprietary network protocol is applied that transmits UDP (User Datagram Protocol) based messages with a constant size of 240 bytes For simplicity reasons we assume the delays due to substation LAN ($t_{CT,LAN}$) and times for the execution of algorithm and PMU data processing ($t_{IT,calc}$) to be negligible, nonetheless, also execution times of complex algorithms or LAN latencies can be accounted for explicitly in the simulation. Finally, the communication network setup is assumed to be incomplete, so that no network route exists at the beginning of the simulation.

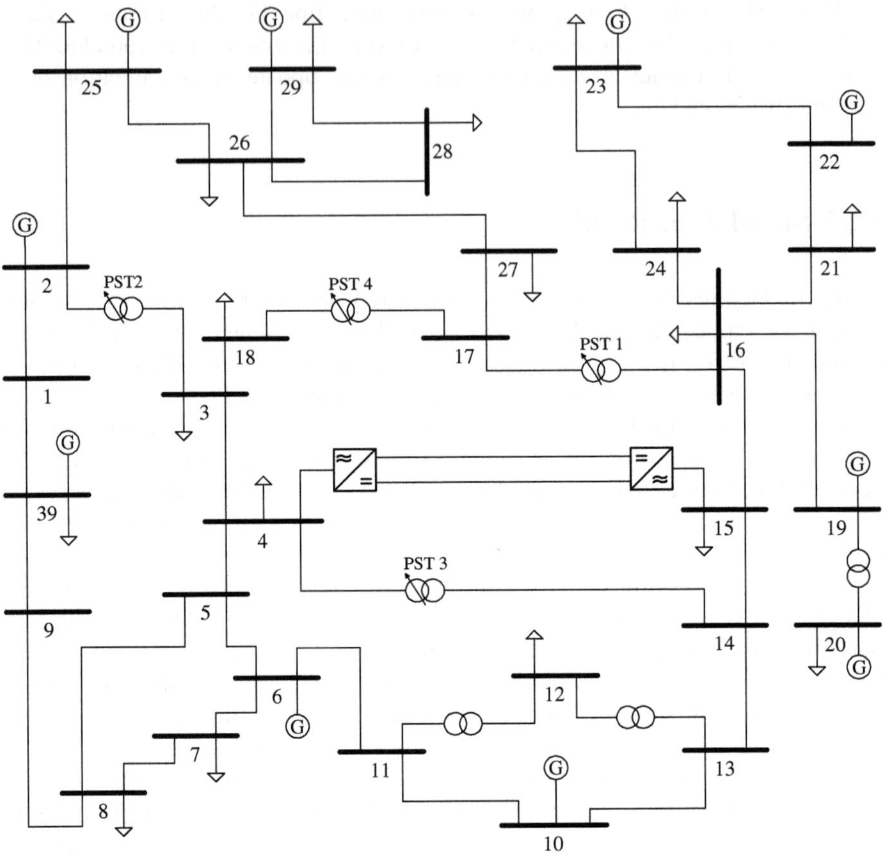

Fig. 19.4 New England Test System (IEEE 39-bus 10-machine system) extended by PSTs and an HVDC transmission line

Figures 19.5 and 19.6 show the simulation results for the general setup as described above, assuming the load at node 15 to be disconnected at $t = 2$ s In particular, the loading of the critical line, the tap position of PST 3 and the IP traffic sent from the SSDPU at node 14 are visualized. Furthermore, the time components that are critical for the real-time performance (compare Fig. 19.2 and Sect. 19.4.2) are indicated. Before the disconnection of the load, the system is in steady state and the loading of line *TL0414* is not critical. Due to the disconnection of the load at $t = 2$ s, the load of the line increases to more than 80 %. The excess of the threshold is determined by the PMU located at node 14. As delays due to LAN communication and PMU data processing are neglected, the SSDPU receives the measurement immediately from the PMU (located at the bay level) and triggers a message to substation 4 in order to report the overload. As the communication network is not initialized at the beginning, this first message is dropped and does not reach the SSDPU at node 4. However, in the course of this failed attempt,

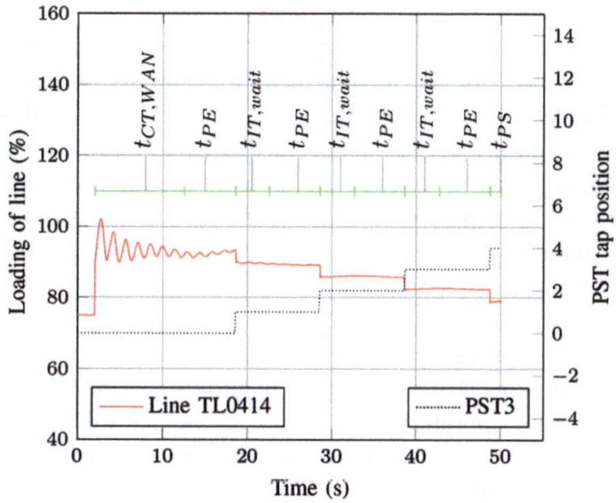

Fig. 19.5 Simulation results: loading of line *TL0414* and tap position of PST 3, based on [2]

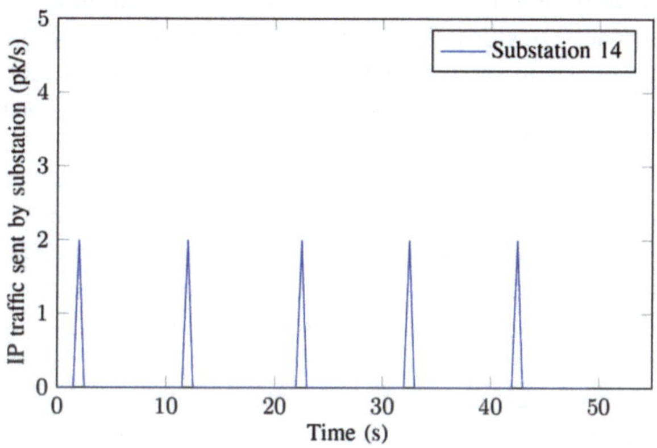

Fig. 19.6 Simulation results: IP traffic sent from substation 14, based on [2]

routes over the WAN are established by routing protocols. Due to the dropped message and the IT logic of reporting at a maximum frequency of 0.1 Hz, a delay of 10 s is caused before the PMU measurement collected at substation 14 is delivered successfully to substation 4 in the second attempt at $t = 12$ s. This delay is indicated in the figure as $t_{CT,WAN}$.

After the second attempt to deliver the PMU measurements has been successful, the decentralized control system at substation 4 initiates a countermeasure by sending a local control command for increasing the tap position of PST 3. As latencies of local communication and execution of the algorithm are neglected, the control command reaches the PST instantaneously, but due to the delays of the mechanics an additional delay of $t_{ps} = 6$ s occurs until the new tap position is successfully set. Thus, the first alleviation of the overload takes place at $t = 18$ s, nonetheless, it is not sufficient to reduce the loading of transmission line *TL0414* below the threshold of 80 %. For this reason, the next PMU measurement is reported from substation 14 to 4 after another $t_{IT,wait} = 4\,\text{s}$ (because of the maximum reporting frequency of 0.1 Hz and the last message having been sent at $t = 12\,\text{s}$). So, the next measurement indicating the overload is received by the decentralized control system at $t = 22\,\text{s}$. Due to this, another increase of the tap position of PST 3 is triggered that takes place—delayed by $t_{ps} = 6\,\text{s}$—at $t = 28\,\text{s}$. Analogously, two more increases of the tap positions are executed at $t = 38$ s. and $t = 48\,\text{s}$, respectively. Finally, the fourth increase of the PST tap position achieves a reduction of the loading below 80 %. After a time of t_{ps} the oscillations following the tap change have decayed below a predefined level and the simulation is stopped after this successful alleviation of the overload by the control system.

The exemplary study case illustrates the basic functionalities of the hybrid simulation. Regarding ICT, it allows for the detailed analysis of the impact of various technologies and logics. Evidently, the maximum reporting rate of SSDPU at node 14 as well as the incomplete setup of the communication network had decisive impact on the performance of the control system. Although these assumptions were a relatively obvious worst case scenario, similar effects are usually not taken into account within simulations, but can have considerable impact. Besides this, the explicit simulation of the ICT system allows for the comparison of different IT and communication technologies. E.g., the impact of different hardware used for the execution of algorithms with relevant execution times, fallback solutions for communication networks, background traffic or different protocols (e.g., using encryption to avoid vulnerability in the context of cyber security) on the real-time performance of WAMPAC applications can be investigated. From a power systems perspective, an interesting capability of the hybrid simulation is that it allows for the detailed comparison of centralized versus decentralized control and protection approaches. E.g., a competing control algorithm for the decentralized control of the PST could be a centralized system analysis performed at the control center. With the hybrid simulation, it can be analogously analyzed how a central collection of system data, state estimation and execution of a centralized optimal power flow for calculation of the ideal tap position of the PST performs compared with the decentralized approach. Further, the simulation allows to investigate the operational benefits by investment in fast controllable primary power system equipment (e.g., FACTS devices or HVDC as a complement or supplement to the slow controlling PST in the study case) in critical network situations and their reliance on the ICT system.

19.6 Outlook and Discussion

In the previous sections, the hybrid simulation design has been discussed and exemplary simulation results have been detailed in order to explain the capability of the hybrid simulation to account for all processing steps and timely impacts at the different levels of power system operation including the ICT. So far the interaction of simulators in a time-synchronized co-simulation has been presented in exemplary case studies in [2, 3]. Below, we summarize and discuss steps for further enhancement and application of the simulator for extended studies on large-scale systems with complex protection and control systems in Sect. 19.6.1. Finally, a summary is given and a conclusion is drawn in Sect. 19.6.2.

19.6.1 Future Developments and Extensions

For future deployment of the hybrid simulation in the context of integrated power system and ICT system analysis, some methodic as well as practical issues like usability, verification and performance need to be addressed. In the following we present enhancements and extensions of the present state hybrid simulation which are currently under development in research unit FOR1511 in order to enable interdisciplinary research in the field of WAMPAC applications with a high level of detail.

19.6.1.1 Scenario Configuration

Usability is a key factor to enable an interdisciplinary research on both power and communication systems. Two major points for this debate will be introduced in the near future, which both are taking into account the specific scenario configuration. First, an automatic scenario-conversion is necessary to enable a repeatable scenario configuration, which can be used without further knowledge on either power or communication system. Here, as scenario generation is power system driven, an easy and automatic ontology matching between communication nodes and power systems primary equipment is necessary to provide enhanced usability.

Secondly, for testing different event scenarios, a malfunction generator will be implemented for both the communication network and power system simulator. By this, failure events during the simulation can be generated following a predefined syntax of event strings and can include events of both power system and communication system, e.g., loss of a transmission line and disrupting a parallel communication link.

19.6.1.2 IT Execution Times

As presented, the execution times of algorithms on substation or control center hardware can be taken into account within the simulation. However, the hybrid simulation relies on adequate information on these execution times which can be significant (e.g., response from a control center may rely on state estimation and security analyses that may sum up to several minutes). Therefore, it is important that realistic estimates on the execution times of the algorithms are provided for the detailed analysis of the applications and a methodology for this estimation would be valuable.

19.6.1.3 Interactive Simulations and Verification of Functionality

It is the objective of the development of the hybrid simulator to enable the interactive simulation of the WAMPAC applications. In the future, applications for monitoring, protection and control purposes like the ones presented in this book and those being under development in FOR1511 will be integrated in the simulation in order to perform joint analyses. The comprehensive simulation of the applications needs to be verified by plausibility checks, comparison to analytical models or other simulations.

19.6.1.4 Distributed Execution

For evaluating the overall simulation performance, distributed execution will be taken into account to speed up the simulation. Due to the fact that the HLA is used in the simulation framework, an execution of the different HLA federates on various workstations is enabled by itself. However, various factors will affect the overall simulation performance and will be detailed in future work.

19.6.2 Summary and Conclusion

In this chapter, we have presented our novel architecture for integrated analyses of power and ICT system based on time-synchronized co-simulation using the IEEE 1516—High Level Architecture. The hybrid simulator design follows the levels of the physical system, e.g., bay, substation and wide area level of power system operation as well as their communication links are explicitly represented in the design. Currently, the simulator is under strong development for performing extended studies on large-scale system and for analyzing the interplay of the WAMPAC applications presented in latest research.

Generally, it can be concluded that the use of the proposed hybrid simulation framework offers particularly benefits for the following fields of investigation:

- analysis of the interplay and interdependencies of several smart grid applications (e.g., interacting or interdependent WAMPAC applications, possibly implemented in different software tools [16]);
- analysis of real-time performance of power system applications, in particular of time critical applications like wide area protection schemes,

- analysis of real-time performance of power system applications, in particular of time critical applications like wide area protection schemes, that require taking into account realistically the data processing, communication latencies and execution times at all involved components of the power and ICT system;
- analysis of emergent effects resulting from the simultaneous action of a large number of autonomous entities, e.g., multi-agent systems or future states of the smart grid with a large number of active and autonomous decision making participants;
- analysis of the impact of ICT technologies, protocols, fallback solutions and overall concepts on real-time performance of applications in power system operation;
- analysis of the mutual interdependencies of the ICT and the power system, e.g., increasing latencies in critical network situations due to increased network traffic, thereby influencing power system protection and control performance;
- analysis of the real-time performance of fast controlling equipment like FACTS devices and HVDC controllers; and
- comparison of different approaches for system management at different levels, e.g., centralized versus decentralized (e.g., as proposed in [17, 18]) protection and control concepts.

Summarizing, the future development of renewable energy source and the rise of smart grids lead to an increasingly dynamic power system operation that becomes more and more dependent on the ICT system. The hybrid simulation framework presented in this chapter enables an integrated analysis of power system, ICT system and various applications at all levels of power system operation. The architecture is designed to allow a high level of details within the simulation by explicitly representing all components of the physical system and specialized simulators for the complex analysis of electromechanical simulation of the power system in discrete time-steps and the event-based simulation of the communication networks. The hybrid simulator will be deployed in the future for investigating the interplay of a variety of power system applications as the ones described in the previous chapters. By enhancing the usability and performance, a comprehensive simulation environment will be provided that enables the analysis of complex systems and facilitates interdisciplinary research in the field of power system operation.

Acknowledgments This work was supported by the German Research Foundation DFG as part of research unit FOR1511 "Protection and Control Systems for Reliable and Secure Operation of Electrical Transmission Systems".

References

1. V. Terzija, G. Valverde, P. Deyu Cai, V. Regulski, J. Madani, S. Fitch, M. Skok, M. Begovic, A. Phadke, Wide-Area Monitoring, Protection, and Control of Future Electric Power Networks. Proceedings of the IEEE **99**(1), 80–93 (2011)
2. S.C. Müller, H. Georg, C. Rehtanz, C. Wietfeld, Hybrid simulation of power systems and ICT for real-time applications, in *Proceedings of 3rd IEEE PES Innovative Smart Grid Technologies (ISGT) Europe Conference*, Berlin, Germany, Oct 2012, pp. 1–7
3. H. Georg, S.C. Müller, C. Rehtanz, C. Wietfeld, A HLA based simulator architecture for co-simulating ICT based power system control and protection systems, in *Proceedings of 3rd IEEE International Conference on Smart Grid Communications (SmartGridComm 2012)*, Tainan City, Taiwan, Nov 2012, pp. 264–269
4. C. Rehtanz, *Autonomous Systems and Intelligent Agents in Power System Control and Operation* (Springer, Berlin and New York, 2003)
5. DIgSILENT, DIgSILENT PowerFactory, April 2012. Available: http://www.digsilent.de/
6. E. Weingartner, H. vom Lehn, K. Wehrle, A performance comparison of recent network simulators, in *IEEE International Conference on Communications 2009*, June 2009, pp. 1–5
7. OPNET Technologies, Inc., OPNET Modeler. Available: http://www.opnet.com
8. IEEE Standard for Information Technology—Protocols for Distributed Interactive Simulations Applications, Entity Information and Interaction. IEEE Std. 1278–1993, 1993
9. IEEE Standard for Modeling and Simulation (M&S) High Level Architecture (HLA)—Framework and Rules, IEEE Std. 1516–2010 (Revision of IEEE Std. 1516–2000), pp. 1–38, 2010
10. IEEE Standard for Modeling and Simulation (M&S) High Level Architecture (HLA)—Federate Interface Specification, IEEE Std. 1516.1-2010 (Revision of IEEE Std. 1516.1-2000), pp. 1–378, 2010
11. IEEE Standard for Modeling and Simulation (M&S) High Level Architecture (HLA)—Object Model Template (OMT) Specification, IEEE Std. 1516.2-2010 (Revision of IEEE Std. 1516.2-2000), pp. 1–112, 2010
12. K. Mets, T. Verschueren, C. Develder, T. Vandoorn, L. Vandevelde, Integrated simulation of power and communication networks for smart grid applications, in *16th International Workshop on Computer Aided Modeling and Design of Communication Links and Networks (CAMAD)*, Kyoto, Japan, June 2011, pp. 61–65
13. K. Hopkinson, X. Wang, R. Giovanini, J. Thorp, K. Birman, D. Coury, Epochs: a platform for agent-based electric power and communication simulation built from commercial off-the-shelf components. IEEE Trans. Power Syst. **21**(2), 548–558 (2006)
14. K. Hopkinson, K. Birman, R. Giovanini, D. Coury, X. Wang, J. Thorp, EPOCHS: Integrated Commercial Off-the-Shelf Software for Agent-Based Electric Power and Communication Simulation, in *Proceedings of International Conference on Machine Learning and Cybernetics*, New Orleans, LA, USA, Dec 2003, pp. 1158–1166
15. T. Godfrey, S. Mullen, R. Dugan, C. Rodine, D. Griffith, N. Golmie, Modeling Smart Grid Applications with Co-Simulation, in *First IEEE International Conference on Smart Grid Communications (SmartGridComm 2010)*, Gaithersburg, MD, USA, Oct 2010, pp. 291–296
16. S. C. Müller, A. Kubis, S. Brato, U. Häger, C. Rehtanz, J. Götze, New Applications for Wide-Area Monitoring, Protection and Control, in *3rd IEEE PES Innovative Smart Grid Technologies Europe Conference*, Berlin, Germany, Oct 2012
17. S.C. Müller, U. Häger, C. Rehtanz, H.F. Wedde, Application of Self-Organizing Systems in Power Systems Control, in *Lecture Notes in Computer Science*, vol. 7343 (Springer, Heidelberg, 2012), pp. 320–334
18. U. Häger, S. Lehnhoff, C. Rehtanz, H. F. Wedde, Multi-Agent System for Coordinated Control of Facts Devices, in *Proceedings of 15th International Conference on Intelligent System Applications to Power System (ISAP '09)*, Curitiba, Brazil, Nov 2009, pp. 1–6

Index

U. Häger et al. (eds.), *Monitoring, Control and Protection of Interconnected Power Systems*, Power Systems, DOI: 10.1007/978-3-642-53848-3,

MIX
Papier aus verantwortungsvollen Quellen
Paper from responsible sources
FSC® C105338

If you have any concerns about our products,
you can contact us on
ProductSafety@springernature.com

In case Publisher is established outside the EU,
the EU authorized representative is:
Springer Nature Customer Service Center GmbH
Europaplatz 3, 69115 Heidelberg, Germany

Printed by Libri Plureos GmbH
in Hamburg, Germany